SEARCHING BEHAVIOUR

Chapman and Hall Animal Behaviour Series

Detailed studies of behaviour are important in many areas of physiology, psychology, zoology and agriculture. Each volume in this series will provide a concise and readable account of a topic of fundamental importance and current interest in animal behaviour, at a level appropriate for senior undergraduates and research workers.

Many facets of the study of animal behaviour will be explored and the topics included will reflect the broad scope of the subject. The major areas to be covered will range from behavioural ecology and sociobiology to general behavioural mechanisms and physiological psychology. Each volume will provide a rigorous and balanced view of the subject although authors will be given the freedom to develop material in their own way.

SEARCHING BEHAVIOUR

The behavioural ecology of finding resources

William J Bell

CHAPMAN AND HALL
LONDON · NEW YORK · TOKYO · MELBOURNE · MADRAS

UK Chapman and Hall, 2–6 Boundary Row, London SE1 8HN

USA Chapman and Hall, 29 West 35th Street, New York NY10001

JAPAN Chapman and Hall Japan, Thomson Publishing Japan, Hirakawa-cho Nemoto Building, 7F, 1–11 Hirakawa-cho Chiyoda-ku, Tokyo 102

AUSTRALIA Chapman and Hall Australia, Thomas Nelson Australia, 480 La Trobe Street, PO Box 4725, Melbourne 3000

INDIA Chapman and Hall India, R. Seshadri, 32 Second Main Road. CIT East, Madras 600 035

First edition 1991

Typeset in 11/12pt Bembo by Witwell Ltd., Southport
Printed in Great Britain at the University Press, Cambridge

ISBN 0 412 29210 6

British Library Cataloguing in Publication Data
Bell, William J.
Searching behaviour.
1. Animals. Behaviour. Ecological aspects
I. Title
591.51

ISBN 0–412–29210–6

Library of Congress Cataloging-in-Publication Data
Bell, William J.
Searching behaviour : the behavioural ecology of finding resources
/William J. Bell. — 1st ed.
p. cm. — (Chapman and Hall animal behaviour series)
Includes bibliographical references and index.
ISBN 0–412–29210–6
1. Animal behaviour. 2. Animals—Food. I. Title. II. Series.
QL756.5.B44 1990
591.51—dc20 90–19161
CIP

Dedicated to Clare, for being there
and to
Norman Davis, Bill Harvey and Bill Telfer
for pointing out that good scientists
can also be good people

Contents

PART FIVE METHODOLOGY

Preface

This is a book about proximate mechanisms. Although some theoretical structure is used to introduce the subject, the intent is to offer a comprehensive view of the mechanistic side of searching (or foraging) so as to balance the current emphasis of books on mathematical and functional models. It seems to me that the pendulum needs to swing back to studies of how animals behave, and that maybe in so doing models will become valuable again in driving experimentation.

I have probably included too many examples in this book, and some are even presented in great detail. Hopefully, they provide a complete picture of the kind of animals used, the experimental setup, the kinds of data yielded, and how the data were analysed. I have done this in response to frustrating experiences of reading chapters in behavioural ecology books that provide insufficient information with which to evaluate an author's conclusion.

In preparation for the ecological aspects of this book I attempted to read the major texts in order to view searching behaviour from an ecologist's point of view. My colleagues in ecology are uncertain as to my success in this endeavour, but I gained respect and admiration for books written by E.R. Pianka (*Evolutionary Ecology*), D.H. Morse (*Behavioral Mechanisms in Ecology*), J.F. Wittenberger (*Animal Social Behavior*), R.R. Baker (*Evolutionary Ecology of Animal Migration*), a chapter by P.W. Price in *A New Ecology: Novel Approaches to Interactive Systems*, and a review by J.A. Wiens on patchy environments.

Finally, I am most grateful to several of my present and former students for their many suggestions and assistance (D. Conlon, R. Roggero, K. Sorensen, C. Tortorici, M.K. Tourtellot) and to many colleagues who either provided me with a paragraph or two, assisted with critical reading, or gave me ideas through discussions (H. Browman, R. Collins, R. Holt, C.D. Michener, W.J. O'Brien, T. Seeley, K.A. Sorensen, R. Swihart, S. Yoerg, W. Wcislo, K. Waddington, J. Wenzel). I also appreciate the guidance I received from the publisher, early on from A. Crowden and later from R.C.J. Carling, and I. Jamieson who made recommendations for trimming the book to its optimal size. The co-editors of this book series, D.M. Broom and P.W. Colgan, have been most helpful and patient.

W.J. Bell
Lawrence, Kansas

PART ONE
Introduction

It appears to me . . . that there are enough real complexities in apparently simple things, that we do not need to make them appear more complex than they already are

Heinrich 1983

Searching behaviour is an active movement by which an animal finds or attempts to find resources. It is perhaps the most important kind of behaviour that an animal engages in, because it is the means by which most motile organisms acquire resources such as food, mates, oviposition and nesting sites, refugia, and even new or different habitats. Since resources such as these are absolutely essential for the growth, development, and maintenance of an individual, and for insuring the success of future generations, efficient searching and accurate assessment of resources are crucial to an individual's survival chances and reproductive potential.

1
Theoretical framework

Searching behaviour represents the confluence of three kinds of factors:

1. The characteristics and abilities of an animal, including its perceptual and locomotory skills.
2. External environmental factors determining what resources are available and the risks generated in obtaining them. Conspecifics may affect competition, or heterospecifics may affect the extent to which predators and parasites comprise a major risk category, as well as other organisms that may provide shelter or impose obstacles to search. The external environment also incorporates abiotic influences that may alter the way an animal searches or how successful it is.
3. Internal factors, such as the level of physiological need relative to a certain kind of resource (starvation or sexual readiness), determining what an animal needs at a particular time, and the manner in which that requirement is altered by age, reproductive status or external environmental influences.

The ultimate success of an animal's searching depends to a large extent on:

1. the method it uses to search in relationship to the availability of resources and their spatial and temporal distributions in the environment;
2. its efficiency in locating resources;
3. the ability of a species to adapt to long-term environmental changes and the ability of an individual to respond, perhaps through learning, to short-term environmental changes.

Given that at least part of an individual's searching phenotype has a genetic basis, we should expect natural selection to operate on whatever variability there is in perceptual and motor traits, the information-processing mechanisms that underly behavioural decision-making, and response times. Thus, to some extent evolution should hone and sharpen an animal's ability to locate and identify efficiently the resources it needs.

1.1 ANIMAL ABILITIES (INTERNAL CONSTRAINTS)

Natural selection, ancestry, and experiences determine an animal's characteristics and abilities. These characteristics and abilities in turn can be viewed as the constraints within which an animal's behaviour is confined. Three important variables relate searching behaviour to the morphological, physiological, and behavioural adaptations of a species. The first is motility, since search behaviour implies active movements in quest of resources. The second is the relative amount of time or energy allocated to searching, because this axis to a large extent determines which search mechanisms an animal employs. Third, the range of acceptable resources within a given resource type is an important constraint that also relates to plasticity if an animal's range can change as a result of deprivation. In the following sections these variables are discussed primarily with reference to locating food.

1.1.1 Motility of searcher

All motile species of animals can be classified, qualitatively at least, according to the amount of time they spend moving in efforts to acquire resources. To be really accurate, the classification scheme for any given species would have to be specific to a particular type of resource (food, mate, ovipositional site), as well as to the animal's sex, age and other characteristics. The proportion of an animal's life spent searching for food items varies greatly. Some animals, such as small homeotherms with severe energy demands, spend most of their time searching for food. In these cases the resource is small relative to the size of the consumer. At the other extreme, herbivores such as ungulates spend less time actually searching for food. In fact, some herbivores, such as those that live and feed within fruit, consume only one food item during their entire lives.

Pianka (1966) applied the term sit and wait (S&W) to predators that obtain prey by waiting in ambush, and widely foraging (WF) to those that acquire prey by extensive searching. WF and S&W mechanisms are both ways for sampling the environment, but the environment 'moves past' a waiting animal, whereas the WF animal 'moves through' its environment. Such differences in motility of searching have also led to divergence in other characteristics. For example, S&W organisms generally have low resting energy requirements, and specialize on highly mobile and densely distributed prey, whereas WF organisms have higher resting energy requirements, and specialize on sedentary or sparsely distributed prey.

Within various broad taxa there are species with a S&W tactic and species with a WF tactic: fish that wait for food to float by them in a stream versus fish that actively seek prey in a still pond, spiders that build webs versus those that hunt, birds that attack prey from perches versus birds that attack prey while soaring or cruising. Enders (1975) developed this concept further, recognizing that the time interval during which animals move between prey captures represents a continuum from those species that stop for periods of seconds or minutes to species that stop for periods of days or months. Thus, virtually all animals could be classified on the basis of the period of the cycle between S&W and WF modes. With respect to searching for food, we can classify animals along this continuum: at one extreme are species that seldom move between long periods of waiting, scanning or ambushing, such as web-building spiders and aquatic invertebrates, praying mantids and some kestrels. At the other end of the continuum are species that move nearly continuously, such as sharks and sunfish, and small insectivorous birds and shrews. In between these extremes are species with shorter cycles, alternating frequently between searching and scanning the environment, as in birds such as plovers, coccinellid beetle larvae, and fish such as crappie.

1.1.2 Time spent in searching versus handling resources

Whereas searching is the process of finding a food resource, pursuing includes chasing down or stalking prey, and handling includes subduing, swallowing, digestive pauses, or otherwise processing food. The relative amount of time or energy invested in search, pursuit, and handling relates to motility, morphology and behaviour of predatory animals (Figure 1.1a) (sensu Enders, 1975; Griffiths, 1980). For example, small insectivores such as ovenbirds spend nearly all of their time searching, and the handling and pursuit costs are relatively small (Figure 1.1b). On the other hand, spotted hyaenas are usually surrounded by potential prey, and so search costs are small, whereas handling costs (of bringing down prey) are somewhat larger, and pursuit costs (of chasing prey down) are larger still (Figure 1.1c). S&W predators such as antlions move little and spend most of their time in ambush; they expend considerable energy handling prey (Figure 1.1d). Coccinellid beetle larvae locate their prey by contact; while the pursuit component is negligible, much energy is expended

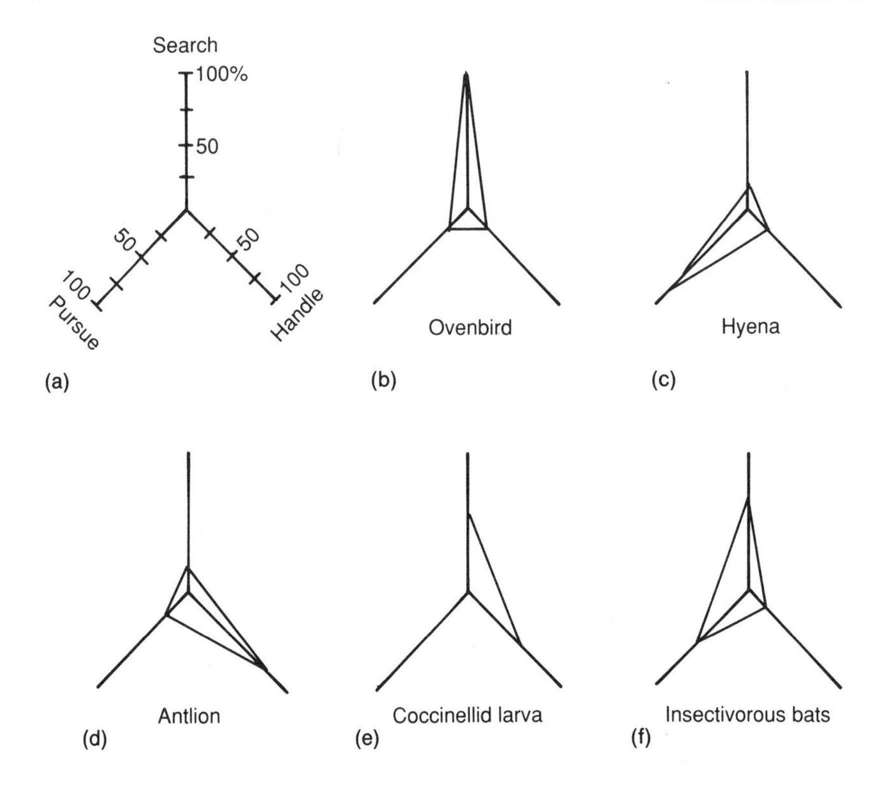

Figure 1.1 Proportion of time spent searching for resources compared to time spent handling (opening, subduing, swallowing) and pursuing (or chasing down) food resources.

searching for patches of aphids, and presumably in handling prey (Figure 1.1e). Insectivorous bats search for and pursue their prey, but probably spend little effort subduing such small insects (Figure 1.1f).

Another important relationship is that animals expending most of their time actively searching tend to consume small prey, whereas animals that must pursue and spend considerable time handling their prey generally consume larger prey relative to their size (Figure 1.2) (Griffiths, 1980). For example, S&W green sunfish take larger prey than do WF bluegill sunfish (Werner and Hall, 1974). Aside from being small, the prey of WF animals is also likely to be relatively defenceless; since searchers expend considerable energy in locating prey, little energy would be available for handling. Animals that consume prey with large handling or pursuit costs, tend to consume prey that are fairly large, because small prey would not be worth chasing or handling.

Schoener (1971) differentiated between time-minimizers and

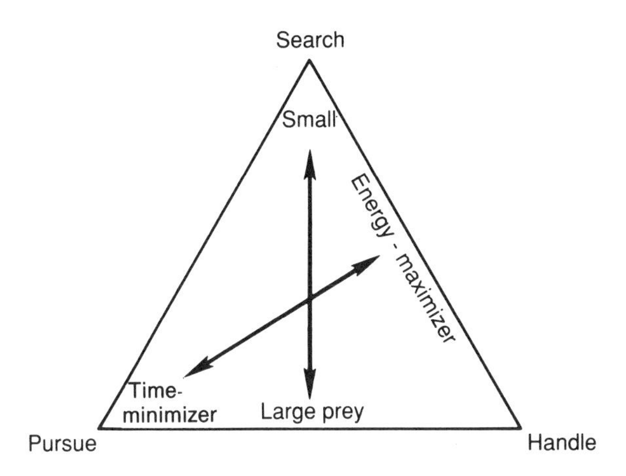

Figure 1.2 Relationships between prey size, time-minimizer and energy-maximizer tendencies, and proportion of time spent searching, handling and pursuing. (After Griffiths, 1980.)

energy-maximizers. Time-minimizers require a certain basic energy input, and once this has been achieved they stop searching for resources and use the remaining time for other activities. Energy-maximizers try to obtain as much of the available energy as possible. Since both sorts of animals are supposed to gather resources at the maximum possible rate while they are gathering food, the simplest way to identify them is by the amount of time they spend searching. Thus shrews, small birds in winter, and larval insects feed nearly all of the time and should be classified as energy-maximizers. An extreme energy-maximizer should consume every potential capturable prey encountered, with a relatively low prey selectivity. Time-minimizers, such as lions, pythons and praying mantids, are generally more selective, restricting their foraging to those prey that will be the easiest or least expensive to catch. A parasite, such as a tick, might represent the extreme of a time-minimizer, waiting for long periods of time for a host to walk by. The searcher–handler–pursuer axis more or less corresponds to an energy-maximizer and time-minimizer continuum (Figure 1.2) (Griffiths, 1980). Thus, energy-maximizing predators feeding on small prey must catch many items to satisfy their energy requirements and consequently spend most of their time searching where their prey are located in abundance. Animals spending considerable energy on handling, such as web-building spiders and ant lion larvae, are probably energy-maximizers. Species investing in pursuit, such as hyaenas and wild dogs, are time-minimizers. They are social, and the killing of large prey provides food for several days, thus releasing time for other activities.

1.1.3 Generalists versus specialists

A third important factor in sorting among search types is the range of resources within a given resource type that is acceptable to a species. A resource type might be food, and an acceptable range might be fruit, with certain fruits being more acceptable than others. This constraint reflects the physiological abilities of an animal initially to recognize a resource as food, to digest and utilize the nutrients in a type of food, and the ability to locate or chase down the food item.

Thus, animals can be classified as generalists or specialists according to the number of different kinds of resources (within any resource type) they use. Just as with the concepts of WF/S&W predators and energy-maximizers/time-minimizers, the categories of generalists/specialists are ends of a continuum, with most species somewhere in between. Although high efficiency in using resources can be achieved only if an animal exploits a few kinds of resources, the extent to which a species can afford to be a specialist depends on the resources that are available and on the number of species using a given set of resources. Generalists often occur where or when they have few competitors, either as a result of low resource availability or because a species becomes limited to a restricted area such as an island. The specialist-generalist characteristics of a species can change in the short term, as when an animal accepts resource A if it cannot find B, or with evolutionary time, as when resource B disappears from the habitat and individuals of future generations become adept at finding resource C. In fact, a substantial body of literature, discussed in other chapters, shows that individuals change position within various resource ranges as their internal needs change. One of the objectives of this book is to elucidate the proximate mechanisms by which individuals are able to shift acceptance criteria in response to changes in resource availability.

1.2 RESOURCE AVAILABILITY AND DISTRIBUTION (EXTERNAL CONSTRAINTS)

The way that resources are organized or become available in time and space constitutes the major environmental constraint on searching success and is an important selective pressure on efficiency in search behaviour. At this point we should attempt to delineate the nature of spatial and temporal resource distributions in the environment, and try to relate the searching movements of animals to these distributions.

1.2.1 Hierarchical levels of the spatial environment

Hassell and Southwood (1978) classify hierarchical levels of how animals perceive their environment, recognizing that 'any framework is bound to be plagued with exceptions and examples of blurred boundaries'. Their classification includes resource items, which are the individual prey, leaves, individuals of the opposite sex, or hosts; patches, which are aggregations of resource items or spatial subunits of the foraging area in which aggregations of food items occur; and habitats, which are clusters of patches (Figure 1.3).

These spatial units provide a framework for searching behaviour such that we can usually distinguish the movements of animals between habitats, patches, and resources from the movement of animals within habitats and patches. Both within and between types of movements are involved in search behaviour, as animals locate resource units and then find resources within these units.

(a) Resources

A resource can be defined as a substance (water), object (potential mate, shelter), or energy source (food) required for normal body maintenance, growth and reproduction (Ricklefs, 1979). Time and space should also be considered resources. Using food as an example, the resource set for a given species can be ascertained as follows. First, observe animals in their natural habitat over the long term, and assemble a catalogue of the actual resources utilized or remnants of them (seeds, peels); second, apply physiological criteria in the laboratory, whereby fecundity, growth and reproduction are measured relative to the utilization of various potential resources.

Distributions Resources can be distributed in various densities and patterns, and these distributions become an important environmental determinant affecting searching success and search behaviour itself. Random distributions occur when each location in space has an equiprobable chance of having a resource in it (Figure 1.4a).However, components of the environment that influence resource distribution are themselves not usually randomly spaced, and so resources are seldom dispersed randomly. Uniform or regular distributions occur when resources are either placed regularly or become so because of regular distributive or selective influences (Figure 1.4b). For example,

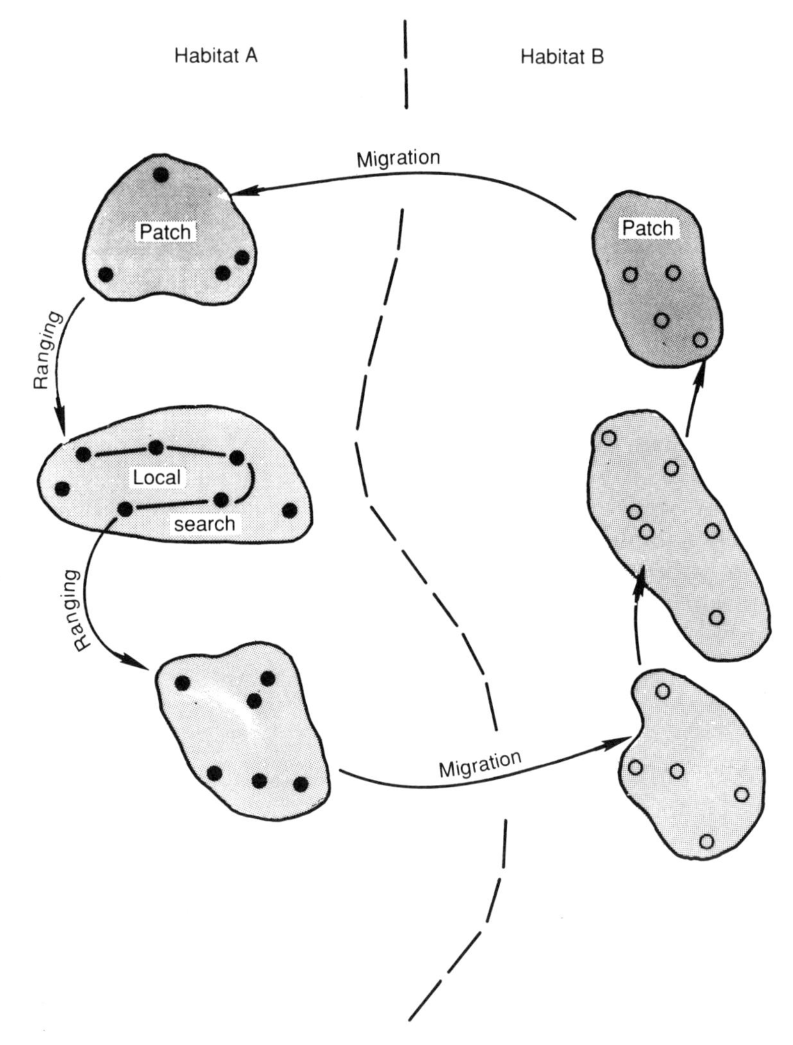

Figure 1.3 Spatial resource units of the environment and types of movements within and between these units. (After Hassell and Southwood,1978.)

individual plants may secrete toxins reducing the viability of nearby plants (allelopathy) or there may be competition for light or other abiotic factors.

Most resources discussed in this book are found in patchy distributions (clumped in space) (Figure 1.4c and d), a result of nonrandom and nonregular seedling establishment, dispersal of seeds or offspring, patterns of oviposition, variations in soil and microclimatic conditions, movement abilities, vegetative growth of plants, or mutual attraction of individuals into aggregations. Because resources are

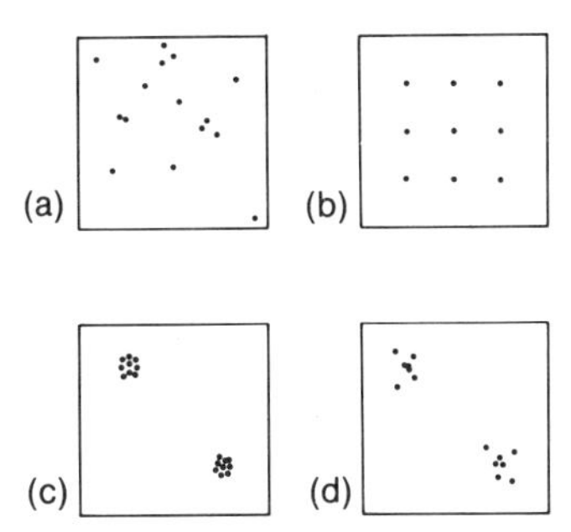

Figure 1.4 Spatial arrangements of resources in patchy and non-patchy habitats. (a) non-patchy, random, (b) non-patchy, regular, (c) patchy, high intensity, (d) patchy, low intensity.

commonly arranged in patches, the search behaviour of animals is designed to locate a patch and then to search within a patch.

Variability If resource units in the environment were really organized as neatly as described above, searching for resources would be a simple task. An animal would merely have to employ the search mechanism that best works on a given kind of distribution. As noted by Bateson (1983), however, 'nature is not necessarily going to package herself conveniently to match our distinctions'. In fact, there are many sources of variability which directly or indirectly affect searching, including differences in or changes in resource units owing to abiotic factors, resource accessibility and usability. Resource availability must therefore be measured not only in terms of abundance relative to the size of the user population and the resource requirements of individuals in the consuming population (Hassell and Southwood, 1978), but also in terms of real accessibility. For example, host plants can be cryptic or hidden by the foliage of other plants, or they may be camouflaged by the odours of nearby plants. If the resource is not accessible, it is essentially unavailable.

(b) Patches

The term patch has been interpreted in various ways (for a review see Wiens, 1976), in part because a patch cannot always be precisely specified. Thus, Hassell (1978) is driven to define a patch as a 'spatial unit of the predator's foraging area – one whose appropriate dimensions are not set by what we perceive, but by the predator's foraging behaviour itself'. Thus, we might define a patch as a local area with a relatively high probability of resource encounter surrounded by areas where the probability of resource encounter is at or near zero. This idea of seeking changes in an animal's behaviour as a means of defining a patch is appealing, because it is less arbitrary than attempting to discriminate between groups of resources according to the human eye. Nevertheless, many investigators have applied some kind of discriminatory function to delineate patch boundaries; for example Kipp (1984) defines a patch in studies of honeybee foraging as all flowers or inflorescences within a population that are closer to at least one other member of that population than they are to other flowers or inflorescences of other populations. This is a case where an animal with keen visual acuity probably treats a group of yellow flowers with few other yellow flowers nearby as a resource unit. It is quite likely that many species of animals know a patch as a particular point in time when resources are encountered at an acceptable rate, rather than as a particular area in space. If animals really do perceive patches in a temporal rather than a spatial sense, the idea of spatial boundaries may be inappropriate and we should be thinking in terms of temporal units. As this book proceeds, the importance of the temporal dimension will become clear.

Patches, or individual resources within a patch, can themselves be spaced randomly, regularly or in clumps, as described above for individual resource distributions. In addition, intensity (Figure 1.4c and d) is a measure sometimes used to quantify differences in density of resources between the interior of a patch and the areas between patches. Homogeneous, dispersed or uniform resources are those with relatively low intensity or a regular or random distribution.

(c) Habitats

A habitat is the kind of place where a species of animal normally lives. The word kind refers to the ecological requirements necessary for normal life, and the word normally leaves room for species that spend considerable amounts of time outside their habitat. Two habitats used by a population may be thousands of miles apart in animals that migrate long distances, or in quite close proximity in absolute distance units. Habitats differ in size and complexity, may be more or less discontinuous, and can differ greatly in the quality of resources or protection afforded to an animal. They can also change over time on a global scale with regard to long-term climatic fluctuations, and on a short-term schedule as in seasonality. The landscape of a habitat may influence search behaviour of foragers in a direct manner by affecting search paths or speed of movement, thereby rendering some resources less accessible than others (Price *et al.*, 1980).

If we look at a habitat as a place where a species of animal normally lives, for a herbivore such as a butterfly, a habitat may simply be a woodland area, perceived according to gross vegetation type; the butterfly may sometimes fly to open fields to collect nectar from flowers, but then it returns and spends most of its time in the woodland. The patch is then the clump of vegetation which provides either the larval or adult food, and the resource items are the leaves or flowers within the patch. The habitat of an aquatic insect is a particular type of water body (fast-moving stream, still pool) within which some patchiness might occur. Habitats of blood-sucking insects may be woodlands, recognized by visual signals, light intensity and humidity. The host animal could be defined as a patch, recognized by odours. The food item is the actual feeding site on the host, localized by a specific temperature range or substances in the blood.

1.2.2 Temporal periodicity

Resource availability can vary in time as well as in space. An analogy to the hierarchical spatial units of habitat, patch, and resource might be a season for a given resource type, within which there is circadian variability (flowers opening in the day and closing at night), and fluctuations during the period of availability (nectar increasing from morning to afternoon and then decreasing toward evening). Thus, the temporal units are time periods which may vary predictably or unpredictably.

The availability of resources may change according to predictable cyclic phenomena correlated with daylength, daily or circadian cycles, tidal cycles, or annual cycles. Resource availability may also be correlated with certain periodicities that are not easily related to abiotic factors. For example, tropical fruits often ripen synchronously on any given tree, but different individual trees of the same species frequently produce fruit at markedly different times. Availability of prey may also be linked to interacting cycles of predator and prey populations, such that the numbers of prey affect the numbers of predators in a frequency-dependent manner.

In temporal distributions, one season can vary considerably from year to year with respect to rainfall, temperature, and wind patterns, as well as in biotic factors such as relative numbers of predators and parasites. Even during one season, day-to-day variation in biotic and abiotic conditions can affect resource availability. An important characteristic of changes in resources, that affects searching success, is the temporal pattern. Price (1984) examines the rate of change and manner of change, and divides temporal changes in resources into the following types:

1. Resources that rapidly increase, such as temperate foliage, to which consumers must respond equally rapidly.
2. Pulsing (ephemeral) resources, such as flowers in temperate regions for which pollinators must be available during the pulse of flowering.
3. Steadily renewed, such as gut contents available to internal parasites, whereby resources decline but are then renewed.
4. Constant resources, which are mostly physical in nature, such as various types of space.
5. Rapidly decreasing resources, such as nesting sites or seeds in the desert, requiring foragers to increase rapidly their searching activity to locate declining resources.

A dynamic example of temporal changes in prey densities and spatial relationships is the distribution of intertidal isopods (*Excirolana linguifrons*) and larvae of sandcrabs (*Emerita analoga*) that are preyed upon by wading birds such as sanderlings (*Crocethia alba*) along shorelines (Figure 1.5) (Myers *et al.*, 1981). Much of the variation in isopods and sandcrab larvae is caused by the sensitivity of marine invertebrates to wave energy, beach profile, and particle size. Mapping the distributions along the beach showed significant differences in 17 of 40 between-week comparisons of isopod density and in nine of 40 crab larvae comparisons. In only two of ten sites did densities remain unchanged for both prey species throughout a five-week sampling period. These kinds of changes are related to rapidly varying physical conditions on the beach, and illustrate the degree and rate of change of prey conditions which sometimes occur.

1.2.3 Relationships between resource distributions and searching

How are spatial and temporal distributions of resources manifested as constraints on searching activity and searching success? First, because of the hierarchical nature of spatial distributions of resources we would expect to find animals initially using gross cues, indicative of certain habitats, then homing in further using patch cues, and finally employing cues from individual resources. Hence, search behaviour at the habitat level can affect search at the patch, and search at the patch can affect search for individual resources. For example, Roitberg *et al.* (1982) showed that the success and experiences of apple maggot flies (*Rhagoletis pomonella*) in a tree can have a marked effect on the behaviour of the fly when it finds a fruit cluster.

Second, the periodic nature of temporal distributions means that animals must be able to restrict search to the most productive times if they are to locate resources efficiently. Controlling the temporal

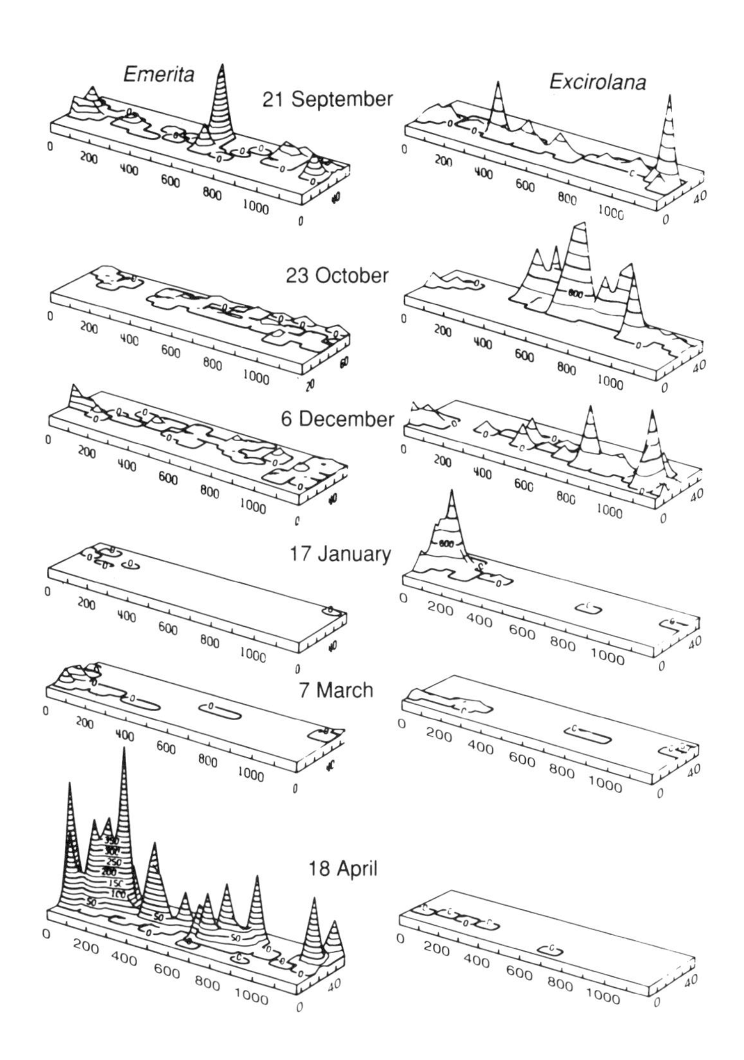

Figure 1.5 Spatial distribution of invertebrate prey of sanderlings (*Crocethi alba*) on an outer coastal beach near Bodega Bay, California. The diagrams are maps of a beach area 1200 × 60 m on six dates in 1977–78. The viewer sees the surface from land looking out to sea. The long axis runs along the beach, while the short axis is cross-tidal. Numbers of crustacean prey are depicted in three-dimensional contour plots (each contour line for *Emerica analoga* equals 600 animals per m^2; for *Excirolana linguifrons* 300 animals per m^2). From Myers *et al.*, 1981. Copyright 1981 Garland Publishing, Inc., reprinted by permission.

nature of search activity often depends on endogenous timing mechanisms or on behavioural mechanisms such as following other individuals that are finding resources. These proximate mechanisms are important for understanding how animals link into the temporal availability of their resources.

Third, we can distinguish between movements among and movements within resource units (i.e. resource, patch, habitat) (Figure 1.3). In fact, movements between all of these units have similar characteristics, such as relying upon environmental cues or topographical contours so as to move relatively straight. Movements within the units are often quite similar also, and can usually be characterized as mechanisms for restricting search to the unit. For locomotory movements and scanning within a patch, we apply the term local search (Jander, 1975). When an animal leaves a patch or a resource and seeks others, regardless of its orientation mechanism (using sun compass, topography, or landmarks), it will be designated as ranging. The term ranging was previously used by Jander (1975) to describe search orientation in the absence of external sensory information. The term migration will be applied to movements between habitats.

1.3 EFFICIENCY AND OPTIMIZATION

'Any organism has a limited amount of time, matter, and energy available to devote to foraging, growth, maintenance, and reproduction' (Pianka, 1988). It is intriguing, therefore, to ask how organisms maximize the use of time, matter, and energy. We can examine this question with respect to the costs and benefits that are incurred in searching behaviour. The benefits include the food, mates, or oviposition sites that are located, and which can be utilized for growth and reproduction. The costs include the energy expended on movement, the risks of predation while engaged in searching, and the time taken away from other activities such as holding territory or protecting nests. Thus, we assume that natural selection would favour searching activities that maximize the difference between searching costs and benefits. The following section introduces the issue of a balance between adaptations promoting efficiency of operation and the constraints within which an animal must operate.

Efficiency is a measure of work accomplished per unit energy consumed, such that efficient search implies locating, accurately assessing, and utilizing resources with the least cost or in the minimum possible time. An examination of a typical study of searching behaviour may provide insight into the importance of efficiency on individual fitness. Individual fitness is defined here as some measure of reproductive success, or number and quality of offspring left for the next generation. In larvae of the coccinellid beetle (*Adalia decempunctata*) which prey on the nettle aphid (*Microlophium evansi*), there is a correlation between the numbers of prey eaten per unit time (a measure of searching efficiency), and the number of eggs successfully oviposited in the right place (a measure of fitness) (Dixon, 1969; Carter and Dixon, 1982). Searching efficiency can be summarized as follows:

1. The female oviposits beside an existing aphid colony. Her search abilities and site preferences translate directly into larval survival, since newly emerged nymphs cannot move very far in their quest for prey.
2. The more efficient are larvae in capturing aphids, the shorter the period for each instar, and thus the faster is the transition to larger adults.
3. The combined foraging efficiency of larvae and adults is proportional to the number of eggs that are ultimately deposited by females at appropriate sites. Thus, other factors being equal, efficient individuals leave the greatest numbers of offspring.

The example of the coccinellid beetle implies that perhaps organisms have been nearly perfectly designed for carrying out the tasks they need to perform. This issue is addressed by optimal foraging theory (OFT), a set of ideas that attempts to formalize the role of natural selection in ultimately determining searching efficiency. It asserts that natural selection is a 'maximizing process and penalizes individuals deviating from the optimal design for their environment' (Krebs *et al.*, 1978). Like 'game theory', OFT is really more of a 'theoretical approach' than a theory *per se* (Stephens and Krebs, 1986), whereby the adequacy of particular hypotheses is tested to account for the evolution of given structures or behavioural traits (Maynard Smith, 1978). The idea is to test a specific mathematical or graphical model, and to accept or reject it. 'An optimality model is a functional account of behaviour and as such does not attempt to describe any particular mechanism' (Krebs *et al.*, 1978). As we shall see, however, many proponents of OFT cannot resist describing the particular proximate mechanism that seems to allow an animal's behaviour to conform to a given model, and these transgressions have occasionally assigned to animals certain unrealistic abilities. One important goal of this book is to elucidate proximate mechanisms that pertain directly to theories set forth by OFT.

A wide variety of studies support the contention that OFT is a valuable tool in stimulating, and in some cases interpreting quantitative studies of adaptation relating to diet selection, territoriality, conflict, cooperative systems, mating patterns, temporal sequences of behaviour, and searching for food and mates (see chapters in Krebs and Davies, 1984; Kamil *et al.*, 1987). The stimulation afforded by OFT to the field of searching behaviour falls into two categories. First, this controversial or at least different approach has encouraged more research on searching behaviour between 1970 and the present date than would have been accomplished if OFT had not been invented (see Stephens and Krebs, 1986, Table 9.1; Gray, 1986, Appendix 1). In fact, without the impetus supplied from OFT there would not have been sufficient material to write this book. Second, OFT has stimulated discussion that flows across taxa, such that insects and birds are being compared and contrasted, and the unifying theories of foraging that are emerging tend to unearth and disclose many isolated studies that might not have otherwise been linked together in a meaningful way.

OFT has been criticized for a number of reasons, some of which have been discussed in detail, and others that stem from inadequate reading or misinterpretation of the facts (see Pierce and Ollason, 1987; Gray, 1987). The remaining criticisms, which point out certain limitations of the approach relative to searching behaviour, or which demonstrate instances in which the approach has failed in specific ways, are worth mentioning briefly, since OFT examples will be used in several chapters of this book.

First, most theories of optimal foraging have generally assumed that the forager possesses complete knowledge of the quality of available patches and its current rate of energy acquisition (i.e. it is omniscient), and it is assumed to have acquired this knowledge at no cost (Orians, 1981). Orians (1981) remarks that only through learning could an animal really fit the OFT scheme, since omniscience is impossible for any animal unless its environment is absolutely stable: animals must learn about resources in the environment. In fact, in most tests of OFT models the animal has had an opportunity to learn the set-up or to practise solving the problem before data are collected. In nature, animals may not have had such opportunities. It would be interesting to discover to what extent animals really know their environment, and how much learning is involved in finding and assessing resources in nature.

Second, most tests of foraging models to date have considered only mean values, and have not even taken the first step of partitioning variance into components due to inter-individual and intra-individual differences (Cheverton *et al.*, 1985). In testing any theory one needs to determine how much deviation from predicted behaviour is compatible with the assumption that the animal is really following the rules specified by the theory. One of the most interesting conclusions we can draw from the coccinellid example mentioned previously is that while averages can be calculated to represent the population under study, the sources of variability in beetle foraging are substantial (although probably typical). Females lay some eggs away from prey, placing a nearly impossible burden upon young larvae to find food. Although the motor pattern of local search seems to be 'hard-wired', starvation modulates this pattern and thereby increases the duration of local search; thus, starved larvae tend to leave a plant sooner than unstarved larvae, and they incur greater risk, gambling on finding better plants in their environment. In addition, response to starvation probably varies in part because of differences in individual thresholds and the way that starvation affects searching behaviour. Because of physiological factors, such as the effects of starvation, individuals change from minute to minute, making it very difficult to define optimal behaviour for any individual and to average across individuals.

Variability can also be adaptive in the long term. Take, for example, patch duration of coccinellid larval searching: some individuals depart a patch too soon, leaving aphids behind when the point of diminishing returns has not as yet been reached, whereas others stay too long, spending unwarranted amounts of time and energy searching for the few remaining prey; the mean performance of the population tends toward the optimal time for searching in a patch relative to the costs of leaving and locating another patch of aphids. However, under conditions of low aphid density in the entire habitat, where patches would be scarce and aphid density within patches would be low, the individuals with traits for staying longer would have a distinct advantage over those leaving quickly. Thus inter-individual variability in searching traits is important in the long term, while selection pressure channeling toward an optimal mean for the average set of conditions is important as well.

Third, Toates and Birke (1982) suggest that '. . . the excitement generated by optimal foraging theories . . . has not produced better understanding of the underlying perceptual and motivational processes. Optimality and decision making are discussed in a way that merely assumes the existence of the necessary causal mechanisms. . .'. Note that Toates and Birke do not argue that causal models must be 'physiological' in the sense of the nuts and bolts of neural details, but that much can be gained by at least making models physiologically viable in that they conform to the reality of physiological evidence. I think this is also the complaint of Heinrich (1983) when he warns that there are already enough real complexities in what animals can really do, and so there is no need to make them appear more complex than they already are.

Finally, as Futuyma (1983) has so succinctly stated, 'The possible adaptive responses that a species may mount to environmental challenges are severely constrained by the properties bequeathed by its ancestors. . .'. Each of the functional systems of an organism has, in this view, severe developmental constraints on its evolutionary malleability (Gould and Lewontin, 1979). Organisms are never as optimally adapted as an engineer might design them to be. Moreover, the differences among species are not necessarily special, optimal adaptations to the particular environments that each species occupies; they may be alternative adaptive solutions to the same environmental problem, and the solution adopted by a species is a consequence of the historical accidents of environment and genetic composition suffered by its remote ancestors.

1.4 TRADE-OFFS: COMPETITION AND RISK

Searching for and utilizing resources must ultimately represent a trade-off in the effort and time expended between those activities and other important fitness-related activities. In other words, an animal searching for a resource could be extremely efficient, but because of an alert predator, could end up dead instead of maximizing reproduction. Risk of predation has obvious consequences of reduction in fitness. Time/energy spent searching may also affect fitness negatively if it is time and energy taken away from other fitness-strengthening activities. Experimental studies of time-budgeting conflicts between searching and other activities in a wide variety of animal species include those of foraging versus vigilance/predation risk (e.g. Caraco *et al.*, 1980a,b), foraging versus territoriality behaviour (e.g. Kacelnik *et al.*, 1981), and mate-finding versus predation risk (e.g. Thornhill, 1984). A major objective of this book is to determine how animals adjust their searching behaviour so as to find a compromise between time expended on various important activities.

The two most important kinds of risks are predation or parasitism, generally by individuals of a different species, and intraspecific competition when the use of a resource by one consumer reduces the amount of resource available to another. An evolutionary response to predation pressure might entail selection for crypticity or mimicry, or the adoption of risk- minimizing tactics such as adjustments in searching periodicity or working in groups. Group effects in particular can confer benefits as long as mutual vigilance allows each individual in a group to spend more time on other behaviours than on vigilance. For example, yellow-bellied marmots (*Marmota flaviventris*) foraging on montane talus slopes (Svendsen, 1974) spend less time watching for predators and yet respond more quickly to a predator when in a group than when alone. Intraspecific competition for the same resource by crowded individuals of the same species can reduce the amount of resource available per individual, or it can simply make searching or resource utilization difficult because of interference between individuals. In sexual behaviour, for example, competition among conspecific males often detracts from time that would be better spent searching for females (Parker, 1978). Large numbers of individuals at a resource can also directly confer benefits, rather than having negative effects, if many individuals can find or use a resource more efficiently than could one individual. Efficiency in hunting prey increases when prey are flushed by other group members (the 'beater' effect) (Rand, 1954). Some species of sea birds encircle a school of fish and drive them ahead of the flock (e.g. Emlen and Ambrose, 1970). Working in groups or packs is also adaptive when predators attempt to overcome the prey defences of large animals. In some species such as bark beetles, too low a density of individuals on a unit of resource may have detrimental consequences to fitness; here selection should favour some degree of aggregation, short of overcrowding. In general terms there is usually an optimal density range of individuals per unit of resource for a given species. An evolutionary response to intraspecific competition might be the adoption of a broader spectrum of resources by some members of the population. For example, a herbivore's fitness may be higher if it expands consumption to include less suitable, but more easily located host plants, rather than deferring all consumption until the most suitable plant is found.

An example of a response to both predation and intraspecific competition is the 'adoption' of a tactic that minimizes risk. Some species may select the most profitable site (e.g. Royama, 1970), finding and remaining on resources that are concentrated, even though they risk competition and predation. Others may 'spread the risk' (e.g. Root and Kareiva, 1984), whereby searching animals are skilled at locating scattered resources, and consequently gain access to resources that are poorly exploited; they may also escape intraspecific competition and the predators that build up in large numbers in dense resource patches. As stated for the 'spreading-the-risk' tactic, when survivorship is variable in space, individual females can gain a selective advantage by spreading offspring among many different independent spatial locations. Thus, by adapting to host sparsity, animals might gain access to resources poorly exploited by other species and simultaneously escape enemies that accumulate in large, dense host patches. Those species that seek out isolated hosts may not be at such a disadvantage as might be expected from the costs of searching.

These points emphasize the need to consider risk factors in

evaluating efficiency of searching behaviour, and suggest that along with selection for efficient search mechanisms there will be selection for trading off search time for risk-minimizing activities or for grouping or other means of reducing risk while searching. Moreover, strategies for reducing predation and intraspecific competition can differ among species, and a single strategy can acquire a dynamic quality supporting aggregation up to a critical number of individuals and then dispersal when the number exceeds that critical number.

1.5 PLASTICITY IN SEARCHING BEHAVIOUR

Over the past decade ethologists have become increasingly aware of the significance of differences in the behaviour of individuals of the same species, and differences in behaviour of one individual at different points in time or under different conditions. Such variability may be continuous, whereby individuals differ in a continuous manner for some behavioral trait such as search speed, or the variability may be discontinuous, whereby some individuals always search slowly and others always search rapidly. Continuous variability is obvious in every data set, and is usually expressed as the deviation from the mean. To a large degree, variation in search behaviour reflects the variability in resource quality, quantity or spatiotemporal distribution.

The terms tactic and strategy are used to delineate the behaviour of an individual and to specify the discontinuous nature of the variability. A searching tactic is defined here as a specific action comprising a group of related and often sequential behaviours, which, when successful, lead to the localization of a resource that increases an individual's fitness. A searching strategy is a 'set of rules stipulating which alternative behavioural pattern will be adopted (or with what probability) in any situation through life' (Dominey, 1984). A strategy then is the overall scheme by which an animal can use the appropriate tactic, depending on its genotype, age, size, or on environmental condition.

Discontinuous variation among individuals in tactical behavioural alternatives has been formulated into the theory of evolutionarily stable strategies (ESS) (Maynard Smith, 1982, 1984). An ESS is a particular strategy, which, if followed by most members of a population, cannot be bettered or invaded by an alternative strategy. If a better alternative evolves, or if environmental pressures change to favour a different strategy, a new strategy may become the ESS. Dominey (1984) suggests that individuals may adopt several tactics probabilistically, e.g. 'guard' with a probability p, 'sneak' with a probability q, or else individuals may be randomly assigned permanently adopted tactics (guard or sneak) with probabilities *p* and *q*, or individuals may adopt a strategy in which the tactic employed depends on the environmental situation, or some characteristic such as age or body size. For example, a pure strategy might be to 'guard' in situation A, and 'sneak' in situation B, or to 'patrol' if aged less than 2.0 years (or if body is small), and 'search actively' if aged more than 2.0 years (or if body size is large). As an individual reaches 2.0 years of age and switches from patrolling to searching, this juncture is termed the switchpoint. In later chapters of this book it will become clear that the overwhelming majority of examples of alternative tactics are derived from environmental cues initiating different sets of events in an equipotent individual.

It is extremely important to distinguish quantitatively between the tactics of individuals in a population under study. Consider the following pitfalls related to animal ontogeny (Caro and Bateson, 1986).

1. 'Young' and 'adult' are typical classes used to discriminate among individuals. However, it is not always easy to distinguish between these classes, and in many species there may be a continuous gradation between age classes, or the age at which individuals switch tactics may depend on local demography brought about by chance events or other factors.
2. There may be more than just two tactics at any given period of measurement. This is shown in scorpionflies of the genera *Hylobittacus* and *Panorpa*, where males of both genera have three distinct ways of respectively securing nuptial prey and mating with females (Thornhill, 1979, 1981).
3. Aging adults in many species are increasingly likely to employ alternative reproductive tactics, such as sneaking copulations, as their competitive ability declines.
4. Phenotypic differences within an individual may not be stable; for example, during the course of one day, individual golden-winged sunbirds (*Nectarinia reichenowi*) switch rapidly between defending nectar-rich flowers and forming dominance hierarchies with no territorial defence (Gill and Wolf, 1975). Differences among individuals in a population also may not always be distinct, or may be difficult to measure.
5. Finally, and most critically, measuring the presence or absence of a behavioural alternative will depend on the precise definition of the functionally equivalent alternatives. For example, continuous tactics are probably quite common in nature (e.g. Parker, 1984), but once they have, for convenience, been arbitrarily divided up into categories such as 'stay for longer than the median stay time'

or 'stay for less than the median stay time', their continuity is immediately obscured.

1.6 SUMMARY AND CONCLUSIONS

The ultimate success of an animal's searching depends to a large extent on

1. the method an animal uses to search in relationship to the availability of resources and their spatial and temporal distributions in the environment,
2. its efficiency in locating resources, and
3. the ability of a species to adapt to long-term changes in resource distributions and the ability of an individual to respond, perhaps through learning, to short-term changes.

Given that at least part of an individual's searching phenotype has a genetic basis, we should expect natural selection to operate on whatever variability there is in perceptual and motor traits, the information-processing mechanisms that underly behavioural decision-making, and response times. Thus, to some extent evolution will hone and sharpen an animal's ability to locate and identify efficiently the resources it needs.

PART TWO

Information for the localization and assessment of resources

A little green worm came crawling over a dewy leaf, lifting two-thirds of his body into the air from time to time and sniffing around,' then proceeding again – for he was measuring, Tom said.

Tom Sawyer, by Mark Twain

If there is a limitation on the amount of time and energy that an animal can allocate to searching for resources, and if resources are not usually distributed in a completely predictable manner, then an animal must search in the right place at the right time to locate resources efficiently. An animal searching for resources really has two distinct problems: it must assess and it must localize the resource. Localization is brought about by orientation, whereby an animal moves towards a resource according to cues associated with the resource itself or with the resource site, or according to information from the abiotic environment. Assessment means that an animal perceives the pattern of an auditory signal, visual pattern, chemical composition, or some kind of special cue characteristic of a particular resource, and based on this information 'decides' whether or not to seek or to utilize the resource.

2

Orientation cues: information for searching

Spatial information used in orientation (review: Bell, 1985) can be delineated into the following categories:

1. external sensory information, perceived from the external environment as an animal searches, typified by perception of visual or chemical cues;
2. internally-derived or internally-stored information, transmitted from proprioreceptors in or near locomotory organs and stored as an individual moves, or acquired through experience learning, or through heredity, characterizing the genotype of the individual or species; and
3. internal stochastic influences or noise.

Two terms should be differentiated at this point: orientation, which is the mechanism involved in the recognition and maintenance of direction, and navigation, which is the mechanism involved in the identification of the direction of a given point in space (Baker, 1978). In orientation, if an animal is displaced laterally, it continues in its original direction as if it had not been displaced. If an animal is navigating toward point X, it compensates for a lateral displacement, and continues to move toward X. For example, capture–displace–release experiments have shown that young birds during their autumn and often their spring migrations continue to perform their migration in the standard compass direction for the deme without compensating for lateral displacements. They are therefore performing these migrations by orientation, not navigation.

.1 ORIENTATION BASED ON EXTERNAL SENSORY NFORMATION

External sensory information can derive from the resources themselves, such as plant odour or female sex pheromone, or they may be abiotic factors, such as wind or water currents, or light from

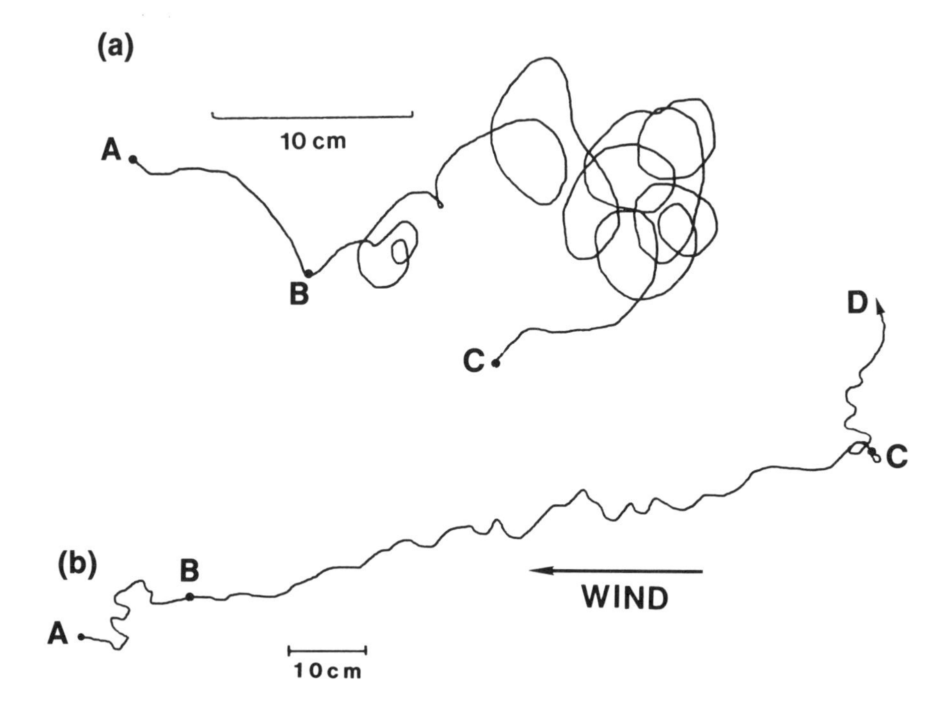

Figure 2.1 (a) Pathway of a male grain beetle (*Trogoderma variabile*) walking in still air (A–B) and after a 2 s puff of female sex phermone at (B). (b) Pathway of a beetle before (A–B), during (B–C) and after (C–D) constant and uniform wind containing sex pheromone. (After Tobin and Bell, 1986.)

the sun or moon. The information may or may not have a directional component.

2.1.1 Non-directional sources

A non-directional cue, such as the call of a predator or a puff of pheromone, from which directional information cannot be extracted, may 'alert' or 'arouse' an animal to the presence of a nearby predator or resource. Arousal may be followed by orientation that improves visual localization or directly promotes localization of a resource or avoidance of a predator. For example, perception of sex pheromone stimulates local search in male grain beetles (*Trogoderma variabile*) (Figure 2.1a) (Tobin and Bell, 1986).

Two principles apply to the use of non-directional external cues. First, the response to non-directional information may initiate a hierarchy of behavioural acts leading toward gathering of more precise information. The response of a variety of crawling, walking, flying and swimming animals detecting a non- directional cue is first to capture the maximum amount of information possible with sensory receptors, then to probe the immediate environment using tactile organs or to investigate visually, and finally to initiate search. This hierarchy allows for a step-wise increase in vulnerability to predation that is proportional to the precision of the information perceived. An animal can generally remain nearly motionless and often hidden while scanning with chemosensory receptors or tactile organs, and will thus be inconspicuous to predators during the initial phase of information gathering. If sufficient stimulation is provided, it may expose itself to potential predators by initiating local search. For example, in the kelp crab (*Pugettia producta*) individual acts of food searching are hierarchically organized (Zimmer-Faust and Case, 1983): low concentrations of food chemicals trigger leg and chela probing, but not locomotion; locomotory search is initiated only after leg and chela probing, but where food is not contacted. Interestingly, starvation has no effect on the hierarchy.

The second principle emphasizes that even though an animal cannot obtain directional information at one point in time from this kind of stimulus, the resulting locomotory pattern and perception of the stimulus over several points in time may enable an animal to find a resource in the immediate vicinity of the site of stimulation. For example, the flatworm (*Dugesia neumani*) increases its rate of change in direction (relative to a control solution) when it absorbs fatty acids from the water (Mason, 1975). The flatworm's turning rate increases linearly with increasing chain length, from a low with pentanoic acid (205.6°/min) to a high with decanoic acid (384.7°/min), correlating with the relative solubility and expected distribution pattern of fatty acids downstream from a prey source. This means that a flatworm would be most likely to localize the region of the least soluble fatty acid, which would normally be found at the site of a resource.

.1.2 Directional sources

Distant orientation, based on directional information sources, is accomplished by moving along a stimulus trail, up a stimulus gradient, or toward a visual object, or by moving up a current which may or may not carry a stimulus. In many species the efficiency of searching is improved through specialized adaptations that amplify or generate directional information as with sonar in bats for locating moths (e.g. Griffin, 1958), through social facilitiation as in trail pheromones of foraging ants (review: Wilson, 1971), or by special scanning mechanisms (see Chapter 3).

(a) Biotic cues

Abiotic directional information occurs in different forms. Trails are commonly laid down and later used to locate food and mates. For example, snails, slugs, limpets and chitons use trail following to locate conspecifics (e.g. Chase *et al.*, 1978), their home site (e.g. Rollo and Wellington, 1981) and previously utilized food sources (Croll and Chase, 1980). Another form of directional information is that provided by directional light and sound waves and odour gradients (review: Bell and Tobin, 1982). A chemical gradient can be deciphered so as to localize the source either by comparing the intensities between two sensory organs at one point in time (spatial comparison), or by comparing intensities from one point in time to another (temporal comparison). The direction toward a source of sound can be detected by comparing the difference in arrival of the stimulus at two sensory organs or the difference in stimulus intensity.

Air and water currents carrying a resource stimulus, such as food odour or pheromones, provide stable directional vectors commonly used by animals searching for food, mates and habitats. Males of many insect species walk or fly upwind in a zigzag pattern when they detect female sex pheromone (reviews: Bell, 1984; Cardé, 1984). Male channel catfish (*Ictalurus punctatus*) orient upstream to odour secreted by adult females; when the odour is removed from the current, they follow the remaining odour focus as it moves across the tank (Kleerekoper, 1972). Mississippi fishermen know about this phenomenon, and place mature female catfish in a cage in the river to attract male catfish. Unless an animal is in physical contact with the substratum, the mechanism for orienting upwind or up a water current usually involves perception of the changing visual inputs during movement. For example, moths flying up a pheromone plume maintain the moving ground pattern from aft to rear, and are thus able to fly upwind without directly perceiving the wind direction (review: Baker, 1985). The same is true for freshwater fish moving upstream in clear water, where they can see the ground pattern 'moving' beneath them.

Five general principles pertain to the use of directional orientational information. First, the perception of directional information may cause an animal to switch over from its initial searching mode based on non-directional sensory or internally-derived information, to directional information to control its orientation. Male grain beetles (*Trogoderma variabile*) switch from sex pheromone-stimulated local search to upwind orientation when a wind current is supplied (Tobin and Bell, 1986) (Figure 2.1b). If the the wind current is removed the beetle switches back to local search. Wolves switch over from chemical cues used at long distances from prey to visual cues when they close in on prey at shorter distances (Mech, 1970).

Second, the type of information an animal uses may be dependent upon the context or the immediate environmental conditions. For example, the grasshopper mouse (*Onychomys leucogaster*), a predator of insects, relies on audition to locate moving prey and on olfaction and vision to locate stationary prey (Langley, 1983). The coyote (*Canis canis latrans*) employs vision when light is sufficient, and olfaction and audition in other environmental contexts (Wells and Lehner, 1978).

Third, up to a point, the accuracy of orientation improves with stimulus intensity. This is illustrated by a decrease in turning oscillations and a more precise correspondence between walking direction and sound direction in female crickets (*Gryllus campestris*) with increasing calling song intensity (Schmitz *et al.*, 1982).

Fourth, the precision of directional information varies according to modality, so that animals may switch from less precise to more precise modalities when given the opportunity. Salamanders (*Notophthalmus v. viridescens*) orient toward their home pond by olfaction in the absence of visual cues, but switch to visual cues when such cues are available (Hershey and Forester, 1980). Localizing a resource through olfaction or audition is less precise than by vision, because sound waves, and especially chemical molecules in the medium or substrate, are subject to perturbations as they move. Thus, the information does not arrive as an image, but rather as a temporal pattern of hits on auditory or olfactory sensory cells. On the other hand, because vibrations, and especially chemicals, can move around corners and may be carried in currents, these communication channels have certain advantages over vision: senders and receivers can remain hidden until the signal is properly received, and perhaps answered.

Fifth, most animals integrate directional cues from more than one modality, as in onion maggot flies that find and assess onion plants through olfaction, vision and contact (Harris and Miller, 1982), and salamanders (*Salamandra salamandra*) that are guided largely by the use of visual landmarks, but also by sun and moon compass orientation (Plasa, 1979). Studies dealing with only one modality often inaccurately portray the mechanisms used by animals to locate resources, and they underestimate the real amount of information that may be available to an animal.

(b) *Abiotic cues*

The abiotic environment provides reliable directional information used by many animals in their searching behaviour, especially but not exclusively in homing and migration. These sources include air and water currents, visual landmarks, olfactory, magnetic and auditory cues, the sun, moon, stars, or magnetic fields (reviews: Keeton, 1979; Wallraff, 1981). Directional responses to cues from abiotic sources fall into two general categories: (1) the sun, moon, and star constellations which are discrete objects or point sources according to which an animal can navigate, (2) wind and water currents, which are not derived from a point source, and so animals orient at some angle relative to the current or directly up- or down-current.

Wallraff (1981) lists the following reasons why animals use solar, lunar, and stellar cues for orientation in a terrestrial environment: the sun/moon/stars are prominant landmarks, the long distance of these objects from earth guarantees maintaining a constant angle to them, and theoretically these cues could be used to calculate position on the earth. The disadvantages are that these cues change seasonally with the Earth's revolutions around the sun, daily with the Earth's rotation, and differ at different geographical latitudes. Thus, the information must be integrated with that from biological clocks in order to navigate efficiently and accurately.

2.2 ORIENTATION BASED ON INTERNALLY-STORED OR INTERNALLY-DERIVED INFORMATION

At least some aspects of search orientation can operate in the absence of continuous sensory input from the environment, primarily because spatial cues can actually be stored within the organism. Internally-derived information includes motor commands and proprioceptive signals stemming from motor activity which are coupled to an animal's body position or changes in body position (review: Schone, 1984). Internal information can also be stored genetically or derived from exogenous sources and stored internally (Jander, 1965).

Searching behaviour draws upon information from motor commands or proprioceptors. When an ant forages along a route 30° to the left of the sun, and is displaced for some reason 10° further to the left, it can, in the absence of the sun to guide it, turn back 10° to the right and onto its previous course. The ant has used information derived from its proprioceptors to 'calculate' the error angle, and this information is then 'recalled' for turning back to the original course (Jander, 1957). This kind of information can be used by crabs searching for the entrance to their burrows along a sandy beach: when frightened, fiddler crabs (*Uca rapax*) and ghost crabs (*Ocypode*) (Hagen, 1967) run straight back to their burrow entrance even though their outgoing path was meandering; the return run is based on vector

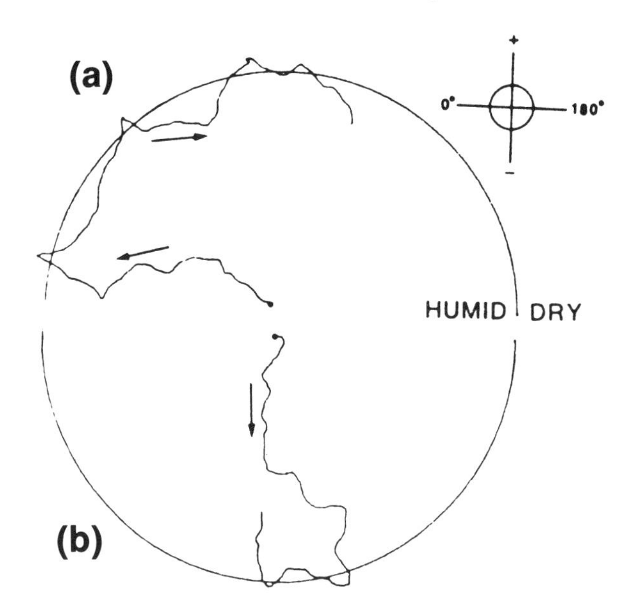

Figure 2.2 Path of an isopod (*Armadillidium nasatum*) inside and outside a simulated circular zone of high humidity. Arrows show direction of movement as the isopod "leaves and returns to the humid zone." (a) and (b) are paths from different individuals. (After Sorensen and Bell, 1986.)

integration of the change in direction and length of each stretch of the outbound path. In other words the crabs perform a fascinating orientiering feat, presumably by keeping track of the distances moved between turns and the turn dimensions, and then calculating the correct angle to turn and the distance to run that leads back to the starting position.

Genetically-stored information allows animals to carry around searching 'instructions' or 'solutions' for solving specific types of problems, just as certain species of animals 'know' how to build their nest without ever having seen one. A simple example is the use of genetically-stored information to solve the problem of remaining within the boundaries of a resource patch. When an isopod (*Armadillidium nasatum*) leaves a humid patch and attempts to relocate it, the external sensory information (the temporal change from humid to dry air) stimulates it to turn and move back into the favourable zone (Figure 2.2) (Sorensen and Bell, 1986). But how does the isopod 'know' how much to turn? Since the turn angle that is usually executed (90° to 180°) is not necessarily correlated with any spatial cue, we conclude that turn angle information is stored genetically and drawn upon when required to complete this kind of manoeuver.

Information used in searching can also be genetically stored as a kind of a 'program'. This seems to be the case for motor patterns of searching that are characterized by sequences of turns (sinusoidal, zigzag) or local search after feeding that decays over time to relatively straight locomotion (review: Bell, 1985). The genetically-stored 'program' would control the sequence of turns, turn dimensions and/or distance or time between these components.

2.3 STOCHASTIC INFLUENCES OR 'NOISE'

Given that how an animal searches is controlled by information stored internally and by information perceived from the internal and external environment, there can be variation embedded in any of these sources of information. Variation could also originate during the execution of the instructions based upon these information sources. Signals produced and stored by the organism, as with motor commands or proprioceptive signals from motor activity, are possible sources of noise, since the signal is produced, stored, and then retrieved; as with gossip, the message is likely to be altered as it passes through neural networks.

Variation that is built into the locomotory activities of an individual can be directly adaptive in several different ways. For example, zigzags superimposed upon looping patterns or various kinds of stop-and-go locomotory modes may be mechanisms for avoiding attacks by predators. Such manoeuvres are called protean defence mechanisms (for example, see Humphries and Driver, 1970). Orienting in a predictable way can be disadvantageous, and so locomotory patterns tend to be stochastic or 'noisy'. However, Curio (1976) points out that

> the predilection of some predators for non-rectilinearly moving prey poses a problem in relation to protean displays, an anti-predator device of many animals. Protean displays partly involve locomotion along an unpredictably twisted path and thus resemble a potent signal eliciting prey capture.

2.4 SUMMARY AND CONCLUSIONS

Animals localize resources by utilizing a combination of externally- and internally-derived information. Animals may switch from less precise to more precise modalities, or from an initial searching mode, based on non-directional sensory or internally-derived information, to directional external sensory information to control orientation. Most animals integrate directional cues from more than one modality.

At least some aspects of search orientation can operate in the absence of continuous sensory input from the environment, primarily because search information can be stored through heredity or derived from exogenous sources and then stored internally. Searching behaviour draws upon information from motor commands or proprioceptors to guide locomotory patterns. Genetically-stored information allows animals to carry around searching instructions or solutions for solving specific types of problems. Variation that is built into the locomotory activities of an individual can be adaptive in eluding predators or increasing the perceptual scanning field.

3
Scanning mechanisms

An important component of search orientation is scanning, the set of mechanisms by which animals move their receptors and sometimes their bodies or appendages so as to capture information from the environment efficiently. Scanning the environment for resources can be accomplished through physical contact or by visual, chemical, or auditory mechanisms. It can be continuous or periodic, and it can occur primarily during movement or primarily during pauses. Scanning for resources is often accompanied by scanning for predators or competitors, such that an animal may run a few steps while scanning for food and then stop to watch for predators, or it may watch for predators while moving and scan for food when stopped. Of importance, how an animal scans is related to the kind of resource it seeks, its own perceptual abilities and movement patterns, the structure of its appendages, and the perceptual abilities of the resource (if prey). Thus, an animal might scan from some distance if its own visual acuity is keen and if the prey are easily startled by nearby moving objects.

The following principles apply to scanning. First, in most cases where data are sufficient, it appears that most animals scan for resources periodically, a process termed saltatory search (O'Brien *et al.*, 1986). No scanning is done while moving, thus eliminating noisy signals that might be caused by movement, and reducing the problem of discriminating between objects that are moving and objects that are stationary. Thus, various species of animals alternate instances of stopping and scanning with periods of moving. For example, Cody (1968) graphs several species of birds (Figure 3.1), showing these kinds of movement cycles. Most probably the ratio of stop-time to move-distance is a continuum, encompassing all species that move to find resources. Second, the interscan interval may be dependent on whether or not the animal is currently finding resources or on some external environmental factor such as turbidity. Third, the reactive distance, over which an animal responds to a resource, may be dependent upon light levels, turbidity and other external envir-

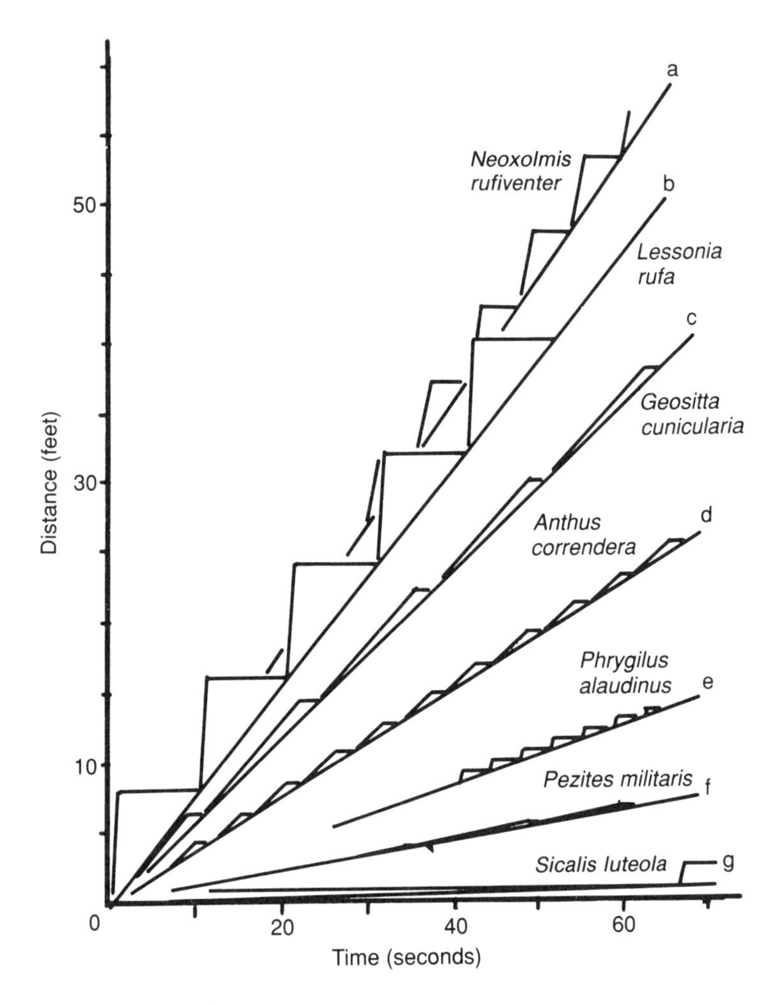

Figure 3.1 Sawtooth curves representing feeding behaviour of several South American Passerine birds from grassland communities. (After Cody, 1968.) Key to species: (a) *Neoxolmis rufiventer*; (b) *Lessonia rufa*; (c) *Geositta cunicularia*; (d) *Anthus corredera*; (e) *Phrygilus alaudinus*; (f) *Pezites militaris*; (g) *Sicalis luteola*.

onmental factors (see Chapter 13), or on hunger and other internal environmental factors (see Chapter 14).

3.1 VISION

Scanning acuity and scanning volumes in visual searchers can be mapped out fairly precisely as compared to animals searching by contact, olfaction, or audition. For many species the investigator need

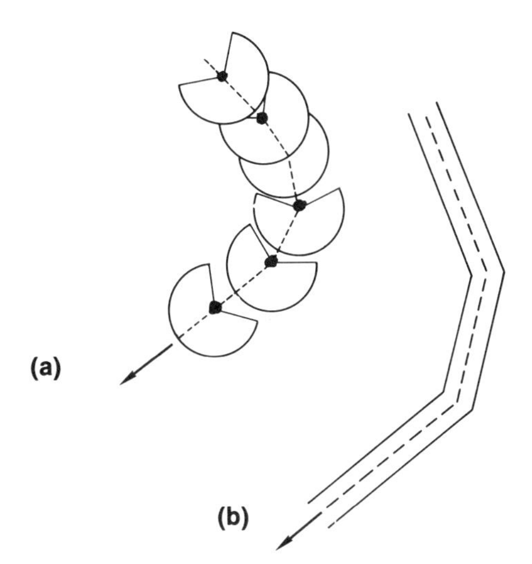

Figure 3.2 Generalized foraging patterns for a bird such as a plover that scans using vision, and a sandpiper that scans using the tactile sense. (a) For the plover, the search area is an approximately constant area (represented as a series of partial discs) with a changeable location. (b) For the sandpiper, the area searched per unit time equals the distance moved in unit time × search path width. Dashed lines represent the path of the bird; solid lines represent the area within which resources can be percieved. Dots in (a) indicate pause positions. (After Pienkowski, 1983.)

only accumulate observations of an animal during stop periods, and measure the distance and direction from the pause position at which an animal moves toward, strikes or picks up a resource.

A particularly good example is the estimation of scanning areas for plovers as the birds pause between runs (Figure 3.2a) (Pienkowski, 1983). The mean length of paces was estimated from cinfilm and from measurements of tracks on mud flats. The angle turned from the arrival position to the direction taken to capture prey was estimated from the films. The angle turned to the left or right was less than 120° in all cases, delineating the limits of perception of the plovers while pausing. From estimates such as these, the scanning area within which prey can be captured was calculated as 0.34 m^2 for the ringed plover (*Charadrius hiaticula*) and 1.03 m^2 for the larger golden plover (*Pluvialis dominica*). These scanning zones can be illustrated as superimposed partial discs representing the sequence of scans (Figure 3.2a). In contrast, birds that locate by contact, such as the sandpiper, have an elongated zone of detection with a width equal approximately to the size of the bill (Figure 3.2b).

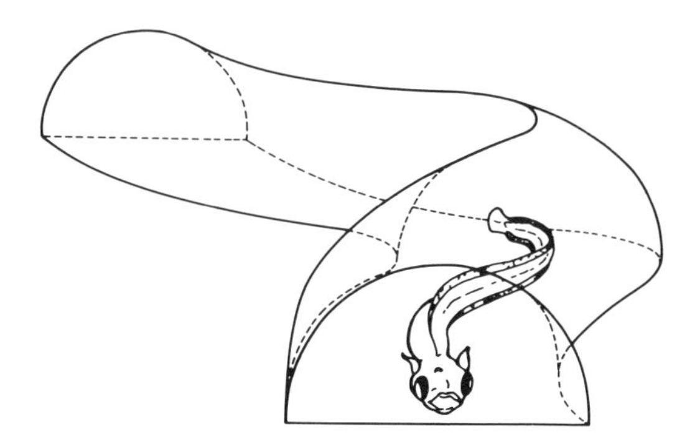

Figure 3.3 'Tube' search by a herring larva while meandering slowly. The hemicylinder represents the reactive sphere of the animal, within which prey can be detected visually. (After Rosenthal and Hempel,1970.)

Perception of prey by herring larvae (*Clupea harengus*) depends on the visual abilities relative to the size and distance of a food item and its visibility (contrast with background, lighting conditions) (Blaxter, 1968). Early investigators believed that the scanning area of the larvae consisted of the 'continuous tunnel' shown in Figure 3.3. However, in light of more recent studies of fish (e.g. O'Brien *et al.*, 1986), the perceptual field may instead be a series of wedges or cones, the apex of each one being the point at which individuals fixate on prey (Figure 3.4).

O'Brien *et al.* (1986) showed that during the scanning pause after a pursuit, white crappie (*Pomoxis annularis*) begin to scan at zero degrees (forward directed) and then scan laterally. Given this mode of scanning, when several prey are present in the visual field, the forward directed prey would be pursued first, simply because it would be the first detected. This prediction is supported by the observations of Dunbrack and Dill (1984) who found a forward directed bulge in reaction distances of coho salmon (*Oncorhynchus kisutch*) and Luecke and O'Brien (1981) who found that bluegill sunfish (*Lepomis macrochirus*) chose the forward directed prey unless off-angle prey were much larger.

The mechanism of scanning in teleost fish seems to be tied to eye movement and focus, which combine to control visual accommodation. This procedure is quite different, but functionally analogous, to head waving in caterpillars or head turning in birds. In fly larvae, where neither the eyes nor the head can be moved effectively, nearly the entire body is moved to the left and then to the right (see below).

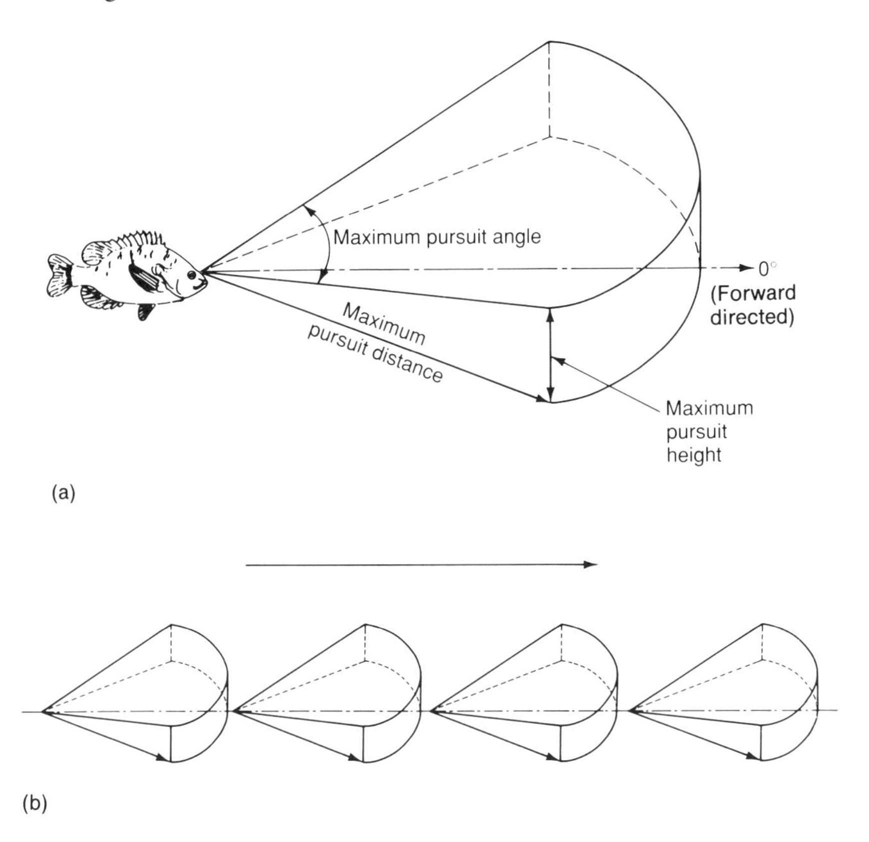

Figure 3.4 Scanning field (a) of a white crappie (*Pomoxis annularis*) consists of a series of wedges or cones as shown in (b). The apex of each cone is the point at which the fish stops and scans for prey. (After O'Brien *et al.*, 1986.)

3.2 MECHANORECEPTION

Some arthropods that locate resources through physical contact scan the environment by periodically casting their bodies or appendages to the left and right of the path. Larvae of certain insect species, such as lady beetles (coccinellidae), nectivorous flies (syrphidae) and lacewings (chrysopidae) stop and 'cast' their body upwards and sideways, thereby increasing the arc within which prey can be contacted laterally to approximately 160° to 220° relative to the forward-going direction (Banks, 1957; Bnsch, 1964; Chandler, 1969) (Figure 3.5). Lepidopterous larvae scan by head waving when stopped at the ends of stems or leaves (Jones, 1977). Presumably head waving allows for both mechanical and visual inspection through an arc of approximately 220° (combining left and right components), and would be

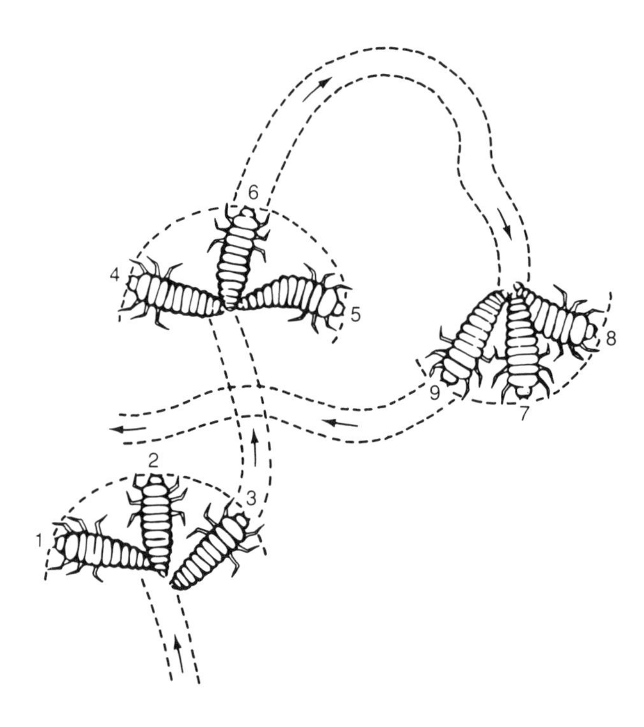

Figure 3.5 Sideways-directed 'casting' movements showing the arc of scaming by a coccinellid larva (*Coccinella septempunctata*). (After Banks, 1957.)

expected to allow the larvae to contact leaves or other structures while searching within a plant.

An interesting example of scanning, which combines vision and mechanoreception, is Hulscher's study (1976) of a captive oystercatcher (*Haematopus ostralegus*). Oystercatchers scan the substrate, and attempt to pierce a cockle located just beneath the substrate surface. In daylight, the bird generally makes single pecking movements when foraging. Each peck consists of one quick movement of the practically closed bill into the substrate. Multiple pecks, consisting of a series of quick up and down movements, occur when the bird remains in about the same place. Localizing prey during daylight is apparently guided by visual stimuli, as in the plovers. The frequently occurring reorientation movements that precede precise pecking into a cockle support this supposition. Mean pecking rate and cockle density were positively correlated, and the yield of pecking, either the number of cockles localized or the number successfully opened, increased up to a density of 150 per m^2. Beyond that density the number of cockles localised decreased, whereas the number opened remained constant. This means that the proportion of localized cockles successfully opened increased with cockle density. At densities of 450 per m^2 nearly one out of every two pecks was successful.

In darkness, cockles are localized by rapid and continuous up and down movements of the bill. A bird takes one or two steps forward, then places its bill into the substrate and walks in a straight line, while the bill makes quick ('sewing') movements in the substrate. This process may proceed for a length of about 150 cm, and if the bird changes direction the bill is retracted from the substrate first and then reinserted after the change of direction.

3.3 CONTACT CHEMORECEPTION AND OLFACTION

Arthropods use their antennae to scan while searching. Tapping the substrate and objects within the scanning range of the antennae extends the effective scanning width considerably. For example, the antennae of the American cockroach (*Periplaneta americana*) provide both tactile and olfactory inputs, scanning the environment in a systematic fashion by tapping the substrate and moving through the air (McCoy, 1984). As shown in Figure 3.6, the two antennae alternate between touching the substratum and sampling air space. In addition to cyclic scanning movements as the cockroach walks, each antenna can be moved independently of the other when the cockroach stops and engages in intensive substratum scanning, or in response to some environmental cue. When the cockroach stops and both antennae operate together, they can scan a volume of air of approximately 80 cm^3 or a surface area of substrate of approximately 60 cm^2.

Local search in houseflies (*Musca domestica*) consists of a rough zigzag path organized into irregular loops (see Figure 17.4b) (White *et al.*, 1984). This tactic, in which the fly swivels its body to the left and to the right, allows it to cover a wider path than if it moved without this irregularity, and so the efficiency of scanning through contact chemoreceptors in the tarsae is improved by the nature of the search path itself. Whereas the body 'width' of the housefly is actually only 3 mm, the effective scanning width increases to approximately 10 mm when the zigzag is superimposed upon the looping path. A fly's successive loops are sufficiently close to one another to allow the fly to scan (by contact with its tarsae) nearly the entire surface area that it traverses.

3.4 AUDITION

The barn owl (*Tyto alba*) can locate and capture its prey by auditory means in complete darkness (Payne, 1971; Konishi, 1973). The owl is

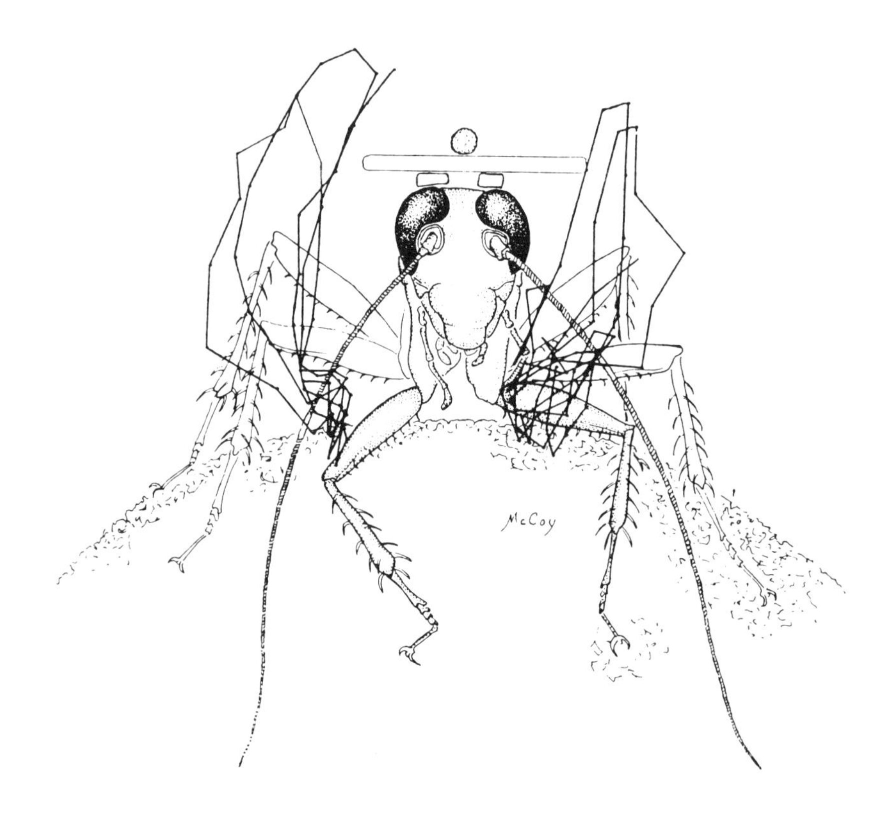

Figure 3.6 Scanning of space and substratum by an American cockroach (*Perplanteta americana*). Lines on each side of the head show the pattern of antennal scanning over several seconds during a pause. From McCoy, 1984, reprinted by permission.

able to locate objects in three-dimensional space, as proven by the ability of one-eared owls to determine the direction but not the distance of potential prey. Removal of the ruff feathers impairs an owl's ability to localize a sound, indicating that the facial ruff may focus the sound like a parabolic antenna.

The eardrum of a frog responds to pressure gradients generated by the phase difference in the sound waves of the external and the internal pathways. This phase difference is a result of the different length of the pathways (Eustachian tubes, and the oral sinuses) leading from one eardrum to the other (Gerhardt and Rheinlaender, 1980). The direction of the sound source can be calculated by the central nervous system based on the difference in stimulation of the right and left tympana (bisensor mechanism). A tree frog approaches a sound source in a zigzag pathway of hops (Rheinlaender *et al.*, 1979). Between hops it moves its head from side to side, apparently getting a bearing on the direction of the sound.

In crickets and related Orthoptera it is the male that produces the calling sound in most species. The tympanal organs of crickets and katydids are located on the forelegs. When the female cricket is aroused by a species-specific song, she emerges from her resting place and scans by turning from side to side. When the direction of the calling male is determined the female begins to walk toward him. The female's walk is interrupted by pauses, which may be followed by a course correction. Although the directional signals are acquired during a pause in some species (e.g. *Scapsipedus marginatus*, *Teleogryllus oceanicus*) (Murphey and Zaretsky, 1972; Bailey and Thomson, 1977), in other species they may be acquired and used in orientation while walking (e.g. *Gryllus campestris*) (Schmitz *et al.*, 1982). The female cricket that orients to sound only when pausing might be designated as a saltatory searcher, whereas the female that orients to sound while continuously moving might be designated as a cruise searcher.

The auditory examples emphasize the distinction between assessment and localization of external cues. The receiving individual assesses the physical properties of the signal (i.e. ratio of pheromone components, sequence of song elements) that are critical for responding to the correct species. Only after the signal has been identified as species specific does the female begin to orient (Zaretsky, 1972). Once localization is initiated, the female continues walking toward the sound even if the song is changed to one that would have been ineffective during the assessment period (Zaretsky, 1972; Weber *et al.*, 1981). That crickets compare signals received by two receivers is demonstrated by the way in which they walk in circles toward the intact side when one of the receivers is damaged (Murphey and Zaretsky, 1972).

3.5 SUMMARY AND CONCLUSIONS

Scanning is the set of mechanisms by which animals capture information from the environment through physical contact or through visual, chemical, or auditory sensory channels. How an animal scans is related to the kind of resource it seeks, its own perceptual abilities and movement patterns, the structure of its appendages (if it has them), and the perceptual abilities of the resource (if prey). Most animals scan for resources periodically, a process termed saltatory search. While the inter-scan interval may be dependent on searching success or on some external environmental factor, the reactive distance, over which an animal can perceive a resource, may be dependent upon light levels, turbidity and other external environmental factors, or on hunger and other internal environmental factors.

4

Initiating factors: when to search

The co-ordination of searching behaviour with the temporal availability of resources is just as important as searching in the right place. This is because certain kinds of resources are required periodically rather than continuously. In addition, resources change over the short term, as in flowers that produce nectar only at certain periods of the day, and over the long term, as in flowers that are available only at certain times of the year.

The temporal organization of searching can be based on:

1. searching rhythm, which may be controlled by internal clocks that are set by perception of changing tides, light/dark periodicity or some other environmental component that changes in either a cyclic or non-cyclic manner;
2. resource stimulus, whereby searching is elicited by the sight, sound, or smell of resources; and
3. physiological state, based on influences such as hunger and thirst, hormonal titre, or number of eggs in the oviducts.

4.1 SEARCHING RHYTHMS

Many animals search according to a schedule that might have a period of a day (circadian rhythms), the lunar cycle (lunar or tidal rhythms), or a year (circannual rhythms), depending on the animal and the type of resource. Interestingly, rhythms of searching behaviour can often be explained in adaptive terms, whereby animals search when it is most profitable to do so (Daan, 1981). These rhythms seem to have evolved either in response to competition for similar food sources or to predation. In other cases searching rhythms may simply reflect changes in resource supply, such as the relative availability of food or mates. In many instances the rhythms are controlled by endogenous mechanisms (e.g. biological clocks) which are ultimately mediated through neuroendocrine systems, and which operate independently of external cues (reviews: Aschoff, 1981; Brady, 1981). Many species exhibit circadian feeding rhythms in the laboratory, even when food is always available, as well as in the field where food availability may fluctuate over the course of the day. The pattern of food searching may also reflect the interaction between an animal's food requirements and changes in food availability in its environment. Thus, the foraging activity of animals can reflect the consequence of selection by predators, and the foraging habits of predators can reflect the timed availability of prey.

Superimposed upon rhythmic factors that initiate search are environmental cues that directly affect searching activity, either positively or negatively. For example, females of the small white butterfly (*Pieris rapae*) develop eggs, but they refrain from ovipositing during overcast weather (Gossard and Jones, 1977); an excess egg load builds up, and then on the first fair day they search for host plants and rapidly oviposit. In some animals, cycles longer than one day are linked to seasonal changes that seem to stimulate searching for certain types of resources. Male African elephants (*Loxodonta africana*) spend most of their time feeding, and show no interest when female family groups come into view. But at six weeks into the rainy season they feed for shorter periods and travel long distances searching for females in estrous (Barnes, 1982). No doubt, reproductive receptivity of females is also linked to the onset of the rainy season, thus synchronizing mate searching with physiological readiness.

Finally, environmental cues may directly affect the timing of searching activity. For example, in some populations of a limpet (*Patella vulgata*), foraging schedules are controlled by intrinsic rhythms, whereas in other populations the limpets seem to respond exclusively to environmental cues by foraging only when covered by sea water during the rising tide (Hartnoll and Wright, 1977).

4.2 RESOURCE STIMULUS

Several examples were outlined in Chapter 2 regarding the initiation of local search by perception of external cues. Recall that many species of male insects begin to search when female sex pheromone is detected. However, such responses to external cues are constrained within certain time windows determined by circadian rhythms and physiological state. Thus, an animal will not (usually) search for food when it is satiated or during its normal inactive periods, even if it perceives resource-specific cues. For example, Schaller (1972) observed lions of the Serengeti laying about during the day, surrounded by tasty ungulates that under different circumstances or times would be potential prey.

Several proximate mechanisms are involved in regulating temporal searching schedules, including direct effects of environmental cues on behaviour and effects on response thresholds. A common proximate mechanism triggering searching behaviour is cyclic changes in response thresholds keyed to circadian rhythms. For example, in the cabbage looper moth (*Trichoplusia ni*) there is a circadian change in male behavioural threshold to sex pheromone, but there is no change in the threshold of antennal receptors (Payne *et al.*, 1970); hence, the change in responsiveness probably occurs in the central nervous system.

4.3 PHYSIOLOGICAL STATE

Searching behaviour may be initiated according to an animal's physiological state. For example, voles of the genus *Microtus* which are normally nocturnal, increase diurnal feeding when the nutritional content of food consumed during the night-time is low (Hansson, 1971) or when lactation of pups necessitates additional energy intake.

Such influences on behaviour are often referred to as changes in an animal's motivational state. Central to the concept of motivation is that an animal's response to perceived environmental stimuli (sometimes called incentives) is influenced by certain internal states. Motivation can therefore be defined as the state of an internal variable which affects an animal's behavioural response to an environmental stimulus. At any given point in time, however, there may be several different, perhaps competing stimuli, and the animal must 'decide' which stimulus to respond to first. In this situation the motivational system must include 'cross evaluation' of unitary variates (drives) in activating choices or decisions among the optional behavioural responses (Jander, unpublished). Sibly and McFarland (1974) argue that when animals are faced with a choice between different behaviours such as eating or drinking, each of which will satisfy a deficit, the response depends upon

1. the level of an animal's internal state relative to the degree of the deficit,
2. the maximum rate at which the behaviour can be performed (since a behaviour that can only be performed slowly will not balance the deficit as quickly as one that can operate at a higher rate), and
3. the relative availability of the resources.

The decision for the animal is to choose the behaviour with the largest value of the product of:

[deficit] × [commodity availability] × [maximum rate of behaviour].

By entering cost functions into the decision-making of animals, such as between drinking and eating in hungry and thirsty doves, Sibly and

McFarland (1974, 1976) conclude that the cost to an animal of incurring additional deficits, when the deficit is already large, is much greater than when the deficit is small. Thus, an animal might be less motivated to eat or drink in proportion to the cost of each activity.

Presumably physiological deficit level, potential rate of intake, and resource availability apply to which kind of resource an animal will search for, but searching motivation is not nearly as clear as motivation dealing with consummatory acts. Quite often when an animal begins to search, the resource required is not within its perceptual sphere, and so it is not the same situation as when the sight of the resource elicits a certain kind of behaviour (Toates and Birke, 1982). In fact, birds begin to search for their overwintering grounds by taking flight on what might entail a 10 000 km journey; squirrels begin to search for acorns hidden months earlier; birds searching for cryptic prey cannot see the prey at the beginning of a search bout, and so there is no prey stimulus to interact directly with the motivational state. Thus, in some cases it seems that a response to a stimulus, dependent on certain deficits, can result from the specification of goal objects within the sphere of an animal (e.g. moving prey, sexual partner). In other cases the goal to be specified might be a geographical location corresponding to a waterhole, a previously visited food site, or a sleeping area; and only when an animal is at the site do consummatory activities occur (e.g. feeding, copulation attempts, or resting). In addition, an animal might not experience competition for behavioural expression between hunger and thirst until it is very near to a site containing either food or water or both.

The following discussion of sea slugs and blow flies relating searching behaviour to hunger and feeding, points out how the proximate mechanisms involved in motivation can be described in terms of neural circuitry and hormone secretion.

Sea slugs (*Pleurobranchaea californica*) have been choice subjects in a series of experiments attempting to delineate the neural mechanisms responsible for behavioural hierarchies (Davis *et al.*, 1977). This mollusc is an opportunistic carnivore, feeding on dead organisms and on the eggs of other gastropods. Behavioural studies reveal a hierarchy of behavioural acts in which escape always suppresses other activities (Figure 4.1): egg laying suppresses feeding, but can occur at the same time as righting; withdrawal of the oral veil due to mechanical stimulation can occur simultaneously with righting or mating, but is suppressed by feeding; when food stimuli are low or when the animal is satiated, withdrawal from tactile stimuli suppresses feeding, such that that the feeding/withdrawal relationship is one of reciprocal inhibition (e.g. Davis *et al.*, 1977; Kovac and Davis, 1980). Suppression of other activities by feeding is adaptive for an

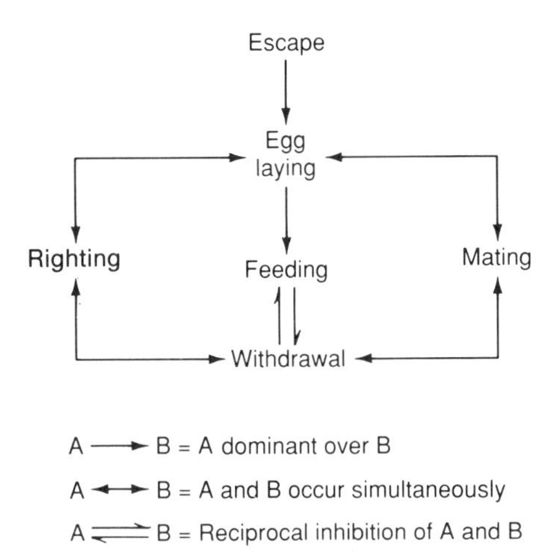

Figure 4.1 Behavioural hierarchy in the sea slug (*Pleurobranchea californica*). (After Davis *et al.*, 1977.)

opportunistic carnivore, and since *P. californica* will eat its own eggs, the inhibition of feeding while egg laying is particularly important. The herbivorous sea slug, *Aplysia*, on the other hand, gives feeding a lower priority and its responsiveness to food shows a circadian fluctuation (Kovac and Davis, 1977). Similar types of behavioural inhibition have commonly been observed in behavioural hierarchies of other animals.

In a series of investigations dealing with hunger, the drive upon which most studies of motivation focus, Dethier (1976, 1982) and his co-workers describe the neural and endocrine circuitry that explains why the blowfly (*Phormia regina*) searches, why it feeds, and why it stops searching and feeding. When a fly is in a nutritionally depleted state it is more active than when satiated, and it engages in random flights until olfactory stimuli orient it to a source of food. Contact of tarsal chemoreceptors with gustatory stimuli causes the brain to send patterned commands to the sucking muscles, and the fly feeds. The resulting ingestion stimulates stretch receptors in the digestive tract, which then send inhibitory signals to the brain, where continuing sensory input is nullified. At the same time, blood-borne factors that are presumed to be hormonal inhibit the locomotor centres. At this point both ingestion and locomotion are terminated. When metabolic deficits again develop, the hormonal titre falls and inhibition of the locomotor centres is released; that is, their thresholds to external sensory stimuli decrease. Concurrently, internal stretch receptors cease firing and the inhibition on central reception of chemosensory input is removed. Analogous data supporting this type of system are available on the feeding system of the grasshopper (*Locusta migratoria*) (Bernays and Chapman, 1974).

It would appear that a scheme relating many of the independent and dependent variables in finding and utilizing resources can be constructed in terms of demonstrated mechanisms, thus reducing the need for intervening variables, such as a drive or a motivational state. According to Dethier (1982), it further appears that in animals like the sea slug and the blowfly

'a hierarchical system of selective potentiation can be constructed that links described behaviour with demonstrated neural and hormonal events'.

It seems as though researchers who work on sea slugs and blowflies are well on their way toward defining the neural circuitry underlying what has been referred to as motivation. McCleery (1983) concludes as follows:

'If phenomena such as learning and motivation are emergent properties of aggregates of neurons, then motivation in higher vertebrates may involve no new principles over similar phenomena in insects or molluscs.'

More complex animals might have more complex circuitry, but the same basic units. Neurophysiologists may never be able to unravel extremely complex systems, however, and so modelling upwards on the basis of known circuitry is the only appropriate alternative.

4.4 SUMMARY AND CONCLUSIONS

Resources and their distributions change over the short term and over the long term, thus requiring the temporal organization of searching activities. This organization can be based on a searching rhythm, stimulation from the presence of resources or resource cues, or physiological state. An animal's physiological state can modify response thresholds to resource cues, thereby increasing or decreasing the likelihood of searching for a given resource.

5

Assessment mechanisms: resource, patch and habitat selection

An animal can assess the quality of a resource unit so as to restrict its search to the best habitats, patches or resources by comparing what it 'knows', its reference, with information perceived at the resource unit. The penalty for inappropriate 'decisions' is severe: eggs fail to hatch when oviposited at unsuitable sites, growth and fecundity are reduced by poor nutrition when herbivores are crowded, nests are flooded when sites close to riverbeds are selected, and offspring may be non-viable or sterile when males mate with females that are too closely or too distantly related.

Although a reference is genetically specified, it can be modified through experience or other factors. Strategies involving certain types of learning allow for alterations in quality assessments relative to global changes in resource unit quality, so that an animal's acceptance criterion shifts according to these changes. Even innate mechanisms are subject to change, as discussed in Chapter 15 regarding deprivation and other internal and external modifiers. Thus, regardless of how a reference develops, it should be modifiable from one time period to another in response to changes in resource quality.

Assessments of habitat, patch and resource are interactive processes, such that a good habitat is defined in part by the quality of patches within it, and a good patch is determined in part by the quality of the resources within it. Proceeding from resource to patch to habitat, the criteria for acceptability may change in predictable ways simply because resources are individual and simple items, whereas patches and habitats are aggregates and complex entities. For example, a patch with very few resources, relative to the density of most patches in the environment, even if it contains high quality resources, may be less valuable to a forager than a patch with more but lower quality resources.

5.1 ASSESSMENT OF RESOURCES

Although there are unique criteria for each resource type and for a given searching species, there are a few principles that apply across taxa. First, many types of resources are distinguishable based upon species identification. Thus a herbivore specializing on grasses will probably never consume oak or hickory leaves. Since animals cannot use taxonomic keys, they probably test resource candidates with one or more sensory modalities to determine if a candidate is the species they need for food, oviposition site or mate. Second, animals often assess the physiological state of a potential food plant, oviposition site or prey item. Mortality or low viability may result from the use of less nutritious plants or hosts, and a predator might be injured if it attacks a healthy prey. Third, the presence of other users or competitors is an important criterion. In some cases, many individuals are actually required to subdue a resource, whereas in other cases groups of individuals in a group interfere with each other in acquiring resources or they simply sharpen the competition.

5.1.1 Oviposition sites

Oviposition site selection is an important factor affecting an individual female's reproductive success, by enabling her to place eggs in sites where offspring have the best chance of survival. Herbivorous insects exemplify many of the factors involved in oviposition site selection. Various species of plants differ in their suitability as larval food plants, in part because of microclimatic differences between sites (Williams, 1981), differences in their physical and chemical constituents (Feeny, 1975), and the parasites and predators associated with particular plants and habitats in which they grow (Wiklund, 1981). If larval success is consistently higher on some plants than on others, and if females are able to locate these 'best' plants, then the species may become increasingly specialized on these plants. Such specialization may also lead to changes in searching behaviour used to locate host plants; more efficient behaviours which lead to rapid location of preferred hosts may evolve, such as discrimination while in flight (Rausher, 1978).

(a) Correct type

The onion maggot fly (*Delia antiqua*) illustrates how the combination of chemicals such as alkyl sulphides and specific visual patterns mediate host-plant finding. Harris and Miller (1982, 1984) presented plant surrogates in various foliar shapes to female *D. antiqua*, and found that most eggs were laid around narrow (4 mm) vertical cylinders. Fewer eggs were laid when the cylinder was reduced in height to less than 2 cm or when the angle between the cylinder and substrate deviated from 90°. Responses to the various forms were based primarily on post-alighting preovipositional behaviours. Females alighting on non-preferred shapes either failed to initiate, did not complete stem runs, or did not probe after completing a stem run (Figure 5.1). Foliage shape, size, and angle may provide *D. antiqua* and other herbivores with information critical to the survival of their offspring. Because *D. antiqua* larvae are subject to desiccation in dry soils (Workman, 1958), survival is better if eggs are laid in moist sand close to food resources. Foliar characteristics stimulating oviposition are relatively uncommon outside the genus *Allium*, and seem to be correlated positively with the developmental stages of plants which may be optimal for larval growth (Perron, 1972). Harris and Miller point out that it is not known if runs over a plant surface allow the CNS to construct a point-by-point image of the entire foliage or fruit form. It is also possible that females perceive only small parts of the form at any given point during a run; if each part stimulates the female to continue, the completed run (rather than a point-by-point image) could stimulate females to perform subsequent preovipositional behaviours.

Female insects are not infallible, however. They sometimes lay eggs on substrates where larvae cannot survive. For example, in a swallowtail butterfly (*Papilio machaon*) in southern Sweden, considerable variation was noted among females in thresholds for acceptance of host plants (Wiklund, 1981). Some females behaved as generalists, laying eggs more or less indiscriminantly, whereas other females were highly selective as long as the optimal host was available. If this phenomenon is a general one in butterfly populations, Wiklund's observation might suggest a mechanism for promoting innovative changes in host plant utilization in new habitats where familiar hosts are absent. Shreeve (1986) suggests that this may happen when females fail to discriminate between host plants because of time and behavioural constraints.

b) *Previous use*

Resources in many instances are ultimately partitioned so that they are effectively divided among competing individuals. The most common mechanism is the use of cues to avoid resources previously exploited by other individuals. A variety of cues are used in making such 'decisions'. For example, the twelve-spotted asparagus beetle

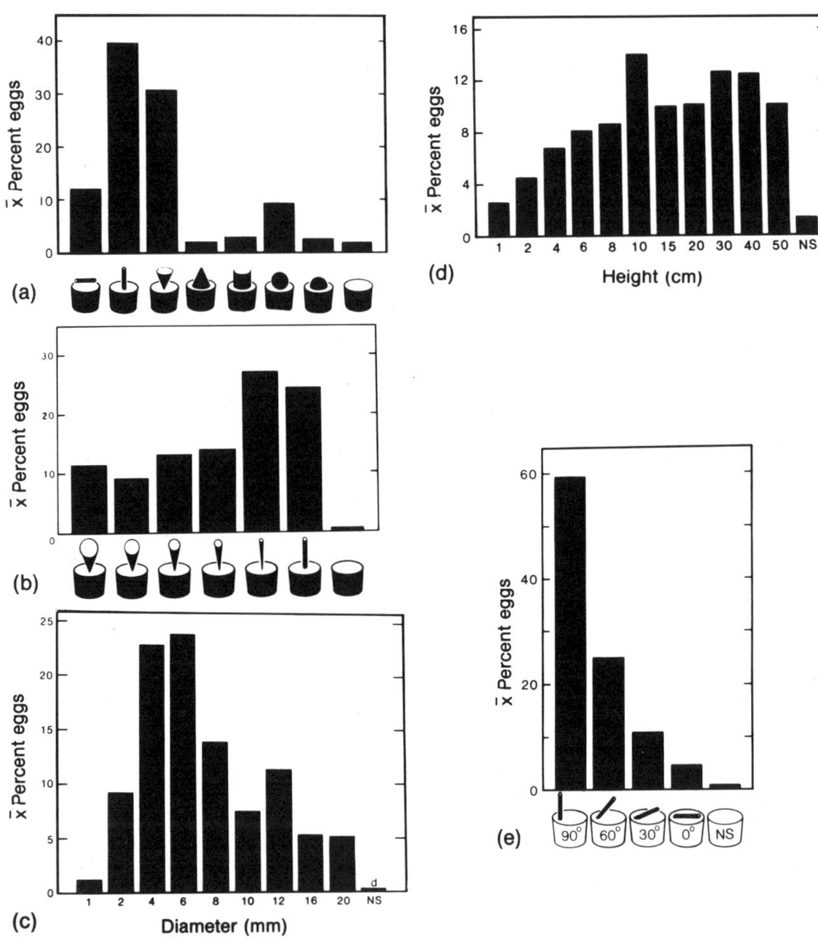

Figure 5.1 Influence of physical dimensions of surrogate plants on the oviposition of female onion maggot flies (*Delia antigus*). (a) Diverse three-dimensional shapes; (b) cone width; (c) cylinder diameter; (d) cylinder height; and (e) cylinder angle. From Harris and Miller, 1984, reprinted by permission.

(*Crioceris duodecimpunctata*) oviposits on foliar or stem tissue of female asparagus plants bearing flowers or berries. Although reproductive plants may have a patchy distribution, each may bare hundreds of berries (van Alphen and Boer, 1980). If a newly eclosed beetle larva locates a berry which is unoccupied, it gnaws a hole into the sepals, enters the berry, and feeds. Most individuals complete their development within a single berry, and a single berry usually can only sustain the development of a single individual. A searching larva may reject a berry if it finds an entrance hole, signalling that the berry is already occupied. In this instance a chemical is exuded by the inhabitant larva, mediating the mechanism for avoiding overcrowding of an exhaustible food resource.

(c) *Physiological state*

Female small white butterflies (*Pieris rapae*) can recognize the physiological state of plants (untreated or treated with fertilizer) and prefer to lay eggs on fertilized plants. Exactly what cues are used by the butterflies is uncertain, but a likely candidate is the greener colour of the fertilized plants, and possibly differences in transpiration rates. Since larvae grow faster and attain larger weights on fertilized plants, oviposition choice seems to be translated into a fitness advantage for the offspring of adults with superior instincts for quality assessment (Myers, 1985).

5.1.2 Food plant

Many vertebrate herbivores consume enormous quantities of relatively low quality plant material in order to extract the nutrients they require. To cope with this problem Coke's hartebeest (*Alcephalus buselaphus cokei*) increases its protein intake through selective grazing (Price, 1978), and moose (*Alces alces*) concentrate on easily digested tender shoots when they are available (Vivas and Saether, 1987). Similarly, red grouse (*Lagopus lagopus*) prefer heather leaves that are high in nitrogen and phosphorus, the two most likely limiting nutrients; discrimination in favour of such food is particularly intense just before egg laying (Moss, 1972; Moss *et al.*, 1972). Sinigrin and other mustard oil glycosides are effective feeding stimulants in larvae of the cabbage butterfly (*Pieris brassicae*) (David and Gardiner, 1962). This species is oligophagous on the cruciferae family, most species of which contain these chemicals. Attempts to find similar host-specific stimuli for some other insect species have not been successful. For example, the oligophagous tobacco hornworm (*Manduca sexta*) which feeds on solanaceous plants, does not respond behaviourally to the most likely plant-specific chemicals, the solanaceous alkaloids (Hanson, 1983).

Learning is often involved in food-plant selection. For example, hummingbirds (*Amazilia rutila*) somehow learn not to visit flowers of *Malvaviscus arboreus* var. *mexicanus* in which the corolla has been punctured and the nectar removed by orchard orioles (Gass and Montgomerie, 1981).

5.1.3 Prey

(a) Correct type

The main criterion for prey selection is usually deciding if the object discovered is the right species. This assessment is analogous to oviposition site and food plant selection, except that predators usually must make their 'decisions' at some distance from their prey, since few prey would be willing to wait about while a predator tastes, smells, probes and otherwise inspects. Predators therefore use indirect cues or cues that can be quickly perceived. For example, Drees (1952) found that salticid spiders responded to prey size, shape, solidity, movement and contrast to background. While omission of one stimulus parameter weakened the response, it did not abolish it. Such response mechanisms have been referred to as stimulus summation, whereby each parameter perceived adds to the total probability of a response.

Many predators prefer that their prey move in an irregular path, which is why fishing lures are so ingeniously constructed to move erratically. Dragonfly larvae (Etienne, 1969), water stick insects (*Ranatra linearis*) (Cloarec, 1969), and pumpkinseed sunfish (*Lepomis gibbosus* (Gandolfi *et al.*, 1968) are ambush predators that respond more readily to prey that zigzag or wriggle than to prey moving in a rectilinear path.

(b) Size

Prey size is an important criterion, because either the catch might not be worth the effort required to complete the capture or the predator may be unfortunate enough to become the prey, or at least be seriously injured. When predators take prey that have dimensions similar to their own, and when size of prey is the major criterion for assessment, some kind of additional screening method is needed. For example, piranhas (*Serrasalmus nattereri*) attack fish within a certain size range of length:width ratios, but that are different from their own ratio (Markl, 1972).

(c) Physiological state

Predators ought not to attack potentially dangerous prey unless the benefit is large. For example, goshawks (*Accipiter gentilis*) tend to capture underweight or injured woodpigeons, which are less alert and also less able to outdistance their predators (Kenward, 1978). Many carnivores such as wolves (*Canis lupis*), grizzly bear (*Ursus arctos*), and spotted hyenas (*Crocuta crocuta*) select older, sick, or debilitated prey over vigorous, healthy prey (Mech and Frenzel, 1971; Cole, 1972; Kruuk, 1972). Their assessment is probably based on movement patterns, rate of movement, or some kind of novel behaviour typical of sick or debilitated individuals.

5.1.4 Potential mate

Female reproductive success is limited mainly by her ability to convert environmental energy into offspring, whereas male reproductive success is limited by the number of matings that can be achieved with different females (Parker, 1978). Choosing an appropriate mate is an important addition to this generalization, and the axiom applies to both males and females. For males, if time, energy or sperm are limiting, mating with appropriate females is crucial, and for females, the number of eggs produced will always be limiting and should be fertilized by appropriate males.

The assessment applied to potential mates should reflect the fitness of the individual and the extent of relatedness between male and female. Fitness can be assessed in a variety of remarkable ways, although why individuals make certain choices is somewhat controversial. Female natterjack toads (*Bufo calamila*) are attracted to deep-voiced males, which on average are the largest (Arak, 1983), and hence most probably the oldest and the most wily in the population. Females may not be able to assess fitness directly, but indirect mechanisms such as this operate in many animal species (review: Partridge and Halliday, 1984).

It is also important that sexual partners be members of the same rather than different species because of problems in chromosome recombination when relatively unrelated individuals mate (e.g. lions and tigers may copulate, but the resulting ligers and tigons are sterile). Sexual partners should also not be too closely related (e.g. brothers and sisters) so as to avoid inbreeding and the expression of lethal recessives. While it is relatively simple for humans to avoid mating with close relatives and/or distantly related species, it is not so simple for other animals. Cues that are used include chemicals, songs, visual patterns, and courtship displays. Which ever sex makes the assessment, and it is usually the female, will assess a courtship signal quantitatively and qualitatively to deterimine if both the pattern and the quantity, ratio, or duration are correct.

There are also various social mechanisms to prevent some potential mating problems. Honey bee queens, for example, fly some distance from the natal nest before accepting drones for mating; meeting up with males of the correct species is accomplished by the secretion of a sex pheromone to which only honey bee drones are attracted.

5.1.5 Nest, roost and hibernation sites

Selection of sites for reproduction, roosting or for hibernation is an important kind of decision in many species. Fitness drops to zero if an individual is preyed upon while roosting or fails to survive through the winter because of a poor choice in hibernation site. Fitness is reduced if a poor decision is made in selecting a site for oviposition, especially if the eggs are vulnerable to attacks by predators or parasites or abiotic factors.

Sites that are suitable for nesting, roosting, or over-wintering are generally discrete areas or structures within the habitat. Each nesting species has a set of criteria that is used to assess potential sites that have been located. Often these criteria are related to the way an animal searches. For instance, the paper wasp (*Polistes metricus*) does not search on the ground for potential sites, but rather around shrubs and trees to which their nests will be attached. Another factor that might guide search is the distance between the natal nest relative to a new nest (Wenzel, unpublished). Overwintered gynes of *Polistes* may return to the natal nest and build onto it, whereas honeybee swarms fly away from the natal nest so as to reduce competition between old and new colonies and to reduce the possibility of inbreeding.

Bivouac site selection by army ants of the genus *Eciton* is probably based on the relative abundance of invertebrate and other prey in the surrounding area (Schneirla, 1971). Nest sites of the ground-nesting sweat bee (*Lasioglossum zephyrum*) are selected according to soil type, in order to ensure a compact matrix and yet allow burrowing, and according to topographical features such as southern exposures in northern climates for optimal solar radiation (Michener, 1974).

Search costs influence nest site (and mate choice) in female pied flycatchers (*Ficedula hypoleuca*) (Alatalo *et al.*, 1988). Females must make their choices quickly owing to their short breeding season and limiting nest sites. However, the quality and height of the nest hole is a major factor influencing their reproductive success, and females must spend at least some time inspecting nest sites already established by males to assess these characteristics. Alatalo *et al.* (1988) allowed females to choose among males that had nest boxes at various heights and with openings of varying sizes, but where the search costs of assessing these resources also differed. Females decisively sacrificed choosiness if the search costs were too high, and they accepted the poorer nest sites. The major sources of nest failure in pinyon jays

(*Gymnorhinus cyanocephalus*) is loss to avian predators such as ravens (*Corvus corax*) and crows (*Corvus brachyrhynchos*), and abandonment after cold and snowy spring weather (Marzluff, 1988). In comparing the relative height at which individual jays nested with the height and fate of the previous nest, it was found that jays changed their nest location if they had previously suffered predation. Experienced jays nested at relatively low heights, enhancing concealment, but nested further out from the trunk early in the season, reducing energetic incubation costs. The birds therefore profit from experiences gained from previous breeding seasons.

Case Study Nest site assessment by honeybees

Nest site localization and assessment by honey bees (*Apis mellifera*) provide a case in which information is gathered by various individuals. Upon this information the colony decides which new nest site a swarm will occupy (Lindauer, 1955, 1961).

When a swarm of bees with a new queen leaves the parental colony it does so on the basis of information gathered by scouts reporting on potential sites at which the new colony might build a nest. Several days before the swarm actually leaves the parental nest, scouts initiate the search for a nest site by exploring the nearby area. When a potential site is discovered, scout bees return to the parental nest and through dancing they communicate information relating the site to other individuals. This dance is similar to that performed by returning foragers which communicate information concerning a food source (see 12.3), except that bees dancing in reference to nest sites solicit food instead of distributing nectar to other bees. In addition, nest site dances may continue for several hours rather than one to two minutes as in food source dances. The dance may stimulate bees to join the swarm as well as communicate information about possible nest sites. Thus by the time a hanging swarm has formed away from the parental nest, several potential sites may have already been reported. Soon after the swarm forms, scout bees begin to dance on the surface of the cluster of bees. Distance and direction to possible nest sites are contained in the dance message, as with foraging dances, and the intensity of the dance relates to the suitability of the site for a new nest. The clustered bees do not head off *en masse* toward any one site, however, until a 'consensus' is reached.

Individual scouts inform clustered bees of a site, recruiting other bees to examine the site. A scout then returns to its chosen site, leading other bees that are probably attracted by a pheromone produced by her Nassanoff's gland. The number of new bees recruited to examine and in turn to dance in support of the site depends in part on the intensity of the dance performed by the original scout and in part on the assessment of the site made by the recruits. The intensity of the dance, relating to the suitability of the site, seems to be based on an innate rather than a learned scale of values, although direct comparison of sites is the only way a bee can make an assessment between sites. Even a dancing bee can be 'persuaded' to change allegiance to an alternative site if she sees another bee dancing more intensely than herself. These is no room for emotional appeals here, since an overestimation of the acceptability of a site would be interpreted as real information! Eventually, this process results in all scout bees dancing in support of the same site. The scouts then perform a series of buzzing runs over the surface of the swarm, and the bees take flight. As the swarm proceeds toward the new site, some bees move out of the swarm in the correct direction, and repeat this maneuver, presumably providing directional information.

In Lindauer's experiments he altered potential sites receiving the attention of scouts, and was thereby able to change the dances of scouts, based on their responses to his alterations. Two identical experimental nesting sites, one of which was close to the natal nest and the other 30 m away, showed that greater support is given for a nesting site that is not too close to the natal site. A site that the bees discovered when the sky was cloudy was at first well supported. When the sun came out, however, the temperature inside the site rose to 40°C, and support fell. An experimentally produced shade cover produced renewed support for the site. Clearly, therefore, bees take into account a number of factors in the assessment of habitat suitability, including shelter from wind and sun, distance from natal site, and odour characteristics.

5.1.6 Ambush and perching sites

Places where animals wait for prey must be concealed, but should also permit fast exits, should not be in areas with too many competitors, and must not expose the ambusher to its own predators.

Schaller (1972) discusses how lions (*Panthera leo*) select ambush sites, often working in groups. One important factor is the height and density of vegetation which influences the ease with which lions can stalk without being detected. Approximately 75% of the wildebeest and zebra in the short-grass plains were captured near at least some cover. The frequency with which Schaller found kills increased as soon as prey moved into the intermediate and tall grasslands. Lions also hunted effectively under cover of woodlands, using thickets for concealment. Approximately 35% of the kills Schaller recorded were made near a river where the dense vegetation and broken terrain enable lions to stalk undetected. One or two lionesses occasionally hid themselves at strategic ambush sites and waited for prey to move close. For example, Schaller observed them waiting concealed in a small patch of grass by a river watching several gazelle at a distance of 50 m. When a male gazelle descended the river embankment, closely followed by a female, the lioness raised herself slowly, only her head above the grass; when the female gazelle passed within 5 m, the lioness rushed.

5.2 ASSESSMENT OF PATCHES AND HABITATS

How animals assess patches and habitats is less clear than their decision-making about individual resources. After all, the larger resource units are more complex and can potentially contain various kinds of resources as well as stress sources.

Patches and habitats are assessed in part by the number and kinds of resources they contain, and so much of the rationale discussed for resource assessment also pertains here. The habitat or patch an animal chooses may depend upon the time and energy available for searching, the probability that additional search will improve the eventual choice, and the degree to which continued search for an even better habitat will change the searcher's fitness. The relative importance of these factors will vary according to the characteristics of the searching animal (i.e. its present energy resources, how fast it can move between areas, and its lifespan), the spatial and temporal characteristics of suitable resource units (large or small, homogeneous or patchy, predictable or unpredictable, temporary or permanent), and the number of competitors and parasites already occupying available sites.

As with resources, the threshold for acceptance depends on the internal state of an organism, which in turn may be modified by external factors. Thus, although not every habitat would necessarily possess all features characteristic of the ideal environment, the main factor is the combined effect of the various features which promotes settling. Certain factors may be absolutely essential; tree-nesting birds, for example, will not find a treeless habitat acceptable regardless of its other merits. Little is known, however, about the exact properties of cues used in the selection process, the number of cues actually employed, or the ways in which they are combined to enable an animal to reach a decision. Sometimes animals must make predictions based on present information as to the future usefulness of a habitat or patch. For example, the number of foliage-dwelling insects found by birds early in the spring may provide erroneous information about the food supply when young have to be fed later in the year.

Although the terms habitat selection and habitat utilization are roughly equivalent, habitat selection should signify the active choice of an area by an individual, whereas habitat utilization may only mean that an individual occupies a given area. For example, although meadow vole (*Microtus pennsylvanicus*) density is correlated with vegetation cover in Wisconsin marshes, this density variation may be due to differential predation in areas of differing cover rather to active selection of areas by individual voles (Getz, 1970). The selection of habitat types usually involves an interplay of genetic ('innate') and experiential influences (Immelmann, 1975). Learned preferences may influence habitat orientation in patchy environments characteristically occupied by frogs of the genus *Rana* (Wiens, 1970, 1972), while the determination of what may or may not be learned could be largely genetic. In other species, such as mice (*Peromyscus maniculatus bairdi*) (Wecker, 1963) and salmon (*Salmo*) (Raleigh, 1971), responses to habitat stimuli may be directly determined by inherited assessment mechanisms. In most cases it is likely that both early experience and innate preference are of importance in choice of habitat. Perhaps innate factors provide the coarse tuning and learned factors the fine tuning for habitat selection.

Individual habitat preferences are strongly influenced by population densities in the various patch types. Other things being equal, individuals would be expected to select habitat types on the basis of fitness determinants, which are a function of both the intrinsic quality of the habitat type and the population density in the habitat. Given small populations, only the highest quality patch types will be occupied, but as populations increase, the quality of that habitat decreases until a point is reached at which some other habitat type has equal potential quality (Fretwell and Lucas, 1970). At this point, animals should expand their range of choices to include additional habitats that would now be of equal quality. In a Wisconsin grassland, for example, grasshopper sparrows (*Ammodramus savannarum*) upon their arrival in the spring, initially occupied breeding habitats with relatively low grass cover and litter depth and tall emergent herbaceous vegetation, whereas savannah sparrows (*Passerculus sandwichensis*) occupied habitat patches with characteristics at the opposite extremes. Individuals arriving later, however, settled in the remaining habitat not already occupied by their species, so that as breeding populations increased in size, the two species converged in patterns of overall habitat occupancy (Wiens, 1973).

While a habitat may be 'selected' according to its contents, the usefulness of a habitat to a given species also depends on factors in addition to the patches and resources within. Temperature, humidity, solar radiation, physical structure, as well as biotic resources required, all contribute to the success or failure of the individual in a given habitat. An animal might be attracted to key stimuli of a given habitat, which are indicators of faunal and/or floral composition. A simple example is the visual difference between open fields and woodlands. Animals may migrate from one to the other using gross visual characters as their cues, which would be a flat line on the horizon for the field and a vegetation silhouette for the woodland. An extreme version of habitat utilization would be if animals arrived in habitats through random movements, and those arriving in habitats with conditions fitting within the acceptable range of a species would survive, and the others would die.

Other individuals may influence habitat selection. For example, even when a male animal has located a suitable habitat according to its criteria, the presence of a territorial male may prevent it from remaining in an area. The conspicuous presence of predators probably acts in a similar way. Stein and Magnuson (1976) showed in laboratory studies that crayfish (*Orconectes propinquus*) selected substrates providing a maximum amount of protection (pebbles rather than sand) when placed with smallmouth bass (*Micropterus dolomieui*) large enough to eat them. Control crayfish showed no preference for one substrate over the other.

5.3 SUMMARY AND CONCLUSIONS

An animal assesses the quality of a resource unit so as to restrict search to the best habitats, patches or resources. It compares the information perceived at the resource unit with its genetically-specified reference. The reference is modifiable from one time period to another in response to internal and external modifiers such as changes in resource quality or period of deprivation.

A habitat is assessed in part by the quality of patches within it, and a good patch is determined in part by the quality of the resources within it. Assessment criteria include: species identification, physiological state, and presence of other users or competitors. Often animals employ an assemblage of cues to assess a resource, and it is the specific combination that will evoke a response.

PART THREE
Search mechanisms

. . . the desire to establish general strategic rules has necessarily sacrificed the consideration of detailed behavioural mechanisms and the constraints to which specific predators and parasitoids are subject, in favour of generalized assumptions

Carter and Dixon 1982

To some extent the search mechanisms an animal uses are matched to resource distribution, such that area-restricting search patterns that would be disastrous for locating scattered resources work well in patches of resources. Some mechanisms operate effectively regardless of resource distribution, as with various learning mechanisms whereby an animal may learn to return to a rewarding patch or an isolated resource. In the following discussion each type of mechanism is discussed with regard to how it works and for which types of resources and resource distributions it proves most effective.

6

Locating patches and distant resources

When searching for a resource patch an animal attempts to localise an assemblage of resources, whereas when searching within a patch it avoids leaving the assemblage until it becomes unprofitable to remain. Interestingly, the problem of finding an isolated resource is similar to that of finding a distant patch because, even though the patch might be larger, initially in both situations the target represents a relatively small proportion of the total environment. The following section considers a few of the strategies used to locate distant resource units; subsequent sections deal with searching for and utilizing resources within these units.

6.1 SEARCHING WHEN NO ENVIRONMENTAL CUES ARE AVAILABLE

When an animal has no information about where resources are located it must attempt to move in such a way so as to optimize its chances of locating resources, but also to reduce the chances of covering areas already searched. Such mechanisms come into play with regard to finding distant resources or resource patches in unknown places and in locating resources that may be lost, as in relocating a nest or burrow entrance. In fact, it is from studies of homing, whereby an animal relocates a previously inhabited site, that several potential search mechanisms are derived. Searching for the home site is entirely analogous to searching for a resource, if in both cases the location of the resource is unknown to the animal. For example, strong fliers among shore birds may range far out to sea in their daily search for food, especially during the breeding season. When a bird moves outside its familiar territory it no longer has known landmarks available, and as outlined below, its search strategy may resemble that of an animal beginning to search for patches of food, mates, oviposition sites, or other kinds of resources. Once a homing bird locates its 'familiar territory' it can use landmarks as guides, and it will be able to locate its site rapidly. Similarly, in many cases animals become familiar with their territory, home range, or defended patch, and in successive searching bouts they can use this information to locate resources efficiently.

Theoretically, there are four possible search strategies to follow when no information is available for finding resources (Figure 6.1):

1. a purely random walk, in which each successive step that the searcher takes is in a direction that is random relative to its previous step;
2. a straight walk of some length in some randomly determined direction;
3. highly systematic strategies, such as an ever-widening spiral, parallel sweeps, or looping from a central pivot point;
4. kinesthetic-input mapping; and
5. combinations and variations of these strategies.

6.1.1 Random walk

In a random walk an animal moves straight ahead for some distance (a 'step'), turns on the spot over an entirely random angle, makes another step and so on (Figure 6.1a). The step length can either be the same each time an animal moves, or each step can be of a length entirely random with respect to the previous step (Brownian movement). If an animal uses a random search strategy, only the probability distribution of its position, not its exact position, can be predicted over time. The disadvantage of a random walk for locating resources is the high probability of often crossing the previous track and of entering areas already searched. Perhaps for this reason there are no examples of purely random movements, even in microorganisms. However, an element of randomness within one of the more deterministic strategies listed below may actually add to the efficiency with which they work (Hoffman, 1983a).

6.1.2 Straight line

When no information is available about the positions of resources in space, 'straight ongoing movement (transecting) is the most efficient strategy, presuming the most likely distribution of resources to be patchy' (Jander, 1975). However, the target is detected, T_d, with a probability of:

$$T_d = \frac{\arcsin(\alpha/r_0)}{180}$$

where
α = detection radius
r_0 = distance between starting point and the resource goal

Thus, the relatively narrow detection zone of the straight move is generally a poor strategy for locating relatively small resource units.

If an animal's search path is sufficiently long, the success of the search effort first increases with increasing directional constancy (e.g. toward a straight line) and then declines. If an animal walks in straight lines, but where the distribution of directions taken is random, the probability of success is maximal when the length of the line spans a distance corresponding roughly to that from which the animal can detect the target. Hoffman (1983a) explains it this way: if the directional constancy of a random search is too low, the animal searches only in the immediate vicinity of the starting point and therefore does not detect the target. If the directional constancy is too high the animal may walk a long way in an inappropriate direction. Hence, most species supplement straight moves by periodic turns to avoid the problem of becoming locked into an unprofitable path direction.

From studies of ranging animals it seems clear that most species are incapable of moving very far in a perfectly straight line without the assistance of some kind of external (topographic or celestial) cue. They tend, therefore, to alternate between left and right turns or else they move relatively straight for some distance and then turn at a definite angle and move another leg. For example, when walking on a flat surface in the absence of chemical cues, the blow fly (*Phormia regina*) moves in a series of short runs with frequent changes in direction (Dethier, 1976). Thrushes (*Turdus merula* and *T. philomelas*) move through a meadow in a roughly linear track, with turn angles uniformly distributed around 0° ± 33° (angular deviation) (Smith, 1974a). The birds tend to alternate left and right turns, with a mean turn angle of 35°. This mechanism is probably more effective in insuring forward movement than would be an attempt to move absolutely straight over long distances. On very few occasions did Smith observe a thrush crossing its own path, suggesting that the tactic works to take the birds into previously unsearched ground.

.1.3 Systematic movement patterns

In systematic patterns the exact position of an animal, and not just the probable distribution of its position, can be predicted. In some cases the equation for the path can be calculated. In the following highly systematic searching patterns the key element is that the distance between 'sweeps' or between successive loops of a spiral path should be approximately equal to twice the detection distance of the searcher. Assuming that the searcher continuously scans left and right, there is minimum overlapping of areas previously searched, but the searcher scans the entire area that it moves through.

Griffin (1983) suggests that homing birds may fly in a large spiral in which the distance between the successive loops of the spiral is approximately twice the distance at which familiar landmarks could be recognized. Through this type of hypothetical behaviour, which might be called spiral search (see Figure 6.1c), familiar territory would eventually be reached if this process were continued long enough. Where a foraging bird is carried far beyond its familiar territory, the approximate total distance, D, of homing by spiral search can be expressed as (Griffin, 1983):

$$D = \frac{\pi D_u^2}{R} + D_f$$

where
R = distance between loops of the spiral
D_u = beeline distance from release site to nearest familiar area
D_f = distance from edge of familiar area to resource (e.g. home)

Actually the most efficient track a bird could follow would be a large spiral with a distance between the loops equal to the diameter of the bird's familiar territory, because it would not have to seek a single point as the simple theory of spiral search assumes, but rather a large area (i.e. the whole expanse of its familiar territory). However, the navigational problem of flying in large spirals would be a difficult one for which to find a sensory basis. Birds often fly in small circles near the point of release, but the large scale spiraling required in this model has never been observed. The task would be seemingly much simpler for an animal such as a walking arthropod which might be able to 'set' its turning rate by systematically taking fewer or shorter steps on one side of the body than on the other. In fact, Jander (personal communication) observed a species of blind termite using radial search to locate their foraging column after becoming disoriented.

Two theoretical systematic search strategies are an expanding square (Figure 6.1b) and parallel sweeps (Figure 6.1d). The expanding square strategy, like the spiral, would be appropriate for locating a target that is moving slowly relative to the speed of the searcher. The searcher begins at the point of last sighting and

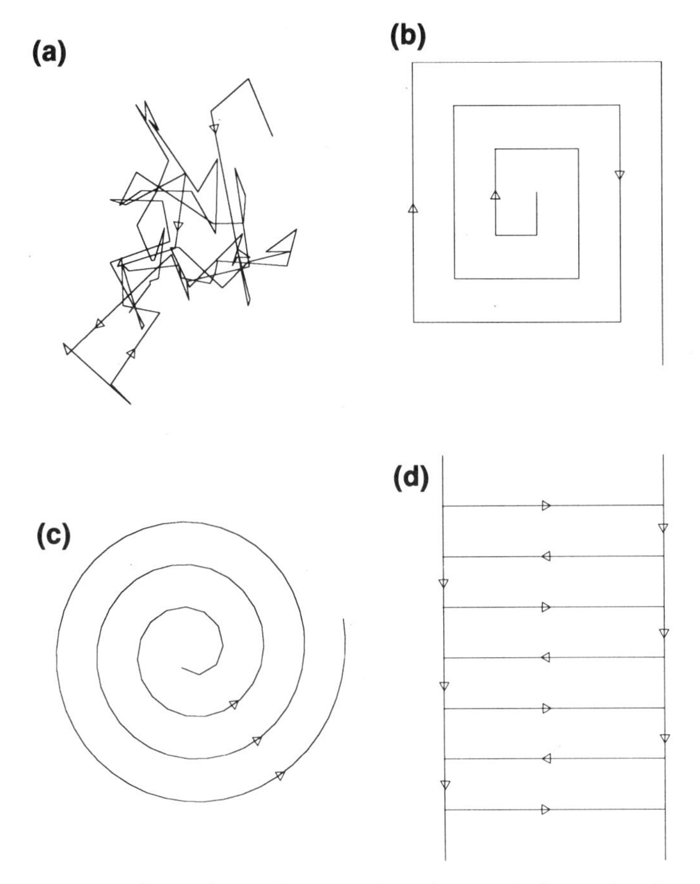

Figure 6.1 Hypothetical search strategies when no information is available as to the location of resources: (a) random walk, (b) expanding square, (c) spiral orientation, (d) parallel sweeps.

initiates an expanding square pathway. In parallel sweeps the searcher moves in straight lines parallel to its preceding path, in which the distance between sweeps equals twice the searcher's detection distance. These search strategies are discussed extensively with regard to search and rescue in humans, surveillance of mineral deposits and fishing areas (review: Haley and Stone, 1980).

6.1.4 Kinesthetic input mapping

Animals can store proprioceptive or other sensory information during orientation and then use this information to determine the direction back home (see Chapter 2). If an animal were extremely good at this technique, it could conceivably search systematically in a variety of possible configurations.

Case Study A desert isopod searching for its burrow

Because of the extreme climatic conditions prevailing in its habitat, the desert isopod (*Hemilepistus reaumuri*) must spend most of the time in its protective burrow. However, because it must leave the shelter to forage and to clean the burrow it has evolved an unusually precise orientation mechanism that enables it to find its way straight back to the burrow (a hole in the ground only 9–12 mm in diameter) (Linsenmair, 1972; Hoffman, 1983a,b). The search strategy of the isopod, comprising both systematic and random elements, includes some of the search mechanisms described in this chapter.

The search behaviour of an isopod guarding the burrow entrance illustrates the initial response of an individual when it loses contact with the burrow. If the guard is disturbed by another isopod or if it moves in the wrong direction and misses the entrance, it must search for it. First, the isopod turns around once or twice on the spot and then begins to move in increasingly wider spirals. Hoffman (1983b) notes that a solution to the search task of the guard might be in the form of an Archimedes' spiral beginning at the point where the entrance of the burrow is most probably located. Such a spiral search, in which the surroundings are explored comprehensively (leaving no gaps), but with minimal overlap, has a pitch twice as large as the maximal detection distance at which the burrow can be perceived by a searching isopod. In a simple spiral search the isopod has approximately a 16% chance of locating its burrow even if it starts just a few centimetres from the burrow entrance. If it misses the entrance and searches in the wrong direction, it cannot survive for more than a few days.

Hoffman (1983) deduced the detection radius (alpha), the greatest distance at which an isopod can detect its burrow, from the following information:

1. The entrance to the burrow is nearly circular; its diameter corresponds roughly to the width of the peraeon of an adult isopod, which facilitates defence of the burrow.
2. The antennae are in constant, rapid movement during a search, with each antenna waving through approximately a 90° sector extending from about 3 mm in front of the head on the midline to approximately 4 mm sideways.
3. *H. reaumuri* pivots continually during the search, oscillating around a vertical axis that passes approximately through the middle of the body. The frequency of oscillation is low in relation to the frequency of antenna movement, but is sufficiently high for a strip with a width approximately half a body length on either side of the path to be traversed by the head.

Thus, the distance, α, from the midpoint of the body to the midpoint of the burrow entrance, at which the forward-pointing antennae tips can contact the edge of the entrance is given by the equation:

$$\alpha = \frac{Db}{2} + \frac{B}{2} + A$$

where
Db = burrow diameter
B = body length
A = antennal length

If the spiral maneuver fails to localize the burrow, the animal adopts a meandering search pattern. This behaviour, which can be elicited by carefully removing the guard from the burrow entrance and displacing it (see Figure 6.2), is independent of the actual position of the animal at the beginning of the search. Remarkably, in less than 400 s after displacement, the isopod returns to its burrow from a distance that is 15 times greater than alpha. In attempting to tease apart the mechanisms used by the isopod, Hoffman has eliminated as possible explanations of this homing behaviour: true navigation, piloting directly to the burrow by using known landmarks, and direct detection of the burrow from the starting point.

The meandering-type search of the isopod has many characteristics of a random walk.

1. It uses the same search strategy in all regions it enters, and therefore needs no information as to the actual location of the burrow entrance.
2. The search strategy is constant in time, and therefore the isopod has only to detect whether it has reached the burrow or not.
3. Once an isopod has chosen a direction, it continues in that direction for a certain distance.

This is the only influence of the preceding parts of the search on the decision of the animal in which direction to search next. Overall, the directions of the isopod's movements are independent, and it does not prefer any absolute direction.

In its search the isopod approaches its starting position sooner and more often than would be expected for a random search (notice the dense accumulation of tracks near the starting position in Figure 6.2). This feature is a fundamental characteristic of the search behaviour of *H. reaumuri* and is not simply a reflection of the direction it takes at the beginning. If an isopod begins its search at a distance of 5–30 cm from its burrow, it returns to it within 800 s with a higher probability than it would if it used a random search strategy; but at the beginning of the search it could have returned even sooner using a random search. Random search with the same directional constancy as shown by *H. reaumuri* would be much less successful than that of the isopod, because the thoroughness with which a given region is searched would be too low. *H. reaumuri* circumvents this problem by returning to the starting point now and then and thus concentrates its search

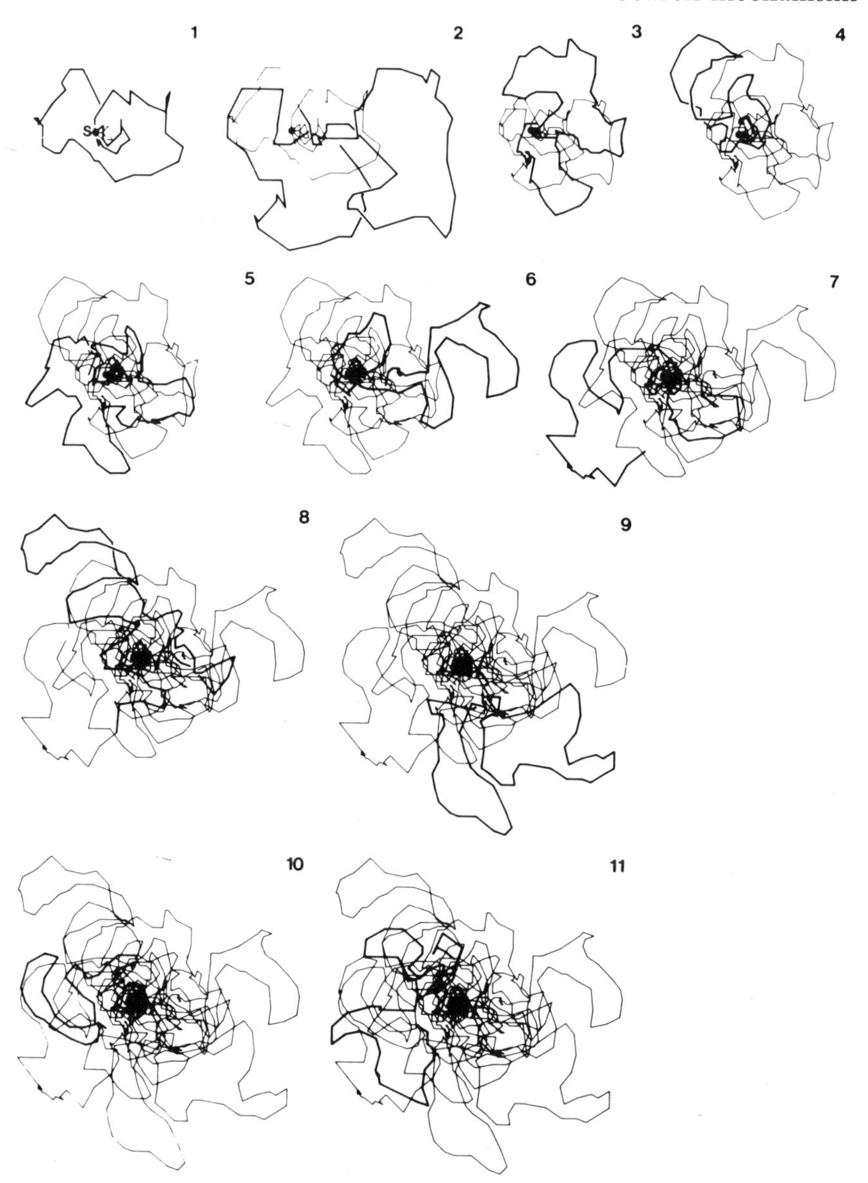

Figure 6.2 Successive stages in search performed by a desert isopod (*Hemilepistus reaumuri*) displaced from its burrow and set on an artificial substrate in unfamiliar surroundings. In each diagram the most recently added part of the track is drawn with a heavier line. Search duration for successive paths: 61, 132, 224, 283, 391, 492, 574, 664, 773, 838, 900 s. Total distance covered equals 14.1 m. Scale bar equals 20 cm for diagrams 1 and 2, and 40 cm for the others. S designates the starting position. From Hoffman, 1983b. Copyright 1983 Springer-Verlag, reprinted by permission.

in one area. The simplest mechanism on which this behaviour could be based would be the ability, demonstrated in other invertebrates, of an animal returning to a starting point by using the stored information about its own previous movements (i.e. kinesthetic input mapping).

6.2 SEARCHING BASED ON ENVIRONMENTAL CUES

6.2.1 Patch-specific cues

Information about potential distant resource patches can be gained through direct sensory input. Insect parasitoids reveal the various ways of using cues from individual resources. As a rule, patch-specific cues do not carry information specifying resource density, and thus would not contribute to the aggregation of foragers in the most profitable patches. However, female endoparasitoids (*Leptopilina clavipes*) of fungivorous Drosophilidae respond to odours of decaying mushrooms only at the stage when the mushrooms are likely to contain host larvae (Vet, 1983). Resources located in association with particular community members may be more easily located because interactions between the two produce attractive sensory signals. For example, the parasitoid, *Cardiochiles nigriceps*, is attracted by the odour of tobacco plant leaf tissue when the leaves are damaged by the parasitoid's host, the tobacco budworm (*Heliothis virescens*) (Vinson, 1975). Herbivores that primarily use vision may locate patches of host plants even at some distance according to the silhouettes of the plants against a homogeneous background. Herbivores using odours may find it easier to locate clumped plants than sparse plants simply because more odour is produced by large numbers of plants. For example, the large milkweed bug (*Oncopeltus fasciatus*) is attracted to host-plant volatiles (Ralph, 1976), and as a result it seldom colonizes small or sparse milkweed patches (Ralph, 1977).

A much overlooked adaptation of animals is the 'turning on' of specific perceptual (usually visual pattern recognition) abilities by specific resource cues. For example, the apple maggot fly (*Rhagoletis pomonella*) which oviposits into apples, flies toward red spheres after searching behaviour is stimulated by apple odour (Prokopy, 1968; Roitberg, 1985). Male sweat bees (*Lasioglossum zephyrum*) and house flies (*Musca domestica*) are attracted to small dark objects when they perceive female sex pheromone (Barrows, 1975). These animals do not respond, or they respond to a lesser degree to the visual pattern if the olfactory cue is absent. This mechanism could increase the efficiency of resource finding in a patch of resources, and yet orientation toward the same visual patterns in an inappropriate context would not be likely to occur.

Case Study Aardwolves in search of termites

The diet of the aardwolf (*Proteles cristatus*) in the Serengeti shows a high selectivity for one species of termite, *Trinervitermes bettonianus*. They find the termites during ranging bouts which begin during the last 3 hours of daylight and continue through the night. Kruuk and Sands (1972) observed aardwolves foraging quietly over grasslands at a speed of 2 to 3 km per hour, with the head bent slightly downwards, and ears moving forwards and sideways. The aardwolves followed irregular paths, with frequent changes of direction unrelated to that of the wind direction. Patches of *T. bettonianus* foragers were apparently encountered either fortuitously or, more commonly, after the aardwolf turned sharply from whatever direction it happened to be walking in. Having located the termites, the aardwolf licked them up from the ground with its broad tongue, pushing the front part of its snout into the vegetation.

The evidence suggests that aardwolves employ ranging without the use of cues to localise patches of termites, and when within close range they switch to sensory cues provided by the termites. While aardwolves may use vision at close distance under good light conditions, it would seem that they locate termites primarily through auditory cues, based on the following evidence.

1. The obvious and striking movements of the ears, and the lack of head and muzzle movements which in hyaenas and dogs seem characteristic of sniffing.
2. In no instance in over 200 recorded observations did an aardwolf turn downwind to find termites. When it made a turn, the aardwolf cocked its large ears and slowly approached the patch of termites.
3. Concentrations of *T. bettonianus* generated a noise which could be heard by humans from a distance of 15 – 30 cm. The detection distance, that is the distance between the sudden turn and the termite concentration, was up to 2 m but most commonly between 0.5 and 1.0 m.
4. The aardwolf immediately ceased foraging when a rain shower began. The sound of raindrops striking the grass would effectively blanket that of termite activity, further suggesting the importance of sound in foraging.

Whenever a feeding site just abandoned by an aardwolf was examined closely, it was found that large numbers of termites were still 'milling around' in dense concentrations. The question arises as to why the predator should leave such an apparent source of food to search for another one. The answer seems to be that soldiers in galleries of the nest system become active when agitated workers abandon their foraging and stream back into nest entrances. The soldiers join their colony mates above ground, but it is only when the last workers are on their way down that soldiers join the fleeing column in any numbers. The secretions of soldiers include various terpenoids such as pinenes, limonene, and terpinlolene, which are probably distasteful to vertebrates such as lizards, as well as a defence against ants. If these secretions were distasteful to the aardwolf, it would probably continue feeding until the proportion of distasteful soldiers increased to an intolerable level, at which point it would stop and leave, even though many termites still remained.

6.2.2 Topographical cues

Animals often use biotic topographical features of the environment to direct their searching when cues from resources cannot be detected. For example, Braconid parasitoids (*Diaeretiella rapae*) employ a combination of cues to search for cruciferous plants on which their cabbage aphid host (*Brevicoryne brassicae*) feeds (Ayal, 1987). Common crucifer species bear large basal leaves from which a high branched inflorescence emerges (Figure 6.3). The aphid colonies are clustered on top of flower- or fruit-bearing branches. Honeydew secreted by aphids drips down and contaminates the upper surfaces of the leaves, and then washes along the branch and stem and is trapped in leaf nodes. The larger the colony, the longer and more concentrated will be the chemical trail leading from it. Parasitoid females are attracted by the plant odour and colour, and alight on the leaves (Figure 6.3, step 1). If the female encounters a contaminated area on the leaf, it flies upwards (step 2), and lands on one of the inflorescence branches. The branch may be contaminated by a colony living on its tip, and if so, the wasp moves upwards along the contaminated section (step 3), and may locate the colony (step 4). If the branch section on which the wasp alights is not contaminated, the female walks downward towards the plant base (step 5). If any of the colonies on the inflorescence are large enough to contaminate the whole section of the lower branch up to the main stalk, the wasp is bound to encounter the contaminated trail on its way down, and will then respond by moving upward (step 6) following the trail to the colony. If this does not occur, the wasp may fly to another branch or to another plant.

Predatory mites, such as *Amblyseius potentillae*, search the edges of leaves for their prey (Figure 6.4a, b and c) (Sabelis and Dicke, 1985). If the mite invades a leaf from the underside, it usually follows the midrib, before climbing onto the upperside of the leaf (Figure 6.4a). If a mite invades from the upperside of a leaf, it usually follows the edges and not the midrib which is less pronounced dorsally. That the mites follow edges is shown in Figure 6.4b and c where a hole has been cut from the centre of a leaf or the edges have been trimmed to a sharp angle. On a much larger scale, adult small white butterflies (*Pieris rapae*) follow natural and man-made topographical structures as they range from one area to another (Figure 6.4d, e and f) (Baker,

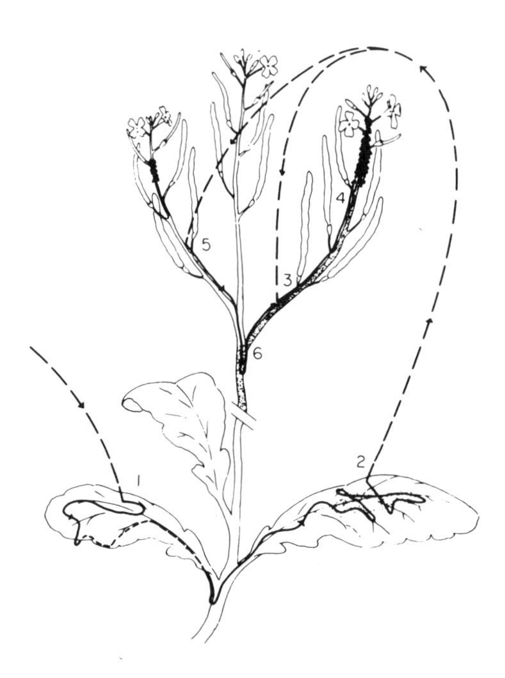

Figure 6.3 Typical track of a female parasitoid wasp (*Diaeretiella rapae*) searching for aphids on a crucifer plant. Shaded areas on plant indicate contamination from aphid honeydew. Aphids are represented by black dots. Dashed line represents movements of the parasitoid in a sequence indicated by numbers. From Ayal, 1987. Copyright 1987 British Ecological Society, reprinted by permission.

1978). The butterflies follow roads, hedges, and lines of telephone poles, and they orient toward distant objects. As shown in Figure 6.4e and f, the butterflies use different cues depending upon wind conditions and probably other abiotic factors.

Cree indians move from one patch to another (sheltered areas along shorelines) when hunting for smaller species such as muskrat (Winterhalder, 1981a). However, when Cree are hunting larger animals, such as moose or caribou that may move from patch to patch, the hunters orient according to tracks left by the prey. As a result the indians adopt an 'interstitial' foraging pattern, following creeks and lakes of a drainage pattern, and moving along the matrix which surrounds the most productive patches. When tracks of prey are located, the hunters follow them to the patch within which the prey will probably be feeding or resting.

The wolf (*Canis lupis*) provides a good example of a predator that captures relatively large prey by searching in groups rather than

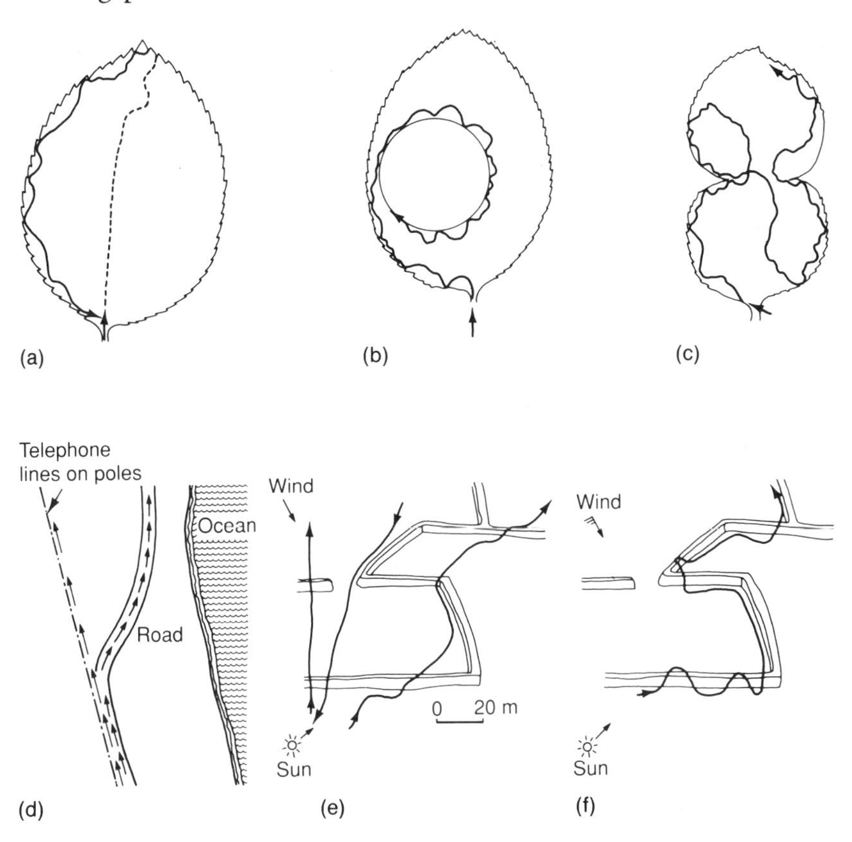

Figure 6.4 Examples of topographical search. (a, b and c) Edge-oriented walking pattern of a predatory mite (*Amblyseius potentillae*) on a rose leaf. Solid lines in (a) represent paths on the upperside of the leaf, whereas the dotted lines are the paths on the lower side of the leaves. (b and c) show that the cue is any edge rather than just the outer edge of the leaf. (After Sabelis and Dicke, 1985.) (d, e and f) Topographical orientation of butterflies (*Pieris rapae*). (d) Orientation to 'leading lines', such as a road or telephone lines. When a leading line coincides with the preferred sun orientation direction, butterflies delegate the maintenance of their direction to following the leading lines. The use of topographical cues ceases if the deviation between the leading line and the preferred direction becomes too large. In light winds, butterflies may fly straight across open spaces (e), whereas in strong winds (f) they fly in the shelter of leading lines. (After Baker, 1978.)

searching as individuals. Their strategy also includes many of the mechanisms that have been discussed thus far: ranging, topographical search, odour-stimulated upwind orientation, and visual stalking. Before a pack of wolves can bring down a prey, the prey must be localized, either by direct scenting or by tracking, both of which occur after a period of ranging (Mech, 1966, 1970). Foraging wolves orient in a relatively straight line, often using topographic contours of the land, and then switch to an upwind direction as a guide, when they perceive the odour of prey. Interestingly, the spotted hyaena (*Crocuta crocuta*) never approaches prey or carcasses from downwind, although its social encounters are usually initiated with one individual approaching another from downwind (Kruuk, 1972). For wolves, direct scenting is the most common method employed, whereby they detect the scent of a prey from approximately 300 to 500 m downwind. By knowing the wind direction and by watching from an aircraft for moose that were upwind of the trail of a pack of wolves, Mech was able to predict where the wolves should catch the scent as they moved across the wind along topographical landmarks. When that point was reached, the wolves suddenly stopped, pointed stiffly upwind, assembled nose-to-nose, wagged tails for 10 – 15 s, and then veered straight upwind toward the moose. The tracking strategy is used by wolves to find prey if, during ranging, a recent track of a suitable prey is encountered. In both cases the initial cue is olfactory, and in either case stalking is initiated once the prey is localized visually. Stalking is a restrained approach, which allows a predator to close the gap before the prey is alerted to the attackers' presence.

The shore crab is the only animal I know of which actually creates a topographical cue during its searching behaviour. It is a bit like Hansel and Gretel, in that a crab leaves its burrow carrying sand, walks straight away from the entrance, and as it goes it drops the sand in small piles. It can then use these markers to guide its return trip. The sand lines may also assist the crab in reducing overlap in successive searching bouts.

6.3 SUMMARY AND CONCLUSIONS

When searching for a resource patch an animal attempts to localize an assemblage of resources, whereas when searching within a patch it avoids leaving until it becomes unprofitable to remain. When an animal has no information about the location of resources in the environment, it moves so as to optimize its chances of locating resources and to reduce the chances of revisiting. Searching efficiency improves when animals have access to sensory cues. With patch-specific cues, information about potential distant resource patches can be gained through sensory input. Resources located in association with particular community members may be more easily located because interactions between the two produce attractive sensory signals.

7

Restricting search to a patch

Although the stimuli that evoke local search and the exact nature of the motor pattern that is exhibited may differ from species to species, local search in most cases seems uniquely designed for finding other resources in a patch once the first one has been utilized. For this reason, when local search is stimulated by resource utilization it is sometimes referred to as 'success-motivated search' (Vinson, 1977). After resource utilization releases a local search program, the motor pattern then readjusts over time until the motor pattern exhibited prior to resource utilization is reached. Resource-specific cues, such as food odours or sex pheromones, may also evoke local search. When resource-specific cues stimulate local search, the animal may eventually habituate to these stimuli and leave the patch. Regardless of the proximate mechanism, transition from local search to ranging or some other emigration mechanism not only ensures that an animal will search intensively at the appropriate time and place, but also that it will terminate search and emigrate before too much energy is expended searching in a depleting patch. Dixon (1959) summarized the adaptive significance of this strategy where resources are aggregated as being advantageous for an animal to search an area more thoroughly where it has already encountered host or prey; by doing so its chance of finding others is increased. However, this tactic would fail miserably if resources were sparsely distributed.

There are basically four ways to restrict search to a profitable patch (Figure 7.1a, b, c, d and e):

1. looping or spiralling motor patterns, in which the animal maintains a turn bias to the left or right which slowly readjusts to straighter locomotion by reducing turn bias; or zigzag patterns in which the animal alternates moves to the left and right, then readjusts toward straighter locomotion by decreasing the frequency of alternation;
2. decreasing movelength (distance walked before stopping to scan) after finding a resource, and increasing movelength when resources are not being encountered;
3. leaving a resource in a different direction than the arrival direction; and

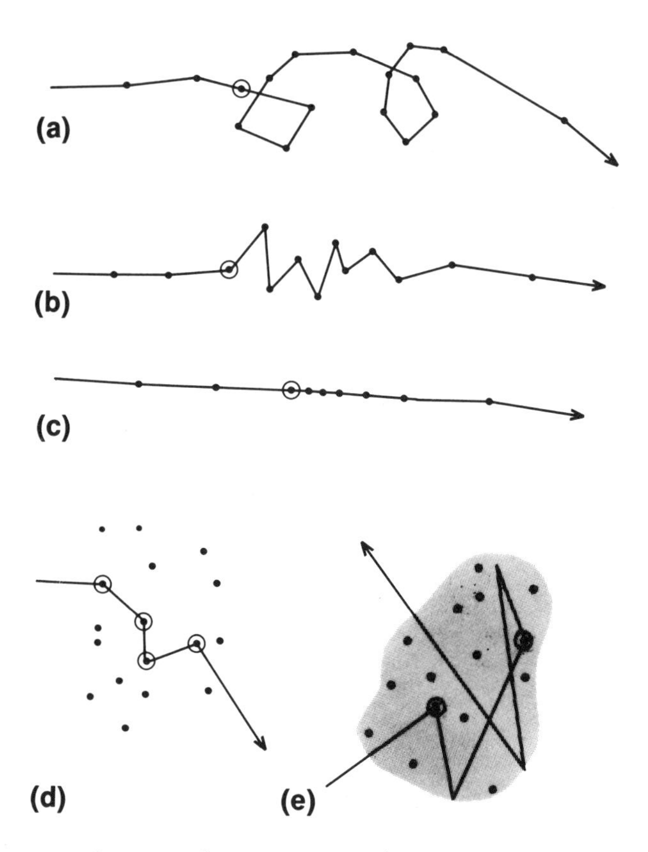

Figure 7.1 Mechanisms for restricting the area of search. (a) Periodic increases in turn bias, generating looping. (b) Turn direction alternation, generating zigzags. (c) Adjustments in move length between stops. (d) Change in arrival-departure directions. (e) Patch-edge orientation. Circled dots indicate resources that have been utilized.

4. turning back into a patch when the edge of a patch is detected.

It is important to point out that the patterns discussed above and illustrated in Figure 7.1 are not mutually exclusive. For example, patterns a, b, d, and e could all be modulated by changing movelength, or pattern e might include zigzagging or spiralling within the patch boundaries.

Local search patterns improve resource localization by maintaining the animal within a restricted zone as it searches the area in a thorough manner, and by widening the perceptual scanning field (discussed in detail below).

7.1 LOOPING/SPIRALLING OR ZIGZAG MOTOR PATTERNS

Local search in many species is characterized by a high turning rate, strong turn bias, and low locomotory rate, all of which decay over time toward a low turning rate, weak turn bias, and high locomotory rate (Figure 7.2; see also Figure 17.7). Motor pattern transitions of this kind, typically forming paths with spiral configurations, occur after a prey is ingested by an insect predator (review: Bell, 1985). Similar patterns occur in parasitoid wasps after oviposition (Hafez, 1961), in flies after ingestion of drops of carbohydrate (e.g. Dethier, 1957), and in predatory mites (Sabelis *et al.*, 1984) and parasitoids (Beevers *et al.*, 1981) when resource or patch cues are perceived. In some species of birds, turn bias is maintained for a few hops after a prey item is found and ingested, although not long enough to execute a spiral (e.g. Smith, 1974a).

On the basis of experiments with carrion crows finding prey arranged in various densities, Tinbergen *et al.* (1967) and Croze (1970) argued that local search was responsible for the birds locating more prey in dense but not in scattered arrays. After finding one prey item, crows searched in the immediate vicinity before moving on, thus increasing the likelihood of capturing nearby prey. In his study of European thrushes (*Turdus merula* and *T. philomelas*) Smith (1974a,b) showed that the birds made larger turns and maintained a turn bias to the left or right (rather than alternating left and right, as they usually do) after capturing an earthworm in a meadow. Observing house sparrows (*Passer domesticus*) feeding on mealworms distributed on a grid, Barnard (1978) recorded decreases in the rate of hopping and in directionality following each worm capture.

Ingestion of a prey item by stickleback fish (*Gasterosteus aculeatus*) is followed immediately by local search, with a lower rate of locomotion and an increase in 'searching movements' (Beukema, 1968; Thomas, 1974). However, area-avoidance behaviour, characterized by an increase in linear displacement and a decrease in searching movements, occurs if the encountered item is rejected. Area-avoidance behaviour is stronger when a fish rejects the prey item on the first encounter than when it makes repetitive encounters before finally rejecting it. It is also stronger when the previous prey was rejected than when the previous prey was accepted. This influence of previous experience wanes over time, dissipating entirely after 11 minutes in the absence of further encounters.

Odours of host, prey, or individuals of the opposite sex can also 'arrest' animals in resource patches by releasing local search. For example, the predatory mite (*Phytoseiulus persimilis*) initiates local

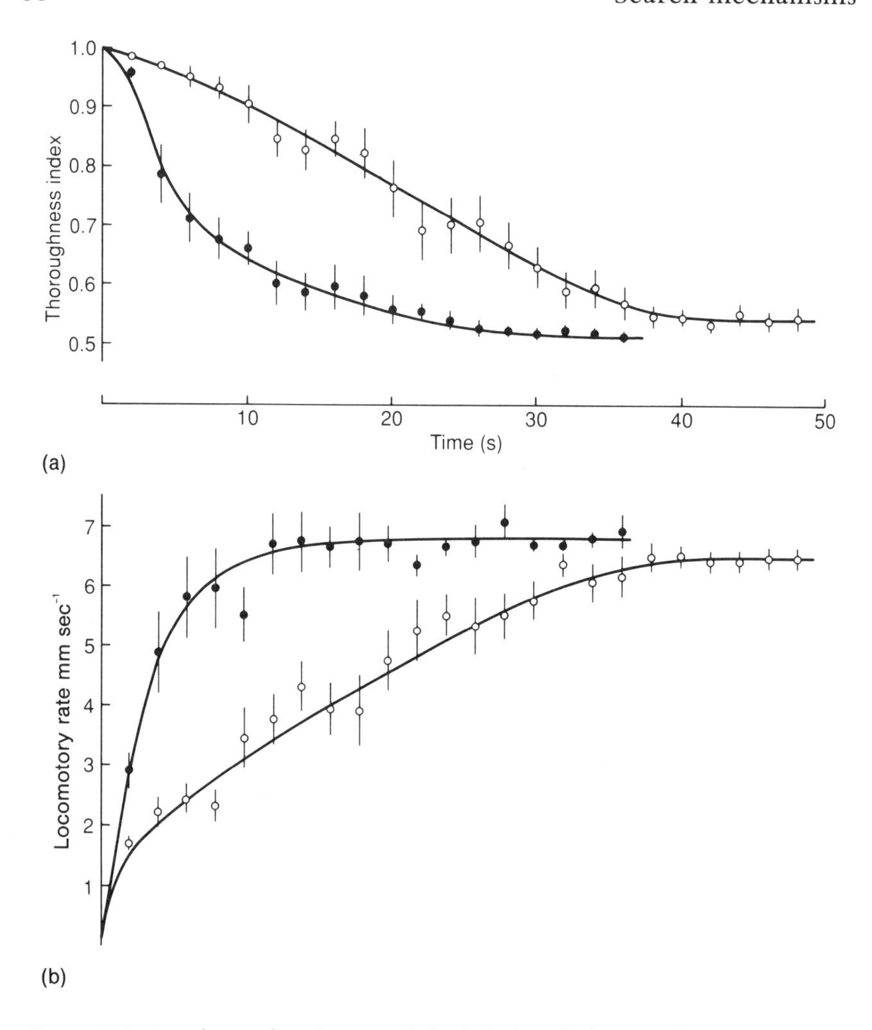

Figure 7.2 Local search in larvae of the lady beetle (*Coccinella septempunctata*) deprived of food for 5 hours (solid dots) or 25 hours (open dots). Vertical lines are standard errors. (a) Changes in 'thoroughness' of searching and (b) changes in locomotory rate, after a beetle has completed feeding on a first instar aphid. (After Carter and Dixon, 1982.)

search when it perceives odour from bean leaves infested with its prey (Sabelis *et al.*, 1984), and males of the grain beetle (*Trogoderma variabile*) respond in a similar manner when given a puff of female sex pheromone (Tobin and Bell, 1986) (refer to Figure 2.1a). As females of a geometrid moth (*Cidaria albulata*) approach a host plant, the plant odour stimulates increased turning frequency, so that the flight pattern describes tighter and tighter loops and circles (Douwes, 1968).

Contact with a resource followed by loss of contact can stimulate local search. Larvae of the cabbage butterfly (*Pieris rapae*) exhibit local search after host-plant contact. Here, restriction of the area searched is due to an increase in turn alternation to the left and right and a decrease in locomotory rate (Jones, 1977). Turn bias appears not to be a component of local search in caterpillars, although directionality decreases when they zigzag. In a similar case, local search is stimulated in the male German cockroach (*Blattella germanica*) if it loses contact with a female during courtship (Schal *et al.*, 1983).

An interesting way in which a chemical can exert its effect is to override positive phototaxis and/or negative geotaxis. In host-finding behaviour of the parasitoid *Trichogramma pretiosum* the presence of a chemical extracted from the scales of its moth host, the corn earworm (*Heliothis zea*), decreased the period before the parasitoid contacted the first egg, and increased the total time spent on an odour-contaminated surface (Beevers, 1981). The chemical also inhibits flight, a major factor in determining when a parasitoid leaves a patch. Thus, host odour perception, combined with the intensified local search pattern stimulated by oviposition into the host (Waage, 1978), provides a dual mechanism for keeping the parasitoid within a patch.

7.2 PATCH-EDGE RECOGNITION

Two forms of directional information which may restrict search to a resource patch are temporal-olfactory and spatial-visual perception of a resource patch border (Figure 7.1e). The cues that operate in patch-edge recognition differ according to the perceptual abilities of a species and the characteristics of the patch border.

Patch-edge orientation may be controlled by temporal olfactory information processing, restricting the greenhouse isopod (*Armadillidium nasatum*) and the yellow mealworm beetle (*Tenebrio molitor*) within a simulated resource patch of high humidity surrounded by an environment of low humidity (refer to Figure 2.2) (Havukkala and Kennedy, 1984; Sorensen and Bell, 1986). Although the edge of the patch is recognized by temporal processing, which triggers the turn back into the patch, the dimension of the turn must be specified by internally-stored information. This is so because there is no way that an animal can 'calculate' the turn required on the basis of on/off temporal information perceived as it walks from the inside to the outside of a patch. Even unicellular organisms such as Paramecium change course by turns exceeding 90° in response to a change from a favourable to an unfavourable stimulus (Dryl, 1973).

Parasitoid wasps, *Venturia canescens*, and predatory mites, *Phytoseiulus persimilis*, remain within a patch of host odour by execut-

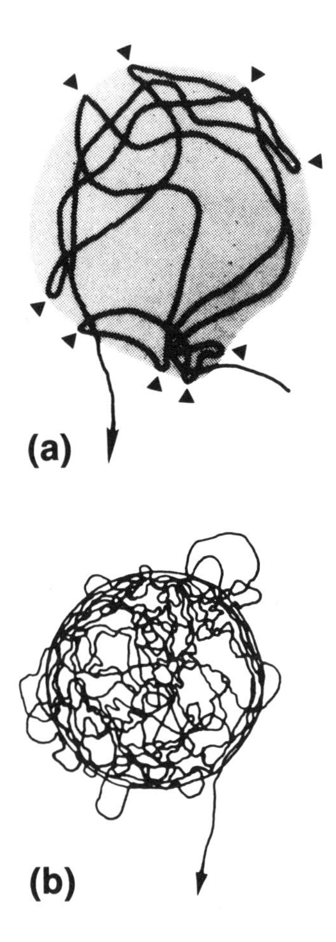

Figure 7.3 (a) Path of a female parasitoid (*Venturia canescens*) on a glass plate bearing a patch of host secretion. Stippling marks the edge of the patch; arrows indicate turns made at the patch edge. (After Waage, 1978.) (b) Path of a sugar-deprived blowfly (*Phormia regina*) on a homogeneous 5-cm patch of 1.0 M sucrose. From Nelson, 1977. Copyright Pergamon Press, reprinted by permission.

ing large dimension turns at the border of the patch (Figure 7.3a) (Waage, 1978; Sabelis *et al.*, 1984). Waage (1978) demonstrated that *V. canescens* moves through a patch, detects prey through olfaction and then oviposits into the prey. A wasp within a patch responded to patch odour by probing, decreasing locomotion, and making sharp turns back into the patch whenever the patch edge was encountered. The parasitoid *Cardiochiles nigriceps* displays a similar behaviour on patches contaminated with the odour of its host, the tobacco budworm (*Heliothis virescens*) (Strand and Vinson, 1982): when the edge of an odour-contaminated patch is encountered, the parasitoid briefly stops walking, begins to antennate the patch surface, and then enters the patch. Both locomotory rate and turning rate increase. After entering a patch, the wasp turns sharply back toward the odour-contaminated zone if it again encounters the edge.

Modalities other than olfaction are also involved in patch-edge recognition. Gustatory receptors in the tarsi of flies (*Phormia regina* and *Drosophila melanogaster*) (Nelson, 1977; Mayor *et al.*, 1987), detect the edges of homogeneous sucrose patches painted onto the substratum (Figure 7.3b). Visual cues are also used. Large dimension turns at the border of milkweed patches by milkweed beetles (*Tetraopes tetraophthalmus* and *T. femoratus*) (Lawrence, 1982), and turns back into a dark zone after entering a lighted zone by cockroach nymphs (*Blaberus craniifer*) (Bell *et al.*, 1983), seem to be mediated by visual information at patch edges. *Rhagoletis pomonella*, a temperate zone tephritid fly whose larvae are endoparasitic on fruit of hawthorn (*Crataegus*), probably uses visual cues to remain within a profitable tree. While searching for fruit within trees adult females hop from leaf to leaf and fly in short loops. If a fly oviposits, it then flies in a local search pattern, and if a fly fails to locate any fruit in trees with fruit clusters, it leaves the tree after a fixed searching time (e.g. Roitberg *et al.*, 1982).

.3 VARIABLE MOVE LENGTHS

When scanning is accomplished primarily during pauses, the ratio of time spent moving versus scanning can vary considerably from very long pauses in animals that actively search for a good perch or ambush site and then wait for prey (e.g. web-building spiders), to short pauses in animals that stop briefly to scan and then move to another site (e.g. plovers). This idea of scanning during pauses has been referred to as saltatory search (O'Brien *et al.*, 1986), which functions to improve efficiency of two different types of foraging tactics. First, animals that seek resources in patches can remain in high quality patches and move through low quality patches quickly by using shorter moves in productive patches and longer moves in unproductive patches. Move length can be defined as the distance moved between scanning inputs and, as implied, some animals apparently do not perceive or at least do not respond to resources while they are between scanning pauses. Second, searching behaviour of species that require individual and often isolated resource items has a search cycle composed of alternating periods of stops and runs; the distance between stops determines how much overlap there will be in scanning from one point to the next.

As discussed in Chapter 3, plovers forage by visual scanning (see Figure 3.2a). They scan an area while pausing, peck at a prey item, and then run to another site. By not moving at all after taking the most desirable prey, and by moving shorter distances after taking an acceptable prey (1.14 paces) than after taking no prey (4.62 paces), the ringed plover tends to concentrate its feeding activity on the areas of highest quality or most abundant food (Pienkowski, 1983). Bumblebees have a similar strategy, as shown by measuring distances moved in rich (screened) and poor patches (unscreened) of clover (Heinrich, 1979b). Figure 7.4a and b show that move lengths were longer in poor than in rich patches. Schmid-Hempel and Schmid-Hempel (1986) studying the same problem in honey bees found that honey bees are sensitive to both the resource quality and patch size.

Changing locomotory rate can have the same effect on the regulation of the distance moved between scanning stops as changing move length. For example, searching of larval anchovy (*Engraulis mordax*) was significantly slower in patches of aggregated *Gymnodinium splendens* than elsewhere (Hunter and Thomas, 1974). Vivas and Saether (1987) showed that foraging intake of moose (*Alces alces*) varies with sapling density. Whereas the number of visits made by moose to low- and high-density plots was approximately constant, they remained longer in high-density plots. Innes and Mabey (1964) found a similar movement pattern for cattle, which ranged in an unpredictable manner along routes between clumps of their selected food species. The explanation seems to be that cattle move more slowly in profitable patches than in less profitable ones.

In the white crappie (*Pomoxis annularis*) and probably other planktivorous fish, the foraging cycle consists of stationary pauses during which scanning occurs, and moves during which no scanning occurs. If scanning is successful, the fish pursues the prey, and if unsuccessful, the fish moves and then positions itself again for a scanning pause (see Figure 3.4a and b) (Evans and O'Brien, 1986; O'Brien *et al.*, 1986). Table 7.1 shows relative times for components of the foraging cycle when crappie prey upon small and large zooplankton. Note that scanning time for large prey is only half that

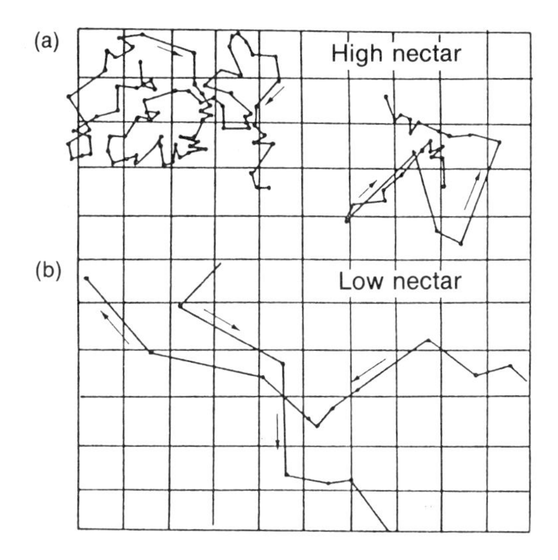

Figure 7.4 Representative search paths of four worker bumblebees (*Bombus terricola*) in clover patches. Clover patch in (a) had been screened to exclude all foragers for two days prior to the experiment; clover patch in (b) had been left available to all foragers (after Heinrich, 1979b). Each dot represents a flower head visited. Lines do not represent actual flight paths. Grid lines are 15 cm apart

Table 7.1 Time budget for the white crappie, *Pomoxis annularis*, feeding on small and large zooplankton at 20°C (from Evans and O'Brien, 1986)

	Successful scanning (s)	*Unsuccessful scanning (s)*
Small zooplankton		
Scan time	1.3	1.8
Pursuit	0.85 (5 cm)	
Run		1.1 (6 cm)
Total time:	2.15	2.9
Large zooplankton		
Scan time	0.7	1.2
Pursuit	1.5 (15 cm)	
Run		0.81 (12 cm)
Total time	2.2	2.0

for small prey, as though the fish adjusts the scanning period according to the time required to fixate on prey.

7.4 CHANGES IN ARRIVAL–DEPARTURE DIRECTIONS

One way that an animal can remain in a rich patch and spend less time in a poor patch is to move straight ahead when the sampled resources are poor and move in some other direction when the sampled resources are rich. Motor patterns of animals foraging within a patch of like resources have been intensively investigated in nectivorous insects and birds visiting flowers. Bees obtain and use information pertaining to resource quality to specify directionality in their orientation, the net effect of which is to remain in food-rich areas and to move through less rewarding areas. This pattern of orientation is functional when nectar sources occur in dense patches.

Pyke (1978) predicted, using optimal foraging theory, that bees should stay in a patch of highly rewarding flowers, pass quickly through a patch of low-reward flowers, and not forage at all in a patch of empty flowers. In testing this theory, Pyke found that indeed the mean angular deviation between the arrival and departure flight direction of bumblebees (*Bombus flavifrons*) increased with the number of flowers visited on an inflorescence, and he developed a rather complex set of departure rules to explain these results:

1. the mean angular change in direction from arrival to departure should be 0°,
2. the animal should alternate right- and left-hand turns between visits, and
3. the width of the visual scanning sector, within which the next flower to be visited is found, should increase as the amount of food obtained at a resource point increases.

As summarized by Heinrich (1983), however, to accomplish this feat a bee must not only measure rewards at each flower visited, but it must remember the direction it came from each time it visits a flower before computing a new departure angle; it must then remember the rewards of previously visited flowers as a basis of comparison, as well as keep track of previous departure directions in order to alternate departures first to the right and then to the left, or vice versa. A more parsimonious proximate explanation is that the more the bee turns on the flower while feeding, the more random its take-off direction will be relative to the approach direction. In fact, while OFT was perhaps useful in generating a 'functional' model of behaviour, it led to a totally inappropriate hypothesis of the proximate mechanism involved. As shown in Figure 7.4a and b, B. *terricola* workers tend to move mainly straight ahead in patches of clover with low nectar yield, and to turn in various directions relative to the arrival direction upon leaving clover heads that had been screened to increase nectar yield (Heinrich, 1979b). More recently, Schmid-Hempel (1984) and Kipp (1987) provided direct evidence for a decay in the angular correlation between arrival and departure direction with time on a flower.

The following points summarize within-patch foraging movements of bees:

1. a bee lands on a flower and engages in probing and searching movements which may vary according to reward,
2. it moves to the nearest edge of the flower in the approximate direction of that assumed while probing the last-visited floret, and
3. flies to the nearest flower which is in the centre of its visual field.

Thus, the directionality of the interfloral track differs from a straight line as a function of the amount of change in direction evoked by its movements on individual flowers. Kipp (1984) argues that these rules emphasize forager 'decisions' relating to the present conditions rather than attempts to forecast future moves relative to previous ones. An implicit assumption of optimal foraging models, that the animal 'knows' or can estimate the magnitude of the parameters which appear in the equations, means that an animal must be able to gather information relevant to these parameters, must 'remember' this information for use at a later time, and must translate all of the available information into estimates of the parameter values (Pyke *et al.*, 1977). Pyke's (1978) basic assumption may be appropriate in order to describe the economic utility of bee foraging paths. However, to assume that bees are cognizant of the parameter values, and then to seek proximal mechanisms invoking such cognizance to explain bees' turning behaviour fails the test of parsimony. This is so because it is not necessary for animals to know anything about the parameters appearing in the mathematical descriptions of their behaviour; it is only necessary that the animals 'appear' as if they know. In other words, we must make the fine distinction between model and mechanism parameters when considering descriptions versus explanations of foraging behaviour.

.5 SUMMARY AND CONCLUSIONS

Local search patterns, which are stimulated by resource utilization or perception of resource cues (food odours or sex pheromones), improve resource localization by restricting an animal as it searches a patch thoroughly, and by widening the perceptual scanning field.

After a local search program is initiated, the motor pattern then readjusts over time until the pattern exhibited prior to resource utilization is reached. This transition not only ensures that an animal will search intensively at the appropriate time and place, but also that it will terminate search and depart a depleting patch.

8

Foraging in the most profitable patches and leaving when profitability declines

If an animal finds a resource in a patch, it may be able to gauge the worth of the patch based on its initial merits and then as it changes over time. Thus, the first important concept in studies of animals searching in depletable patches is that patches differ in quality before the animal begins to utilize resources. Secondly, resource availability within a patch decreases (resource depression). This decrease may be brought about because of an animal's own foraging activity, because a resource remains within the sensory field of the animal but has no further value (i.e. when females are already mated or when the available hosts have been oviposited in), or because resources become less vulnerable to attack and/or exploitation as a result of previous use (i.e. when trees produce toxins rendering leaves less palatable than before herbivory occurred). Obviously, the concepts of initial patch quality and patch depletion are not distinct, since at some point in time even a high quality patch turns into a lower quality patch as the resources are depleted.

Other things being equal, a profitable patch can usually be defined as one with a relatively high initial resource density, since high density usually translates into a greater resource procurement per unit time than does low density. In fact, Royama (1970) defined patch profitability in terms of capture rate. The idea is that a forager in an environment with a set of patches and a certain amount of foraging time allocates this time so as to forage in the most profitable patches (with the highest capture rate). Many studies point out that most species spend their time in patches with the highest density of resources. For example, Figure 8.1a and b depicts preferences of bats (*Pipistrellus pipistrellus*) for patches with the highest densities of flying insects both in time and in space (Racey and Swift, 1985). At densities of less than 300 insects per 1000 m^3 air, pipistrelles did not remain in an area to forage, but moved away within 1 minute of their arrival.

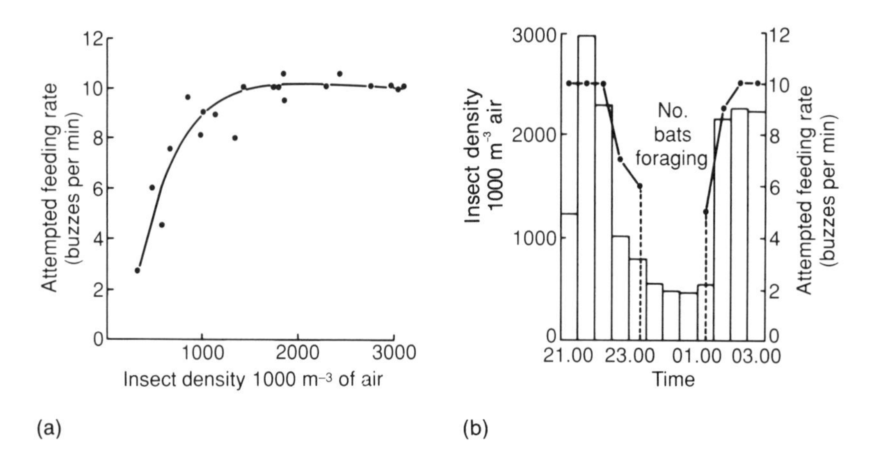

Figure 8.1 Effects of prey density and temporal availability on foraging of individual *Pipistrellus pipistrellus* bats. (a) Attempted feeding rate by individual bats relative to prey density in various foraging areas. (b) Insect density (columns) and attempted feeding rate (solid dots) of lactating female bats measured at 30 min intervals through the night in one foraging area. From Racey and Swift, 1985. Copyright British Ecological Society, reprinted by permission.

At higher densities, the rate of feeding attempts increased proportionally with insect density, and then began to level off with a maximum rate of 10 buzzes per minute at approximately 1250 insects per 1000 m^3 (Figure 8.1a). The feeding rate varied throughout the night when female bats were lactating (Figure 8.1b). It was maximal for the first 1.5 hours, but then declined as insect density fell and the bats became satiated. Then a period followed when no bats foraged, coinciding with both the lowest insect density and the time when females returned to the roost, presumably to suckle their young. The rate of feeding then increased sharply to reach a maximum at the time of the predawn insect peak.

Patch depletion, the second important point concerning patch quality, could come about by direct use of the resources, as in a coccinellid beetle ingesting aphids in an aggregation, or when females are successively mated, or as plants become oviposited upon. Depletion can also occur indirectly, as when females take evasive action because of a predator and are thereby unavailable to males. The consequence of resource depression is that an animal arrives in a patch and initially locates resources at a rapid rate. As resources are depleted, the rate of acquisition decreases.

Leaving a patch or equivalent foraging area and searching for new patches because of resource decline is well documented among herbivores and carnivores searching for food, parasitoids searching for hosts, and males searching for females. Consider a coccinellid beetle searching on a plant for aphids. It makes sense that a beetle will remain in a patch if there is a good chance of locating more aphids, but it should leave if the chances are slim. This is because the beetle should not squander all of the energy gained from its patch visit by staying too long in search of declining numbers of prey. Although most species seem to follow the rule of leaving a patch before the point of diminishing returns, the proximate mechanisms by which this feat is accomplished differ markedly depending on an animal's abilities. In the following section various proximate mechanisms are discussed, from simple patch-departure mechanisms to mechanisms requiring counting, sampling between patches, calculating, averaging, and remembering.

8.1 SIMPLE PATCH-DEPARTURE MECHANISMS

It is clear that the local search mechanisms described in Chapter 7 can delimit the movements of animals to productive resource patches. It is just as important to realize, however, that each of the local search mechanisms discussed provides a proximate means for the animal to leave a patch. Moreover, not all animals have to register capture rate or tell the time in order to leave a patch before the point of diminishing returns.

If an animal switches to local search after it utilizes a resource, and if searching in a patch is simply the repetition of this time-dependent change in the motor pattern over and over again, this simple mechanism could indirectly determine when an animal will leave a patch in relation to resource decline. For example, Carter and Dixon (1982) carried out experiments with larvae of a coccinellid beetle (*Coccinella septempunctata*) foraging in an apparatus in which the larvae moved up an artificial stem onto circular horizontal discs containing aphids. Each time a beetle larva ingested an aphid, its locomotory rate decreased and its turning rate increased, followed by a decay in both measures toward the prefeeding values over a period of 30 to 40 s (refer to Figure 7.2). Larvae left the patch when most of the aphids were eaten. The proximate mechanism for leaving the patch seems to be that the relatively straight emigration pattern, to which the larvae averaged 30 to 40 s after feeding, decreased the probability of contacting aphids and subsequently increased the probability of leaving the patch. As might be expected, the time spent in a patch was greater when the initial density of aphids was 0.66 per cm^2 as compared to 0.16 per cm^2.

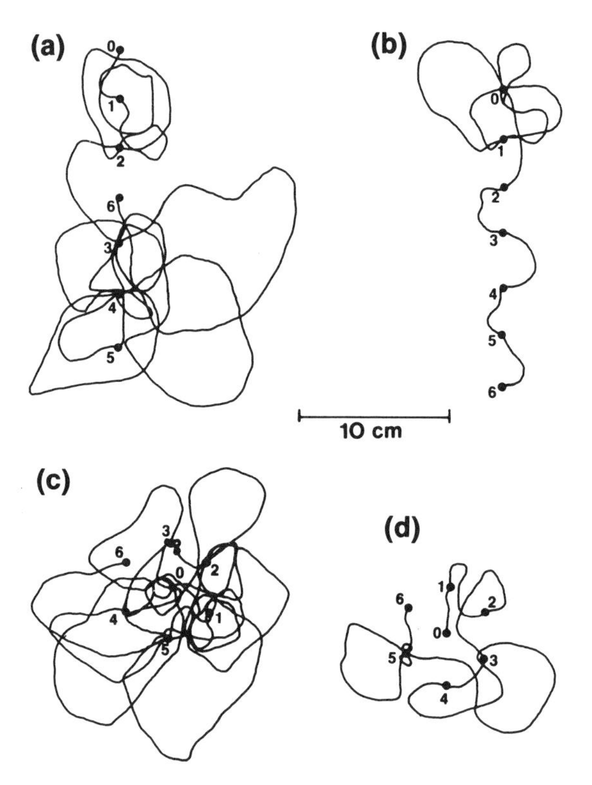

Figure 8.2 Videotaped search paths of houseflies (*Musca domestica*) in patches of sucrose drops. Sucrose drops are positioned in a row (a and b), or in a hexagon (c and d). Drops are 3 cm apart. Numbers show order in which drops were found. (After Fromm and Bell, 1987.)

Searching for sucrose drops in patches has been studied in the housefly (*Musca domestica*) (Murdie and Hassell, 1973; Fromm and Bell, 1987). Sucrose drops simulate honeydew deposits on leaves beneath aphid colonies, upon which *M. domestica* and other flies commonly feed (Downes and Dahlem, 1987). Fromm and Bell (1987) introduced flies into a patch of 1 μl sucrose drops (0.25 M) arranged in a row, by leading the fly to a drop at the end of the row. The search paths (Figure 8.2a and b) were highly variable because the flies did not always discover the drops sequentially and also because they often encountered a previously visited drop. Revisiting of this sort is a phenomenon that would not occur in animals like the lady beetle or house sparrow which consume the entire resource. The important point here is that the turning rate increases and the locomotory rate decreases immediately after ingesting each drop, as the fly locates successive drops over a 4 min period (Figure 8.3) (Fromme and

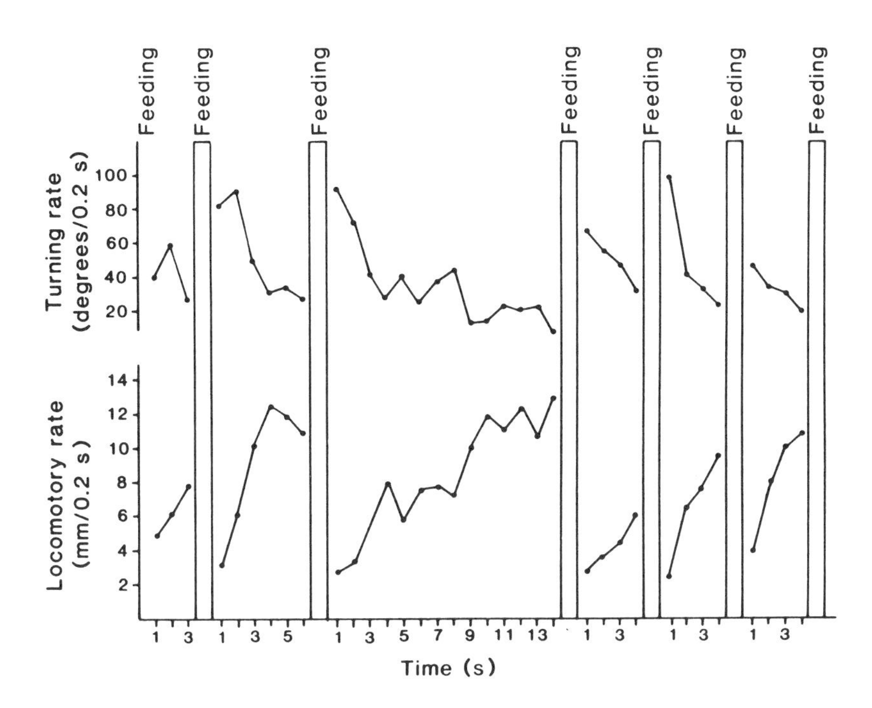

Figure 8.3 Turning rate and locomotory rate between feeding bouts for an individual housefly (*Musca domestica*) searching in a patch of sucrose drops. Columns indicate feeding periods. (After Fromm and Bell, 1987.)

Bell, 1987). If satiation is not a factor, the fly repeats its searching tactic over and over again, stimulated at each drop by the ingestion of sucrose. As with lady beetle larvae, if a drop is not found within a certain time, which would be likely to occur as the drops are used up, the motor pattern reverts to relatively straight and rapid movement, reducing the chances of encounters with sucrose drops and leading the fly out of the patch (Figure 8.4).

There is substantial evidence that parasitoids, and perhaps other animals, leave a patch because they become habituated to odours released by prey or hosts within a patch (see Chapter 8 for details). For example, patch times for a parasitoid wasp (*Cardiochiles nigriceps*) in odour-contaminated patches decrease on successive visits (Table 8.1), suggesting that patch abandonment is caused by habituation to

Table 8.1 Average duration of successive patch visits by the parasitoid *Cardiochiles nigriceps* (from Strand and Vinson, 1982)

Visit	*Time**
1st	349.4 (314.2)
2nd	133.5 (74.9)
3rd	37.9 (27.5)
4th (after termination of 3rd visit and held for 1 h)	317.0 (202.6)

*Mean values with standard deviations in parentheses.

the host odour (Strand and Vinson, 1982). Although this mechanism seems to lack precision, it reduces the chances of a parasitoid remaining in a patch so long that it begins to encounter hosts which it has already parasitized. When combined with additional mechanisms, such as the way that oviposition lowers the olfactory threshold to host odour (Waage, 1979), the required precision can be attained. Time away from host odour would diminish habituation and the parasitoids would again be stimulated to oviposit. Thus, habituation to host odours, in combination with threshold changes in response to resource utilization, can act as a 'timer' to govern the duration of a patch visit. Habituation in this case is analogous to satiation in the flies, and the change in threshold caused by oviposition in the wasp is analogous to the change in locomotory pattern after feeding in the flies (Figure 8.4).

.2 COUNTING AND KEEPING TRACK OF TIME TO)ECIDE WHEN TO LEAVE A PATCH

In the following discussion we solve the basic problem of whether or not animals in nature or in semi-natural conditions can perceive changes in patch quality, and if they alter their behaviour accordingly. In the field, an investigator could either look for patches of different quality or monitor patches as they change in quality over time, or see if animals search mainly in the best patches and if they treat patches differently as they sense a change in overall patch quality. In the laboratory the idea would be to manipulate patch quality and to observe changes in the behaviour of test animals.

It is extremely difficult to obtain data from the field that clearly reveal the mechanisms by which animals decide when to leave a

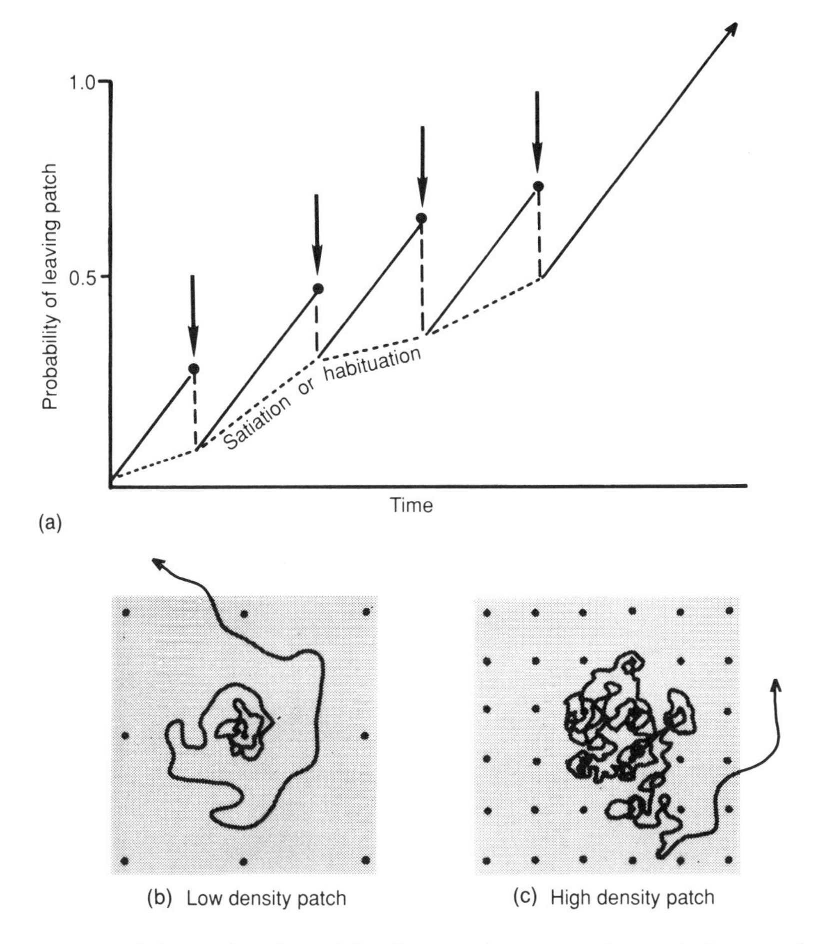

Figure 8.4 (a) Graphical model of a proximate patch-restriction mechanism. Arrows represent points in time at which an animal utilizes a resource (e.g. oviposits into a host, feeds on a resource). The probability of leaving a patch increases because an animal becomes satiated or habituated, and decreases because of resource utilization. (Based in part on Waage, 1979.) (b and c) In low density resource patches an animal is likely to 'miss' resources, and because its local search is not 'reset', it tends to leave the patch; in the high density patch an animal is more likely to encounter resources, and because its local search is reset, it tends to remain in the patch.

patch. A study by Gibb (1962) illustrates this point clearly. Larvae of a pine cone borer (*Ernarmonia conicolana*) eat the seeds and then overwinter beneath the cuticle of ripening pine cones. Blue tits (*Parus caeruleus*) and coal tits (*Parus ater*) prey upon the larvae in autumn and winter, and in so doing they leave unmistakable traces on the cones, which Gibb used to estimate the number of larvae eaten. Dissection of the cones allowed an estimate of the number of larvae surviving.

Gibb (1962) compared intensities of prey both seasonally and between neighbouring localities. Intensity is defined as the number of larvae per 100 cones from an area, whereas density is the number of cones per $m^2 \times$ intensity per cone. Each cone is considered to be equivalent to a patch. Gibb was able to compare foraging of the birds in what would seem to be ideal situations for an investigator: changes in intensity over time and differences in patches at the same point in time. For the change over time, in 1955–56 when both cones and larvae were abundant, predation was clearly intensity-dependent: tits' predation increased linearly over a range of intensities from 10 to 65 larvae per 100 cones. In 1956–57, when pine cones were scarce but larval intensities were still high, predation was very low at the lowest larval intensities, increased to 50% with an intensity of 25 larvae per 100 cones, and remained at this level even in areas with intensities up to 105 larvae per 100 cones. As for differences in patches at the same point in time, tits experienced both high and low intensity in neighbouring localities during the course of a single day. The birds seemed to remove approximately the same number of larvae from cones independent of the number of larvae per cone in a locality (Gibb, 1958). In fact, in one particular locality with an average of 26 larvae per 100 cones, a single tree was found with 256 larvae per 100 cones. The tits' predation at this tree was significantly lower than might be expected, suggesting that the birds were using the average for the area to decide how many larvae to take per cone at this tree.

Gibb (1962) reasoned that perhaps the birds learn how many larvae to expect in a given locality, and that they use this reference to determine how many larvae to remove from a cone. It seemed as though the birds were leaving a cone after a fixed catch or that they were hunting by expectation. If the average for the environment was 10 larvae per cone, the bird would count to 10 and then leave, even though a particular site might have more larvae per cone than the average. Gibb (1962) concluded that

> such predation by expectation . . . represents a high achievement by the birds, implying not only that they learn how many larvae to expect in different localities, but more remarkably that they slacken their search when the expected number of larvae have been taken from the cones which bear the traces of previous attacks.

This explanation has been termed the number-expectation hypothesis, wherein a forager expects to take a certain number of items per patch, counts the items it uses, and then leaves when it obtains the expected number.

Gibb's conclusions spawned a series of 'fake pinecone' laboratory studies in which the experimental variables could be carefully controlled, rather than being dependent upon the whims of nature. Simons and Alcock (1971) designed an experiment in which white-crowned sparrows (*Zonotrichia leucophyrys*) learned to search at one of four possible 'cones' (blocks of wood with food hidden in six small holes). The sparrows sampled blocks by looking in one or two holes, and only continued searching the block if food was located. Thus, they used a simple rule of testing a few holes, and then continuing to sample if rewarded. In these experiments the birds did not form an expectation for number of prey they would remove. Simons and Alcock, criticizing the number-expectation hypothesis, point out that the mechanism proposed by Gibb would only work if most patches contained a standard number of prey; otherwise a bird would not be able to form a clear-cut expectation to use in its assessment. Krebs *et al.* (1974) raised another issue, noting that few instances of hunting by expectation had actually been observed, and that Gibb only examined cones after the tits had searched through them, using beak scars as an indicator of bird predation. If several birds had searched the same cone, each hunting with a different expectation, and if they did not avoid cones already inspected by others, the signs would not be indicative of predation rates per bird.

Krebs (1973) put forward an alternative theory, that birds spend a fixed amount of time on each patch, rather than eating a fixed number of prey. The birds in Gibb's study would then have taken proportionally fewer larvae from abnormally high-density cones if they spent a fixed time on each cone. This idea has been coined the time-expectation hypothesis, predicting that a forager expects to spend a certain amount of time in each patch; when that time is completed, regardless of patch quality, the forager leaves.

In still another series of artificial pinecone studies, Krebs *et al.* (1974) investigated hand-raised black-capped chickadees (*Parus atricapillus*) searching on artificial trees in an aviary. Artificial pine cones were placed on the trees with pieces of mealworm inserted into holes in the cones to achieve various densities. During preliminary training sessions over a 4-day period the cones contained one mealworm piece per cone, and then in the subsequent testing sessions the environment was switched to high density cones or high density trees. This protocol neatly replicated the presumed conditions of Gibb's (1962) study, but had one advantage over Gibb's in that the behaviour of the birds could be monitored. The results clearly rejected hunting by number expectation, because the birds did not

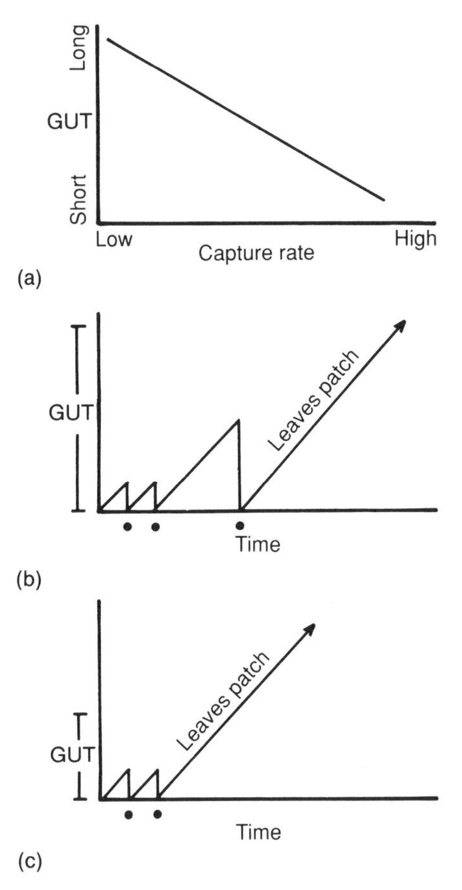

Figure 8.5 Graphical model showing how capture rate might set giving-up-time (GUT). (a) At a low capture rate, GUT should be relatively long, whereas at a high capture rate, it should be relatively short. This allows an animal to adjust to changes in resource density in patches in the habitat, as shown in (b) and (c). Dots indicate resource captures.

adhere to the rigid expectation which might have been gained through the early experience of finding one mealworm piece per cone, and because the birds found more mealworm pieces as soon as more were available. The data neither ruled out nor supported hunting by time expectation, since the time spent on high density cones or groups of cones was not significantly different from the time spent on cones of low density.

Krebs *et al.* (1974) also applied the above data set to a threshold capture rate model, in which the threshold rate is made equal to the average rate for the habitat, such that this rate increases and giving-up time (GUT) decreases as the average patch density in the habitat increases. GUT, the period between utilizing the last resource taken in a patch and leaving the patch, is 'set' according to the average capture rate for the environment (Figure 8.5a), such that a short GUT would be appropriate for an environment with high density patches, and a long GUT would be appropriate for an environment with low density patches (Figure 8.5b and c). Such differences in GUTs would be adaptive because an animal should search longer before giving up and leaving a patch in a poor environment than in a rich environment. As compared to the models previously discussed, the threshold capture rate model would theoretically allow an animal to leave a patch before the point of diminishing returns, but not so soon as to leave many resources behind. Analysis of data from chickadee foraging tests (Krebs *et al.*, 1974) showed that the mean GUT was longer in low-density environments, where capture rate was lower (2.8 prey per minute) than in high-density environments, where capture rate was higher (3.5 prey per minute). These results confirm the predictions of Charnov in that the capture rate threshold seemed to be adjusted to the average for the environment. When the performances of individual birds are examined, all of them failed to develop a number expectation; some seemed to follow the time expectation criterion by spending more time at high- than at low-density patches, but others did the reverse; all birds except for one individual conformed to the threshold rate model prediction.

The most fascinating foraging model is Charnov's (1976) marginal value theorem which attempts to relate the energy spent moving between resource patches to the time that an animal should spend in a patch. To maximize its rate of gain of resources from patches under conditions when resources are being depleted, we might expect a predator to remain in a patch for a period of time related to the cost of travel associated with locating the patch or to the probable cost of finding another patch. If a predator spends a certain period of time (T_t s) travelling to a patch and then consumes food according to the gain curve shown in Figure 8.6a (curved line), where on the curve of diminishing returns should the predator leave the patch and travel to the next one? The marginal value theorem states that in order to maximize energy per unit time the predator should leave the patch at point T_{opt} which gives the steepest slope of the line AB (solid line in the figure). The line must both touch the gain curve and intersect the x-axis at the travel time, since these are assumed to be fixed by the environment.

If travel time increases (patches are further apart) the optimal time to stay in a patch also increases (Figure 8.6b): the longer it takes to travel, the lower the value of moving, and so a predator should remain longer in the current patch. If patches vary in quality, the

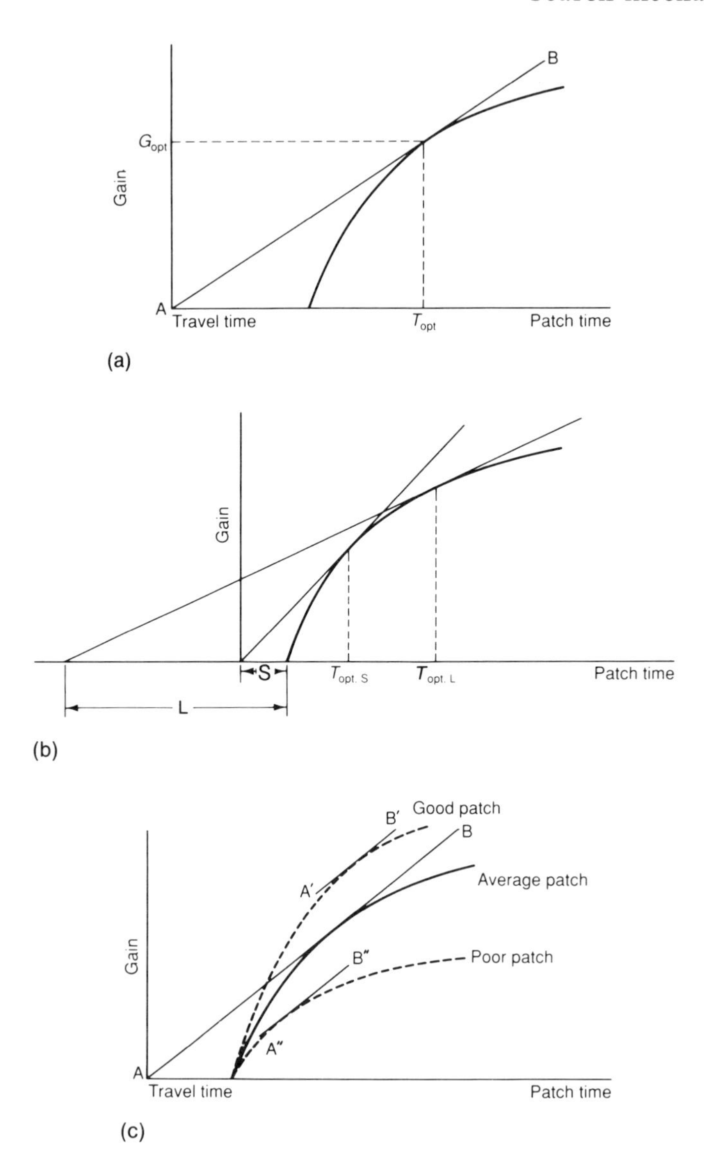

Figure 8.6 The marginal value model. (a) An example of a gain curve arising from resource exploitation, and the tangent (AB) represents the optimal patch resident time T_{opt}. (b) As travel time increases, so does T_{opt}. If travel time is short (S), patch time should be short, $T_{opt,S}$, and if travel time is long (L), patch time should be long, $T_{opt,L}$. (c) When there are several different patch types, each should be exploited until the gain drops to the average for the environment (line AB). (After Krebs and McCleery, 1984.)

model predicts that each patch should be exploited until the gain rate within the patch drops to the average for the environment (Figure 8.6c).

One way of testing the marginal value theorem is to measure the time between the predator's last capture in a patch and the moment it leaves the patch (GUT). As mentioned previously, a predator should stay longer in a poor environment because it can deplete the current patch further before it is worth spending energy in travelling to another patch. Some tests of the marginal value theorem are found in Chapter 12.

Waage (1978) provides a detailed study of searching behaviour of a parasitoid (*Venturia canescens*) which oviposits in pupae of its host, a meal moth (*Plodia interpunctella*) usually found in fallen fruit, and delineates several important factors which interact to determine patch time. The wasps searched in patches of wheat middlings in which pupae were arranged in various densities. A wasp moved through the patch, locating its prey through olfaction and ovipositing into the prey, and then after a certain period of time left the patch. Recall from Chapter 7.2 that parasitoid wasps within a patch respond to patch odour by probing, decreasing locomotion, and by sharp turns back from the patch edge whenever it is encountered. Gradual habituation to the odour results in an eventual abandonment of the patch. Two factors are important for understanding the proximate mechanisms underlying patch times in this wasp:

1. increased concentration of patch odour (which might result from high host density) increases the duration of visitation, indicating that habituation at higher chemical concentrations takes longer than at low concentrations; and
2. oviposition also increases the duration of patch time, probably by decreasing the rate of habituation to odour.

Waage (1979) applied to his study the models discussed previously. Two hypotheses were tested:

1. the hypothesis that each oviposition of the wasp confers a fixed increment of time on to the duration of a patch visit (additive hypothesis), and
2. a threshold rate hypothesis, predicting a variable increment which is dependent on the time since the last oviposition and independent of the total number of ovipositions previous to that one. In other words, as the time between ovipositions in a series gets shorter, the associated increment of patch time per oviposition decreases from the GUT to zero.

To discriminate between an additive and a threshold rate hypothesis,

Table 8.2 Time spent by the parasitoid *Venturia canescens* in a patch of hosts (from Waage, 1979)

	(A) *1 oviposition upon entering patch* *(s)*	*(B)* *5 oviposition upon entering patch* *(s)*	*(C)* *5 ovipositions over first 15 min on patch* *(s)*
Time* until off patch for 1 min	26.81 (10.59)	22.79 (6.23)	34.92 (8.70)

*Results are mean values with standard deviations in parentheses.

A vs C, $P < 0.02$; B vs C, $P < 0.053$; A vs B, $P < 0.99$.

naïve wasps were used in the following experiment. Over patches (5.5 cm in diameter) of host-contaminated middlings a screen could be raised or lowered to prevent or allow oviposition. Three different regimes were tested in which a wasp was allowed to oviposit

(A) once upon entering a patch (this required approximately 1 minute),
(B) five times consecutively upon entering a patch (this required approximately 3 minutes), and
(C) five times upon entering the patch, but with ovipositions spaced at 3 minute intervals over a period of 15 minutes.

If the additive hypothesis is correct, then (B) should equal (C), and both (B) and (C) should exceed (A). The threshold model would predict that patch times for (B) are less than (C), provided that the GUT was greater than 3 minutes, and if not, then (B) should be greater than or equal to (C); and patch times for (B) and (A) should be equal, because of the similar timing of the last oviposition (3 and 1 minutes, respectively). The data shown in Table 8.2 support the threshold model, in that the timing of ovipositions appears to be a better determinant of patch time than the total number of ovipositions.

Based on the results described above, Waage (1979) developed a simple quantitative model based on the idea that patch time is determined by '. . . the interaction of two incremental processes, the response to the contact chemical stimulus and the response to oviposition, and one decremental process, the waning of the response to the patch edge . . . through habituation':

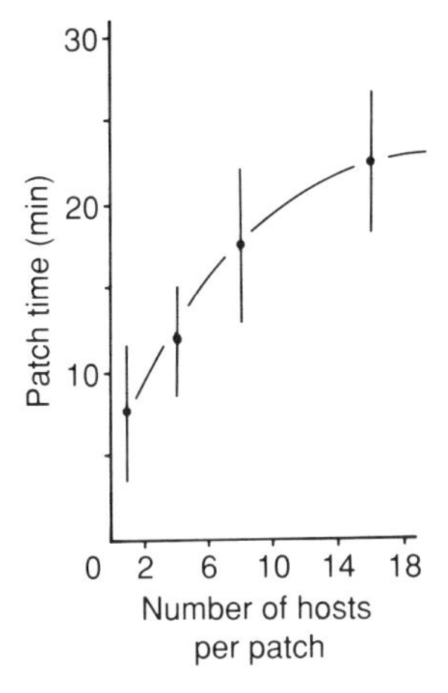

Figure 8.7 Average durations of first visits by female parasitoid wasps (*Venturia canescens*) to patches of different density of hosts. Vertical lines are 95% confidence intervals. (After Waage, 1979.)

$$T = \frac{aP + \Sigma I_t - I^*}{b}$$

where P = number of hosts on patch, a = constant relating host density to chemical concentration, b = decay rate of responsiveness, r^* = level of responsiveness of wasps entering the patch (decaying to r and patch abandonment) and I = ovipositions on patch affects decay to r^* by adding increment of responsiveness, I, to value of r.

The model predicts short patch times at low host density and long patch times at high host density. It also predicts highly variable GUTs on patches of different densities. The rate of oviposition over the last few minutes on a patch would be expected, however, to be similar for a range of patch densities. At very low densities this terminal oviposition rate would drop below that for higher densities, and persistence on patches at these densities is caused primarily by the response to the patch stimulus.

In experiments to test the model, patches were prepared with host densities of 1, 4, 8, and 16. As expected, patch duration increased with host density (Figure 8.7), whereas the GUTs measured for different density patches were highly variable both within and between patch densities. Thus a threshold rate model based on equal GUTs for all densities is not applicable. In fact, the observed GUT in most replicates was shorter than some previous interval between ovipositions on the patch, which is impossible in a threshold rate model.

In a final series of experiments, duration of patch visits and ovipositions were recorded in a multi-patch habitat containing two depend on previous patch experience; following the first visit to a patch, the probabilities of discovering new patches of 1, 4, or 8 hosts, having just exploited a patch of any of these densities, did not differ significantly. Waage (1979) notes, however, that in nature where patches may be separated by longer distances than in the laboratory, and where more dense patches may give off more volatiles than less dense ones, wasps might orient to those patches providing a greater odour stimulus by flying upwind toward them. Nevertheless, what might appear to be patch selection seems instead to be variability in the duration of visits to patches of different density.

In summary, some of the predictions of a basic threshold rate model are satisfied by Waage's (1979) data. The problem with the threshold rate model, however, is that patch leaving is supposed to be based on a specific interval between prey captures (or ovipositions). The contribution of patch stimuli to arrestment would involve only the initial period on the patch before oviposition, equal to or less than the GUT, which is maximally the same for all host densities. This makes the model particularly sensitive to the natural stochasticity of oviposition rates which could result in a forager leaving a patch too soon or too late and thereby reducing efficiency. In Waage's model, however, the contribution to patch time of responses to a patch stimulus (contact chemical) is substantial and tends to fine tune the assessment of patch quality because it increases with host density. Instead of adding a fixed increment to patch time, ovipositions in this model alter a hypothetical level of responsiveness to the patch stimulus. As a result, short-term stochasticity of oviposition rates is effectively damped, and patch leaving is influenced not by the timing of the last oviposition alone, but by the timing of all ovipositions while on the patch. This permits a more accurate assessment of patch density and a more appropriate response by the forager than in the threshold rate model.

Waage's (1979) results tell us that an animal does not have to 'know' at any moment in time the locations and the current profitabilities of patches in a habitat in order to 'compute' the optimal set of patches for exploitation; instead, animals such as insect parasitoids, houseflies, and lady beetle larvae need only have innate responses and need not depend upon previous experience. Their behaviour is modulated by the physiological state, however, as when the arthropods discussed here experience changes in their locomotory pattern in relation to the time period they have been deprived of the resource they seek. Hence, in Chapter 14 it will be pointed out that the change in duration of local search from short to long correlates with period of starvation in flies and coccinellids and with period of deprivation in parasitoids. Unfortunately, while there is an oversupply of patch-leaving models based on unlikely abilities of animals, there are few patch-leaving models based on hunger.

Commenting on expectation models in general, Zach and Falls (1976c) point out that the adaptiveness of hunting by an expectation of any kind would depend not only on how quickly expectations can be formed, but also on how quickly such expectations can be modified in response to changes in prey density. If both of these kinds of modifications were to occur rapidly, hunting by expectation, as used by Gibb (1962) and Krebs (1973), is not a useful concept since a predator would simply allocate its search effort in response to local differences in prey density. On the other hand, if the modifications were to occur too slowly, the animal would still be using information that might be outdated to determine its present expectation. The predator must therefore use a fairly accurate and updatable 'sliding memory window' to compare the current intake with the intake rate in the recent past (Cowie, 1977; Cowie and Krebs, 1979; Ollason, 1980).

8.3 CAN ANIMALS REALLY ESTIMATE CAPTURE RATE?

It is clear from previous discussions that individuals of some species tend to spend most of their time in the most profitable sites, but as we have seen there are several simple mechanisms that could account for this behaviour without animals having to measure capture rate directly while using resources in a patch. However, for those species that can perceive capture rate, such abilities would improve the precision of mechanisms controlling patch-leaving times, especially where patches differ subtly in density of resources.

First, several studies have been reported that deal mainly with diet selection of animals, but which also test to see if animals become more selective as prey density increases. If animals do become more selective under these circumstances, then this change would mean that the animals are aware of the number of prey consumed per unit time. In a field study by Davies (1977), spotted flycatchers consumed more large prey when prey was abundant than when prey was relatively scarce. In another example, bluegill sunfish (*Lepomis macrochirus*) were not selective when hunting in a mixture of small, medium, and large *Daphnia* at low density (20 of each class), but instead consumed all three size classes according to how often each was encountered (Werner and Hall, 1974). The bluegill became more selective at higher prey densities (350 of each class), consuming mainly the largest size prey. At an intermediate density they chose mainly the two largest size classes of prey. In this case the handling times for all three size classes were nearly equal and so profitability, in terms of nutrients ingested per unit time, depended mainly on the relative sizes of the prey captured. Possibly bluegill and flycatchers gaze into their habitats to estimate prey density, and then use this information to determine which size prey to consume, but it is more likely that the change in selectivity means that the proximate mechanism by which fish and birds estimate prey density is perception of capture rate.

Krebs *et al.* (1977) designed an ingenious experiment that eliminates the opportunity for a test animal to estimate directly prey density, while allowing it to estimate indirectly prey density based on capture rate. Great tits were given a chance to select large or small prey types, which were different sized pieces of mealworms presented sequentially on a moving conveyor belt. As mealworm pieces moved by, the bird could choose whether to pick one up or leave it. The encounter rate could be controlled by the rate of conveyor belt movement, such that prey encounter was 0.025 prey per second in one condition and 0.15 prey per second in the other. When the encounter rate for both prey types was low, the birds were not selective. With higher encounter rates, however, the birds selected the larger mealworm pieces. This experiment provides good evidence that prey capture rate is an important proximate mechanism by which an animal assesses prey density.

A quite different approach in testing for perception of capture rate is simulation of foraging behaviour using maze behaviour and operant conditioning (see chapters in Kamil *et al.*, 1987). These approaches permit rigorous control of variables and the application of laboratory technology perfected by experimental psychologists over several decades.

The analogy between foraging behaviour as it occurs in nature, and operant and maze behaviour as it occurs in laboratories has received considerable attention (e.g. Mellgren, 1983; Kamil and Roitblatt, 1985). Maze behaviour relates to foraging in that searching in a maze is essentially movement through space from a nonfood source to a potential food source. Operant behaviour relates to foraging in that the major component is harvesting (procuring) food from a food source. Hence, search and harvesting, the main components of foraging behaviour of many mammals, are represented in the experimental psychology laboratory by mazes and operant chambers, respectively. In the maze procedure an animal runs to a goal box where food is available in a food cup. Here the relevant process is supposed to be 'search', with virtually no emphasis on the procurement process, and in fact most such experiments are carried out using food pellets that an animal can ingest very rapidly. The operant procedure is almost exactly the opposite, since the subject is placed at the food source and is required to procure the available food through the repetitive act of pressing a lever or button; there is no locomotory search component.

In one kind of operant experiment, an animal is allowed to press either of two keys, each of which can deliver a food reward. Each key is associated with a schedule, and the schedules run concurrently. On concurrent variable interval schedules, the responses of an animal to the two keys are in direct proportion to the frequency of reinforcement obtained for those responses. In other words, the relative frequency of responding on an alternative matches the relative frequency of reinforcement on that alternative. For example, if one schedule provides reward twice as often as the other, two-thirds of an animal's time should be spent there (e.g. Rachlin and Green, 1972). Although animals spend most of their time on the schedule offering more frequent rewards, they still make visits to the other schedule to collect any rewards that have accumulated; this is analogous to sampling between patches.

Experimental psychologists are interested in the basis upon which animals are making their 'decisions'. For example, are they matching the proportion of time spent in a given schedule to the proportion of all rewards received in that schedule (Herrnstein and Vaughan, 1980), referred to as the matching law (Herrnstein, 1970)? Or do they adhere to the delay-reduction hypothesis (DRH), in which a stimulus correlated with a great reduction in time to food presentation will be a stronger reinforcer (in choice tests) than one correlated with a lesser reduction in time to food presentation (Fantino, 1969)? Fantino and Abarca (1985) suggest that the DRH may be an appropriate proximate mechanism by which animals maximize the rate of energy intake per unit time. For example, they might use the DRH to decide when to leave a patch currently being used when the marginal capture rate in the patch drops to the average capture rate for the environment as a whole. Animals may be more likely to show sensitivity to delay reduction than to variables such as overall rate of energy intake or rate of reinforcement.

Is the experimental psychology approach sufficiently realistic to allow us to draw conclusions about what animals do in nature? Baum (1983) lists three artificialities of a typical operant conditioning experiment:

1. it occurs inside a small box, which is quite different from anything natural,
2. it occupies a small proportion of an animal's active hours, and
3. it presents food on a schedule that is not natural.

Another problem is that wild animals may seldom have so many identical and consecutive chances to perfect their actions under any set of conditions, since the problems they face are continually changing (Gass, 1985). In operant experiments the subjects are exposed to the experimental conditions thousands of times before data-collecting trials. Thus, in concurrent schedules that are constant, trained animals really do 'know' the worth of the two patches, and so they are as 'omniscient' as an animal can be, as predicted by some OFTs. I would add that simulating travel time as time spent waiting at a window for food to appear may also be less than realistic, given the differences in metabolic costs between standing and searching. However, similar results were obtained by Baum (1982) and Dunn (1982), even though Dunn used a small box simulating travel, and Baum used a larger arena, allowing actual travel during the search.

8.4 SUMMARY AND CONCLUSIONS

Two important concepts regarding patch quality are that patches differ in quality before an animal begins to utilize resources, and that prey availability within a patch decreases over time either because of foraging activity, or because a resource remains within the sensory field of the animal but has no further value, or because resources become less vulnerable to attack and/or exploitation as a result of previous use. An animal does not have to 'know' at any moment in time the locations and the current profitabilities of patches in a habitat in order to 'compute' the optimal set of patches for exploitation; instead, animals such as insect parasitoids, houseflies and lady beetle larvae need only innate responses and need not depend upon previous experience.

Various hypotheses have been formulated to explain how animals 'know' when to leave a depleting patch: number expectation, time expectation and capture rate expectation. Most tests of these hypotheses have been performed in the lab, owing to the difficulty of obtaining data from the field that clearly reveal the mechanisms by which animals decide when to leave a patch. Several studies provide evidence that animals use a threshold capture rate; the threshold rate is made equal to the average rate for the habitat, such that this rate increases and giving-up time (GUT) decreases as the average patch density in the habitat increases. GUT, the period between utilizing the last resource taken in a patch and leaving the patch, is 'set' according to the average capture rate for the environment. Such differences in GUTs would be adaptive because an animal should search longer before giving up and leaving a patch in a poor environment than in a rich environment. Evidence is generally lacking for the other hypotheses.

The adaptiveness of hunting by an expectation of any kind depends on how quickly expectations can be formed, and on how quickly such expectations can be modified in response to changes in resource density. If both of these modifications occur too rapidly, hunting by expectation is not a useful concept since a predator would simply allocate its search effort in response to local differences in resource density. If both of these modifications occur too slowly, the animal would still be using information that might be outdated to determine its present expectation. The predator must therefore use a fairly accurate and updatable 'sliding memory window' to compare the current intake with the intake rate of the recent past. Studies of diet selection and operant conditioning show that animals become more selective as prey density increases. These data indicate that animals are aware of the number of prey consumed per unit time. Perception of capture rate may, therefore, be a proximate mechanism for estimating resource density.

9

When to return to a resource patch

In some cases, a resource is permanently altered or rendered worthless by utilization (e.g. consumed seeds, parasitized eggs, mated females in species where females can mate only once). In other cases the resource can be renewed or recycled, such that it can again provide benefits after a certain period of time. The question here is how long should a forager wait before returning to a patch? Cropping and traplining are two mechanisms that promote revisiting after a strategic period of time. In traplining, a route is learned and then followed during each search bout. In cropping, resources are utilized within the home range or habitat of an individual or a group. Quite often territoriality is involved, whereby animals defend an area containing the resources that they periodically crop.

.1 CROPPING

Cropping is a unique foraging style that functions to remove renewable resources periodically, with an intervening duration sufficient to allow resource renewal. The period between cropping sessions can be regulated by the locomotory pattern of the animal, through learning, or by exchange of social information.

Flocking in birds aids in regulating 'return-time' to previously depleted areas. Return-time can be defined as the elapsed time between successive visits to points within a bounded area searched by a flock, and is of primary importance in the harvesting of renewing resources (Cody, 1971, 1974). A flock can move in such a way as to minimize the variance in the time intervals between successive visits to a particular sector of the habitat and to adjust the mean return time interval to the rate of replenishment. Individuals that feed independently turn more often than do flocking individuals in their search for food and these loners probably waste a considerable amount of time searching where others have previously fed and have already depleted the food.

Cody (1971) observed flocks of birds, consisting mainly of Fringillid species, cropping seeds in the Mohave Desert. This area at the base of

Table 9.1 Data from 14 finch flocks in the Mohave Desert, spring 1968 (from Cody, 1971).

Distance from mountains (ft)	*Path of flock*					
	Mean speed (ft/s)	*Directional probabilities*				*Length of mean straight path (ft)*
		F	*R*	*L*	*B*	
100–800	0.269	0.74	0.11	0.12	0.03	207
1050–1700	0.306	0.71	0.14	0.14	0.01	229
2500–6000	0.365	0.58	0.22	0.15	0.05	143

a mountain range forms a sloping plateau at an altitude of approximately 4000 feet. The vegetation is mainly low shrub species such as *Yucca schidigera*, *Ephedra*, *Larrea* and *Happlopappus*. Flock size, composition, and behaviour varied predictably

1. between years of high and low food density,
2. as the winter season progressed with ever-decreasing (nonrenewing) food supplies, and
3. over a food density gradient that is higher at the wetter mountain base than it is in the desert plain (Cody, 1974).

Take, for example, the gradient from mountain base to desert plain. Because seed abundance and ripening rate varied predictably with distance from the base of the mountain range, Cody grouped data from 14 flocks into three sections according to distance from the mountains. The analysis (Table 9.1) showed that speed increased and the length of straight path sections decreased with distance from the mountains. The directional probabilities indicate that the shortest return times were in the areas most distant from the mountains.

To delineate proximate mechanisms, Cody (1971) followed flocks and mapped their paths by recording turn angles and average speed (Figure 9.1a). Paths are primarily forward-moving, as shown in a typical example in Figure 9.1b. Also, the average straight segment is quite long (Figure 9.1c). A mixed-species flock of approximately 100 birds covers an area of 50 feet × 150 feet and moves in an elongate shape in the direction of its long axis over the desert floor. Within each flock section there is a rolling motion, as each bird flies, lands, and feeds by hopping forward for some distance and, when the end of its section arrives, again flies forward to the front of the section and the process is repeated.

Various species of ants crop the area around the nest, and in some

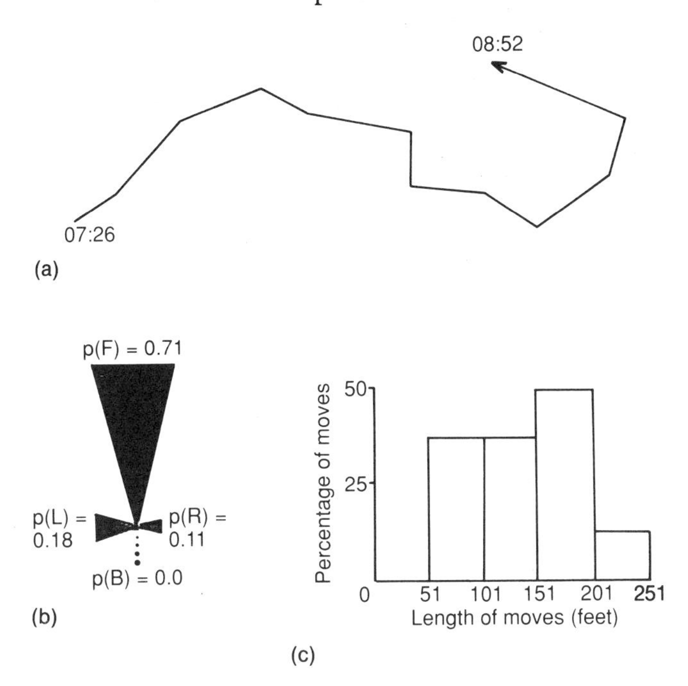

Figure 9.1 (a) Sample path (1620 feet) of a flock of birds foraging in the Mohave Desert. (b) Probabilities of the flock moving forward [p(F)], left [p(L)], right [p(R)], or back [p(B)]. (c) Frequency distribution of lengths of straight moves. Units are in feet. (After Cody, 1971.)

species food abundance determines if they forage as individuals or as groups. The transition is analogous to birds switching to and from flocking. In *Cataglyphis bicolor*, an ant inhabiting arid regions of Northern Africa and Greece, searching is conducted by individual ants collecting dead arthropods in the vicinity of the nest. Each forages in a restricted area which is typically a sector of a circle with the nest as its centre. The entire area around a nest is searched by the combined efforts of many individuals setting out in different directions, each ant searching one sector of the circle (Wehner *et al.*, 1983; Harkness and Maroudas, 1985).

The ants have a very precise vector navigational system (Wehner, 1982) which they use when leaving and returning to the nest. At the end of the outgoing phase they 'know' where they are in relation to the nest, and they return to it in a nearly straight line after finding a prey item (dashed line in Figure 9.2a). They also use a kind of landmark map to specify locations of foraging sites, so that they change their foraging routes after the natural panorama has been modified artificially. Whereas a vertebrate 'cognitive map' specifies

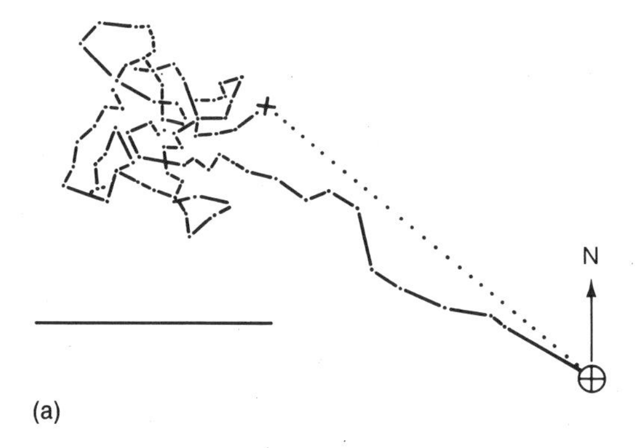

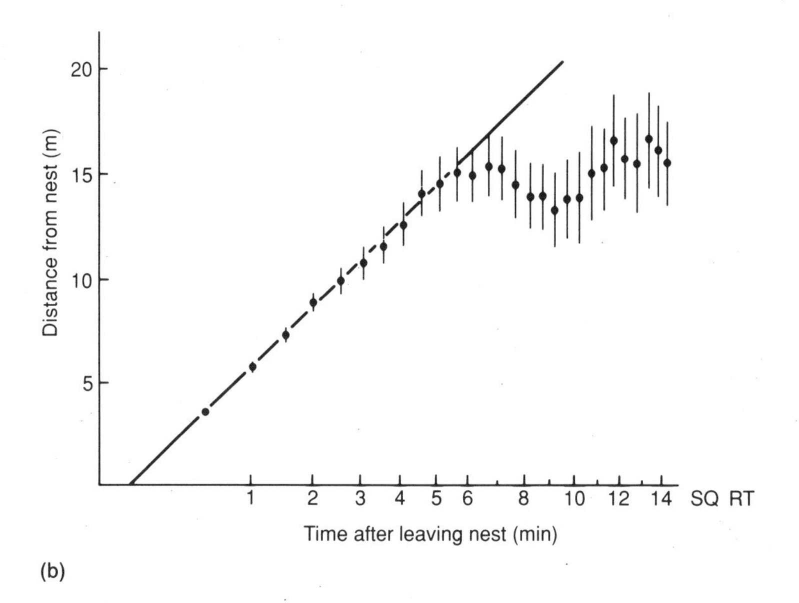

Figure 9.2 (a) One example of an individual ant (*Cataglyphis bicolor*) ranging out from the nest and then switching to local search; dashed line indicates navigation path as an ant returns to the nest. A cross marks the site where a resource was found; horizontal bar equals 5 m. (b) Distance from the nest of desert ants at different times after they leave to forage. (After Harkness and Maroudas, 1985.)

positions of landmarks defined in absolute terms, the ants use a map based on relative distances, consisting of familiar routes passing through a certain area. Thus it is probable that, as stated by Wehner *et al.* (1983)

'. . . each route is defined by a certain sequence of landmark images which the ant expects to appear in succession while proceeding along its foraging or homing paths'.

After walking straight away from the nest a certain distance, an ant begins local search (Figure 9.2a); this is illustrated in Figure 9.2b as the point where the data drift to the right. The convoluted and circling movement pattern, which enhances visual scanning, tends to increase the chances of locating prey from a reactive distance of approximately 10 cm. In subsequent searching bouts an individual ant tends to move toward the same compass direction at which food was previously found. The switch from straight, outgoing locomotion to local search would appear to be controlled by either a timing mechanism or one in which the ant counts its steps. However, the distance walked by the ant in the outgoing phase, if it is set by internally-derived information, must be genetically programmed as an appropriate distance to walk for a given habitat, or perhaps the distance is set through the experience gained on previous foraging trips. Unsuccessful foraging bouts are time-limited, suggesting that the ants may rely on some kind of time expectation, after which they return to the nest. Since spiders and robber flies prey upon foraging *C. bicolor*, predation may have partially shaped the time-limited foraging strategy of this species.

Given that the strategy of *C. bicolor* operates to partition spatially the foraging area around the nest by individual foraging, what is the advantage of site constancy to an individual ant leaving the nest in a particular direction each day? First, since the food density in any particular sector may vary from day to day, independently of other sectors, it will pay a particular ant to revisit a sector that has been profitable. Secondly, given the ant's navigational abilities, site constancy may promote learning of landmarks in a particular area, which in turn would enhance foraging efficiency. The payoff for this search strategy must be measured in terms of nutrients required for the growth and development of a replacement individual. In other words, can a forager replace its own body mass during its own lifetime? In Tunisia (with low food abundance), approximately 50% of foragers return to the nest each day with food items, as compared to 90% in Greece (with high food abundance). However, in both study areas each forager during its lifetime delivers to the nest 17 times its own body weight. The problem of lower seed abundance in Tunisia than in Greece is countered by the larger body size of ants in Tunisia, allowing them to bring back more biomass than can the smaller ants in Greece. Thus the strategy in both places seems to provide greater than the minimum replacement requirement. The ants are not as efficient as they could be, however, since under certain prey-density conditions higher returns could probably be expected by the recruitment of other individuals.

A seed-eating harvester ant (*Veromessor pergandei*) offers an excellent contrast to *C. bicolor*, because it forages both in groups and individually (Bernstein, 1975). This species uses the individual foraging tactic during brief periods of high seed density, whereas when or where it is relatively dry and seeds are sparse, the foragers leave the nest in one continuous column and search for food in a group. Rissing and Wheeler (1976), studying *V. pergandei* in Death Valley, California, also observed this flexible foraging strategy, although they sometimes observed columns foraging even when seed production was high. During group foraging the front of the column fans out to increase its search area. On consecutive days the foraging columns of harvester ants rotate clockwise or counterwisethe direction taken by a factor of 14° to 17°. This presumably prevents the overlap of already searched areas (Figure 9.3). Rissing and Wheeler (1976) observed that when seeds were abundant the columns changed directions to a lesser extent then when seeds were sparse, suggesting that radially changing column directions around the nest are a response to low seed yield on previous days.

Sunbirds and wagtails illustrate how territoriality is often an integrated component of cropping efficiency. Golden-winged sunbirds (*Nectarinia reichenowi*) in central Kenya use a common weed, *Leonotis nepetifolia* (Labiatae), as a nectar source (Gill and Wolf, 1975). Nectar production rates average about 0.70 μl per flower per hour in the morning and decline to a low of 0.25 μl per hour late in the day, providing an average of 4 calories per flower per day. Sunbirds often establish feeding territories encompassing several patches of flowering *L. nepetifolia* in which the resident attempts to restrict the use of resources by other individuals. Since nectar volumes inside a territory are on average two to four times higher than nectar volumes in adjacent undefended areas, territorial defence would seen to be a profitable enterprise. When defence costs become too high relative to the gains, the sunbirds temporarily cease defending their territories. As the number of new flowers opening declines, each bird attempts to increase the size of its territory by incorporating parts or all of adjacent territories. Each territory yields between 13 000 and 14 000 calories from 1500 to 2500 flowers, which as shown in the following calculation correlates well with a sunbird's daily energy requirement.

Territorial pied wagtails (*Motacilla alba*) occupying and defending banks on the River Thames add another dimension to cropping of territories by sometimes allowing intruders onto their territories.

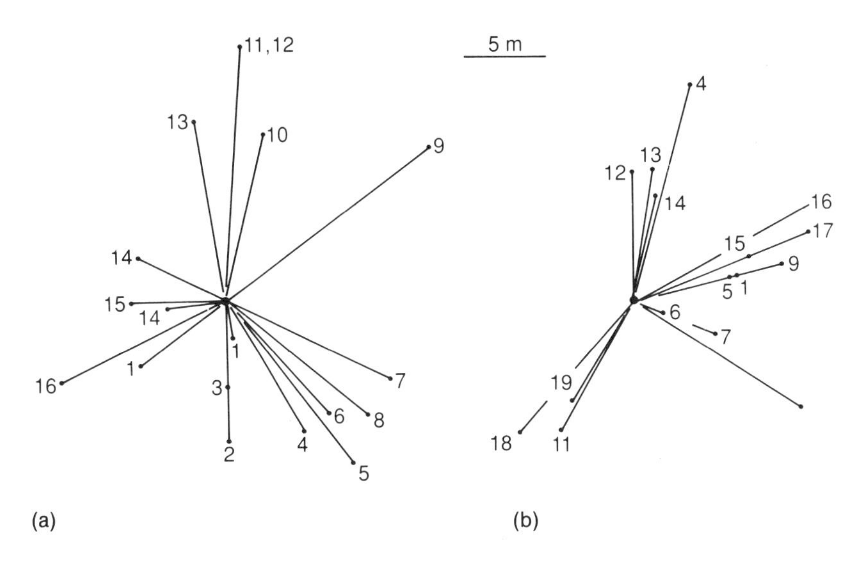

Figure 9.3 Column rotation during consecutive foraging periods of two nests of harvester ants (*Veromessor pergandei*). Lines indicate length of foraging columns, and numbers indicate consecutive foraging periods. In (a) the columns generally rotated counterclockwise, whereas in (b) the pattern was more complicated, but tended to be generally clockwise. During periods 3, 8, and 10 in (b) the ants left the nest in all possible directions; they did not forage to the west and northwest of this nest, possibly due to a lack of plants in these areas. The dot in centre of each array is the nest entrance. (After Rissing and Wheeler, 1976.)

They forage by walking along the water's edge and picking up small insects that wash onto the banks (Davies and Houston, 1981). The insects form a renewing food supply in that after they have been depleted from a stretch of the river edge, it is only a matter of time before further prey are washed up onto the bank. As in the previous examples, feeding efficiency is increased by foraging in a territory systematically, because the available food increases exponentially with time since depletion.

By evicting intruders, territory owners are able to increase the time allowed for resource renewal between successive visits to the same stretch. If other wagtails were allowed to land on the territory, the owner's feeding efficiency would be reduced because it would be visiting stretches soon after they had been depleted by other individuals. Intruders on territories have a lower feeding rate than do owners, because even if they are able to land undetected, they often end up feeding on stretches that have recently been depleted by the owner. Therefore, by exploiting its territory systematically, an owner can make it relatively unprofitable for intruders. The most profitable place for an intruder is just in front of the owner, where the time since the last depletion is greatest. However, any intruder that lands just in front of the owner is easily spotted and evicted. If an intruder lands far away from an owner it can avoid detection, but it will be feeding over a less profitable stretch.

What makes the wagtail example so interesting is that while they usually defend their territories alone, they sometimes allow another individual onto the territory for a period varying from one day to several weeks. These 'satellites' land on territories and attempt to appease the owner with special postures (Zahavi, 1971). Although satellites are costly in that they consume a portion of the available prey, they benefit the owner by helping to defend the territory against other intruders. Thus, if the owner allows a satellite onto its territory, it will enjoy the benefits of assistance with territory defence; but it will suffer the cost of having to visit stretches after a renewal time of only half that which it would achieve if it was alone. Since the rate of intrusions onto a territory increases with food abundance, intruders are common and an owner gains by taking on a satellite because dividing the available food still leaves plenty for both birds. Conversely, the owner does better alone on days when food is scarce; because halving the return time may drastically reduce its feeding rate, and because intruder pressure is low when food is not abundant, assistance with territory defence may not be of much benefit.

9.2 TRAPLINING

Some species avoid revisiting by learning which resources have been utilized previously or by following the same route on each foraging bout. Racey and Swift (1985) mark-recaptured female bats, *Pipistrellus pipistrellus*, and found that each night they moved between foraging areas on a regular route. In fact, several were sighted at the same time and place on successive nights. The bats deviated from their regular flight path to pursue an insect, but then returned to their beat.

Euglossine bees (Janzen, 1971), bumblebees (Heinrich, 1976; Manning, 1956), and *Heliconius* butterflies (Gilbert, 1980) avoid revisiting by learning the spatial position of individual plants and then moving between resource sites in a sequence that is the same as or similar from one foraging bout to the next. The butterflies collect pollen as a protein source for yolk production and to sustain their relatively long lives (review: Gilbert, 1980). In *H. charitona* a large load of pollen contains sufficient nitrogen for the production of five eggs, which is the average daily egg production of one female. The primary cost of obtaining pollen for egg production is foraging expense, because their bright colours and unpalatable taste minimize the risk of predation upon these butterflies. Since pollen sources are spatially constant over many weeks, but unpredictable on a daily basis, traplining is clearly an efficient strategy. Older females are more adept at traplining than are younger or less-experienced ones, and it is possible that young females are stimulated to orient along the same path each day by following older females from communal roosting sites. Males orient in a similar manner when seeking females, and females trapline when ovipositing on Passiflora plants (Williams and Gilbert, 1981). The following case study illustrates several elements of traplining in the cropping of plants in the home range, as well as topographic search and other types of orientation.

Case Study Frugivorous primates cropping through their home ranges

Many frugivorous primates live in small groups that move through the home range, cropping fruit and other plant materials. The spatial distribution and temporal availability of fruit varies considerably among plant species, especially in the tropics. Some plant species fruit synchronously over a short fruiting season as an adaptation for reducing losses to seed predators (Janzen, 1971; Smythe, 1970), while others fruit over a period of weeks or months to reduce intraspecific competition among their seed dispersers (Frankie *et al.*, 1976). The various plant species also fruit at different times of the year, apparently to reduce interspecific competition for seed dispersers (Snow, 1965; Smythe, 1970; Janzen, 1972). As a result, some fruits are fairly evenly distributed while others are clumped in time and space.

The foraging strategy of the orang-utan (*Pongo pygmaeus*) which is primarily a fruit-eater, depends largely on food availability. When fruit is scarce, individuals spend more of their time travelling, presumably in search of food, and under these conditions will feed on less preferred foods such as leaves, young shoots, lianas, flowers, epiphytes and small vertebrates or their eggs (MacKinnon, 1974). Orang-utans typically follow zigzag routes through the forest, feeding at many small food sources or thoroughly searching a small area for a few days before heading off quickly to another region. There is a clear difference between the zigzag local search movements within an area and the kind of movements used when heading to a new region, which is typical of ranging. While travelling to a new area they follow natural features such as streams or ridges, and often take the exact routes (arboreal highways) used previously by other individuals.

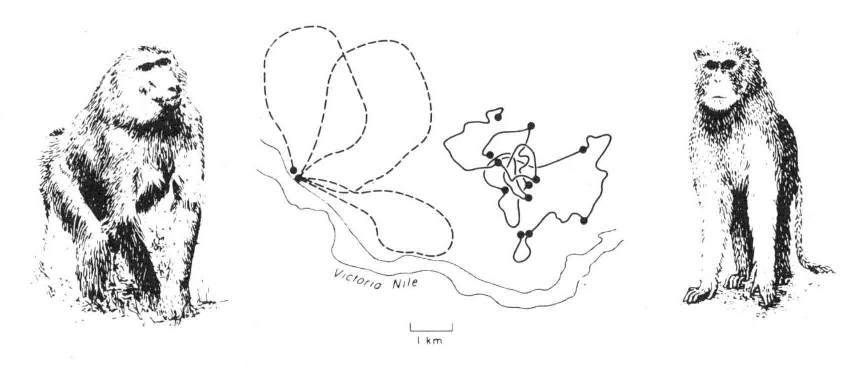

Figure 9.4 Track pattern of a social group of patas monkey (*Erythrocebus patas*) (solid lines) and the savanna baboon (*Papio cynocephalus*) (dashed lines), in relation to areas suitable for roosting and feeding at Chobi, Murchison Falls National Park, Uganda. Solid dots indicate overnight roosting sites. From Baker 1978; copyright R.R. Baker, reprinted by permission.

Howler monkeys (*Alouatta palliata*), travelling high in the canopy on Barro Colorado Island, appear to follow the same pathways through trees repeatedly, rather than at random. The switch from ranging to more localized search is exemplified by a group of *Cercocebus albigena* monkeys in western Uganda which move only 100 to 800 m per day when foraging at a resource patch, but 1100 to 1400 m per day when travelling between patches. This strategy assists in locating large or rare fruiting trees; the monkeys then remain at the site once they have been found (Waser, 1978).

Foraging tracks of the patas monkey (*Erythrocebus patus*) and the savanna baboon (*Papio cynocephalus*) reveal strategies for reducing search of areas already covered (Hall, 1965). The baboon, with a mean troop size of 25 to 30 individuals, occupies areas in southern Africa, foraging in savanna, woodland and forests. Each day the troop forages in a circular-shaped path, 0.5 to 12 km in length, returning at night to the same roosting site (Figure 9.4). Plant material is supplemented by animals in the diet, including relatively large mammals. A baboon troop consistently selects one roosting site, usually near water, within a home range which averages 8.4 km^2. Yellow baboons move in single file when leaving their sleeping grove and heading toward a distant foraging site, but walk in ranks when foraging in open areas (beside and at a distance from each other). They also tend to walk closer together when in areas where vulnerable to predation. The strategy of walking in ranks, while efficient for reducing search of already covered areas, has its limitations. If a troop is too large, those at ends of the rank must walk a long distance when the troop turns in a new direction; also, those individuals out at the ends may not even be in a resource patch upon which those at centre are feeding (Altmann and Altmann, 1970). The patas monkey ranges over grassland and woodland savanna, feeding on plant material, insects and lizards, with daily movements of 0.5 to 1.2 km within a home range of 52 km^2 for a troop of 30 individuals (Figure 9.4). Roosting sites are scattered throughout the range, defining daily movements. Both the forward-moving tactic of the monkey and the daily loops of baboons tend to minimize searching of areas previously covered as the animals crop resources within their ranges.

Groups of *Lemur catta* in Madagascar move from one feeding site to another, covering their home range of 6 to 23 ha within 7 to 10 days (Sussman, 1977). The regularity with which they move around their range maximizes feeding efficiency, and may also keep food supplies at the right growth stage so as to achieve maximum regeneration.

Primates such as the orang-utan have excellent knowledge of local geography, and can move in a direct route to feeding sites, salt-licks, arboreal water holes, or nesting sites (MacKinnon, 1974). Howler monkeys often move in a direct route through the trees for a distance of several hundred metres to a *Platypodium* tree, to eat leaves, or to a fruiting tree (Smith, 1977). This is a good example of animals using their past experiences for increasing efficiency of resource localization. Individual orang-utans have been observed visiting the same site repeatedly, suggesting that they have 'favourite haunts'. Individuals may be able to remember which parts of the forest are productive at different times of the year, even though individual trees do not necessarily fruit every year. Some fruits such as durians and figs have strong odours, and animals such as orang-utans could conceivably use olfactory orientation to locate these trees, as indicated by the direct routes often taken. They might also be attracted to noisy, frugivorous birds congregating at good sites. Kummer (1971) suggests that certain individual hamadryas baboons (*Papio hamadryas*) may act as 'deciders', leading foraging movements of a troop; the deciders are usually older males who can probably remember where food and water are located and where risks are at a minimum. Frugivorous primates probably benefit from information gained through patrolling and exploratory behaviour, and perhaps they can return to trees which they previously noticed were flowering or beginning to fruit. It is not clear from the literature whether or not learned searching information of this kind can be transmitted from individual to individual or from generation to generation. However it is accomplished, these animals are able to home in efficiently on sources of food that are small, rare, and dispersed.

The extreme plasticity in food-searching behaviour of many frugivorous primate species appears to be an adaptation for life in a diverse rainforest flora characterized by complex fruiting cycles. Because of the scattered distribution of food, it is essential for populations of many of these species to be split up into small dispersed foraging units. Only under unusually good fruiting conditions could large bands find enough to eat at the same place. Moreover, they must be able to search over large areas, since different parts of the forest may produce different fruit species which ripen at different times of the year. Group foraging for all of the species discussed here seems to be advantageous by virtue of enhanced communication about resources as well as predators, and preventing overlap in areas searched by individuals. Group foraging would be less costly than foraging by individuals as long as the size of the group does not exceed the capacity of the food patches.

9.3 LEAVING PATCHES EARLY

One way to avoid revisits to resources is simply to leave patches before using all of the resources. This strategy, which would only make sense if resources were plentiful, is exemplified by several species of nectivores which leave flowers after visiting fewer than the available number of nectaries or after visiting fewer than the available flowers on an inflorescence. For example, Kipp (1987) found that honey bees foraging on artificial flowers, each with three florets, tended to probe only two of the three florets before leaving. In those cases where three florets were probed, bees returned to the first floret in 25% of cases. Zimmerman (1979) suggested, using optimal foraging theory, that if an inflorescence has few flowers, then a bee should show high directionality (i.e. small turn angles) in order to avoid revisiting the same inflorescence. If there are many flowers on an inflorescence, the bee could revisit it, since there might still be high reward flowers remaining. However, in order to accomplish this, before a bee departs it must count the number of flowers on an inflorescence, so as to decide whether or not this inflorescence should be revisited or not. Such behaviour seems unlikely. Rufous hummingbirds also visit fewer nectaries on average than are available on a flower (Figure 9.5). For example, *Amazilia rutila* visit a mean of five flowers before leaving *Calliandra*, which has 50 small flowers on a small compact inflorescence (Montgomerie, 1979). This is true for several other species of flowers with various numbers of nectaries (Gass and Montgomerie, 1981). Although this mechanism works to avoid revisits, its use suggests that these animals lack an ability to keep track systematically of which flowers have been used and which have not.

9.4 SUMMARY AND CONCLUSIONS

Cropping, traplining and leaving a patch early are search mechanisms for reducing the chances that an animal will revisit a depleted resource site.

Cropping allows an animal to utilize periodically renewable resources. The period between cropping sessions, which can be regulated either by the locomotory pattern of the animal or through learning or exchange of social information, should be of sufficient

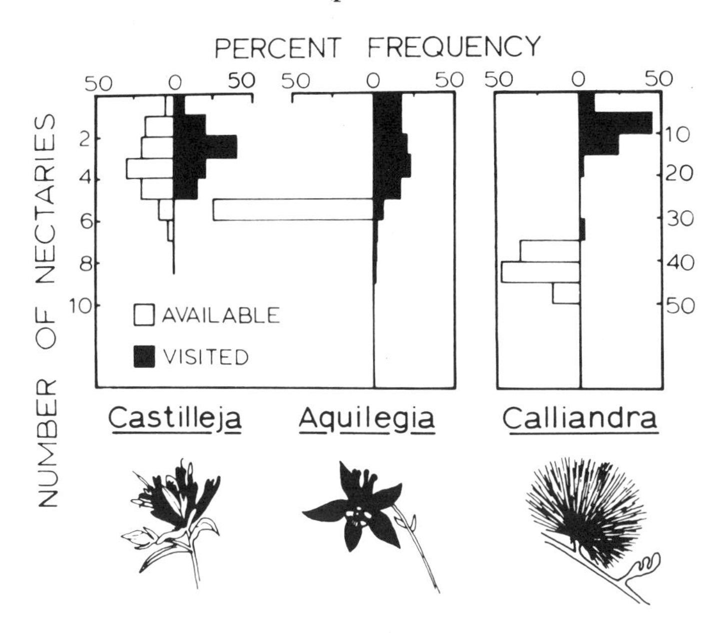

Figure 9.5 Relationship between the number of nectar sources available and the number visited by hummingbirds for three flower species. Foraging data are from rufous hummingbird visits to *Castilleja miniata* inflorescences and *Aquilegia formosa* flowers, and from *Amazilla rutila* visits to *Calliandra* inflorescences. From Gass and Montgomery, 1981. Copyright Garland Publishing, Inc., reprinted by permission.

duration to allow resource renewal. Cropping may be performed by groups to ensure that each individual has access to resources. Individuals may forage in different sectors so as to partition the foraging area spatially, or a group of individuals may forage together. Food abundance may lead to switching between an individual foraging tactic during brief periods of high food density, and group foraging when or where density is low. Individuals sometimes establish feeding territories which restrict the use of cropping areas by other individuals. However, when the costs of defence become too high relative to the gains, defence of territories may cease or a territorial holder may allow another individual onto the territory to help defend the territory against intruders.

In traplining, an animal moves between resource sites in a sequence that is the same as or similar to one foraging bout to the next.

Leaving a patch early is one way that an animal can avoid revisits to previously utilized resources.

10
Learning to forage efficiently

If a species has the ability to profit from experience, it can improve its foraging efficiency by responding appropriately to changing environments and by capitalizing on information gained through sampling resource patches. Experience may affect subsequent choices among resources or the way that an animal searches within a resource unit.

10.1 CONDITIONING

Conditioning, which is the most common type of learning in searching behaviour, can be delineated according to the temporal position of conditioning in the stimulus-response chain that is modified by experience (Papaj and Rausher, 1983):

1. conditioning of response, when the probability of an animal exhibiting a certain response increases or decreases with repeated presentation of stimuli; and
2. conditioning of perception, when the probability of perceiving a stimulus complex increases as a result of repeated presentations of stimuli.

In brief, an animal 'learns to see' the item of search through conditioning of perception, and 'learns to prefer' the item through conditioning of response.

It should be stressed that learning does not always occur, even in species that can learn, nor does it occur in all species of animals. Inflexible species-specific responses sometimes exist which prevent the learning of certain tasks, or there may be marked differences between species in the tasks that are learned or the ways in which they are learned. Moreover, learning is not always an asset, and sometimes it can even be a liability. If an animal learns to solve a search problem in a certain way, but then that solution fails to solve future search problems, it might be better for that kind of animal to approach each problem independently of previous experience. Animals that are not keen learners probably use resources in which distributions change often, so that learning one distribution is not an asset for seeking others, or where learning how to handle one type of resource will actually be detrimental for handling another type.

10.1.1 Conditioning of response

Animals commonly learn that one kind of resource (e.g. species, colour) of a given resource type (e.g. flower, fruit) is more profitable than another kind, and they subsequently 'specialize' on the more profitable resource. In other cases the animal may focus on the place or type of place where resources were found previously, and limit its time spent at sites where resources were not found.

(a) *Flower constancy*

Flower constancy in nectivores is one of the clearest examples of conditioning of response. That nectivores often visit a certain kind of flower (or small spectrum of those available) more often than other available ones has been observed commonly since Darwin (1876) first made notes on this subject.

Lewis (1986) used the small white butterfly (*Pieris rapae*) to determine if being constant to certain types of flowers might be a mechanism to reduce the time and energy required for learning how to extract nutrients from different flower forms. She first tested to see if butterflies learn to handle flowers through experience. Individually marked butterflies were observed in flight cages containing flowers on which they could forage. When given either bird's foot trefoil (*Lotus corniculatus*) or vetch (*Vicia cracca*) for 24 hours, a significant number of butterflies chose the species they had foraged on previously. In other tests, the time elapsed between landing and finding the nectar in a flower, termed 'discovery time', decreased with successive attempts of most butterflies on both bluebells (*Campanula rotundifolia*) and on trefoil. The initial times varied among individual butterflies, but the discovery times of most of them followed a classic learning curve. To determine if learning a second plant species interferes with flower constancy, marked butterflies were divided into two groups, and individuals of both groups were given bluebells until a minimum of five and a maximum of ten successive discovery times did not exceed 3 seconds each (the experimental criterion). Butterflies from group I were then given trefoil to learn until discovery times did not exceed 2 seconds each. Individuals of group II were given no flowers for 20 minutes, the maximum time required by individuals of group I to reach the criterion on trefoil. Individuals of both groups were then tested on bluebells. Butterflies in group I had to relearn bluebells, whereas those without flowers during the same time period did not have to relearn. Final learning times and initial test times on the two rounds of bluebells were significantly different for group I butterflies but not for group II. These results suggest that interference occurred, and this potentially translates into an energetic cost, as the butterfly depends on nectar for at least half of its energy budget. The cost of learning nectar extraction probably exceeds other costs associated with specialization, as for example in increased travel time between flowers.

These costs doubtless vary among habitats differing in resource concentration and among insects in various physiological states. In fact, flower constancy is no longer advantageous under conditions of a changing resource environment (Heinrich, 1976; Waddington, 1983a). Rather, a higher average reward can be attained by employing a 'mixed strategy', that is, foraging with some frequency at more than one floral species. Thus, nectivores avoid problems of variable resources by 'majoring' and 'minoring', whereby they concentrate on one species, but they keep sampling one or two other flowers as a type of bet hedging. The sampling allows them to detect changes in profitabilities of specific flowers that they might have concentrated on, either because the yield of the flower is less certain than previously (variance has increased), or because the mean profitability has dropped.

Major and minor foraging specializations in bumblebees were demonstrated by Heinrich (1976) and Oster and Heinrich (1976). The bees first sample the resources available, and then, as established above for butterflies, they learn which species is best, based on handling time and reward, and forage on that one particular species (Figure 10.1) (Heinrich, 1979c). However, profitability changes over time, and so minoring on a few other species permits tracking of fluctuating resources. Majoring and minoring is more efficient than random foraging or complete constancy, since it reduces costs of switching between flowers, but allows foragers to track changes in resource availability.

(b) *Avoiding uncertainty*

A series of experiments by Real (1981) demonstrates that individual bumblebees (*Bombus sandersoni*) and paper wasps (*Vespula vulgaris*) learn to distinguish between constant and variable nectar rewards. The area around a wasp or bumblebee colony was enclosed by mosquito netting, and an artificial patch of flowers was established within the

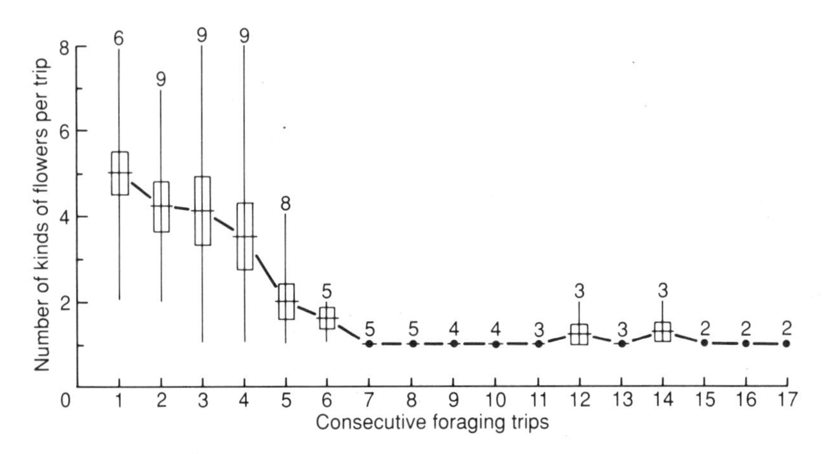

Figure 10.1 Number of different kinds of flowers per foraging trip in bumblebees (*Bombus vagans*). On their first trip the bees visited an average of five flower species. By their seventh foraging trip, all bees were only visiting jewelweed (*Impatiens biflora*), the most rewarding and abundant flower. Numerals indicate sample size. Boxes enclose one standard error on each side of the mean. Vertical lines indicate the range. From Heinrich, 1979c. Copyright Ecological Society of America, reprinted by permission.

enclosure. Artificial flowers (yellow and blue cardboard squares) were randomly placed beneath nectar wells drilled in a clear sheet of Plexiglas, and the volume of nectar reward per artificial flower was manipulated to generate different degrees of variability within a colour type. The first bees tested showed a significant preference for yellow. However, when blue flowers were kept constant (2 μl in each) and yellow made variable (6 μl in every third flower and none in the remainder), individual foragers avoided the variable yellow flowers even though the expected average reward was equal in yellow and blue. In control runs with the variable floral type reversed (yellow being constant), individual foragers then showed an equally strong avoidance of blue flowers.

In order to determine if bees were avoiding uncertainty rather than merely avoiding empty flowers, an additional set of experiments was run with some nectar in every flower. The constant floral type contained 2 μl of nectar in every flower while the variable type contained 5 μl in every third flower and 0.5 μl in the remainder. Again, the foragers preferred the constant floral type; however, the preference was not as strong as when there were empty flowers.

Analysis of the sequence of visits made by individual foragers seem to indicate that bees learned to distinguish the variable floral type during a foraging bout. Similar experiments with wasps showed that they preferred the constant type when flowers were of equal quality. For the same magnitude of variability, however, the bumblebees' avoidance of the variable floral type was more pronounced than was the wasps' response. Waddington *et al.* (1981), using another species of bumblebee (*B. edwardsii*), also found that they preferably fed from artificial flowers yielding the same reward on each visit (low variance) rather than from flowers yielding variable rewards (higher variance), even though the long-term expectation (the mean) of reward was the same at each type of flower.

(c) *Energetic efficiency*

Waddington and Holden (1979) suggest that a bee achieves maximum caloric intake by visiting the flower on each move which maximizes the ratio of energetic gain to foraging cost. Marden and Waddington (1981) further stress that where flower species grow intermingled (rather than in clumps), the flower constancy strategy is not likely to yield the maximum caloric intake for the bee. That is, a bee would have to pass by suitable flowers of the nonfavoured species to reach a flower of the favoured species. Under these circumstances, bees should use an energy optimization mechanism rather than flower constancy.

Marden and Waddington (1981) designed two laboratory experiments to record honeybee flower visitation sequences on two equally rewarding flower types. The goal was to determine which model, flower constancy or energy optimization, is the better predictor of honeybee floral choice behaviour. Two cases were examined:

1. where a bee achieves equal energetic efficiency using either strategy, and
2. where energetic efficiency is decreased by using the flower constancy strategy.

Whereas the flower constancy theory predicts that honeybees will always visit a single flower type, the energy optimization model predicts that in the special case of equally rewarding flowers, bees will always visit the closest flower. In experiment I, where blue and yellow flowers were equidistant, the bees maintained flower constancy. Hence, where all possible strategies were energetically equal, and there was no added cost for a bee in maintaining flower constancy, no bees chose flowers randomly to colour. In experiment II, the colour types were alternated between being the close or the distant flower. A strategy of flower constancy in this situation would be energetically less rewarding than the strategy of always visiting the closest flower. In experiment II only one of the ten bees showed flower constancy, while others visited the close flower on most flights. It would seem that bees are flower constant when other strategies are energetically equivalent, but switch to alternative strategies when it is energetically favourable to do so.

Because of the various problems or exceptions to flower constancy discussed above, Waddington (1983) stresses the importance of presenting and analysing the floral-visitation-sequence (F-V-S), rather than assuming constancy or some other pattern. The important part of the term is 'sequence', for it is the actual sequence or order of visits to flowers that most completely describes a bee's behaviour. The best description of the F-V-S is a quantitative determination of the probabilities associated with the possible transition between the plant species on successive interplant flights (Waddington, 1983). The probability matrix is the basis of the theory of Markov chains, and as Straw (1972) has indicated, the movements of pollinators between flowers 'fall naturally into the category of phenomena amenable to description as Markov processes'. Markov models of pollinator activity have been developed by Straw and Bobisud and Neuhaus (1975) for studying the effects of pollinator foraging patterns on the maintenance of intermingled simultaneously blooming species of plants.

(d) Cues involved in learning

Bumblebees with no field experience do not work on flowers as efficiently as do experienced individuals (Laverty, 1980). Inexperienced individuals land on incorrect parts of flowers (buds or outer surfaces of lateral petals), probe incorrect sites, and collect nectar without pollinating flowers. Both the time spent per flower type and the number of tactical errors decline as the bees become more experienced.

The stimuli used by nectivores for learning flowers include flower morphology, odour, and colour. In some cases nectivores are predisposed toward the use of one kind of cue or another, or at least it seems that prior to gaining experience they operate on instinctive mechanisms. For example, although inexperienced honeybees also have to learn the exact locations of rewards within flowers, some of their initial responses are instinctively directed to certain areas of flowers (Weaver, 1957; Daumer, 1958). Another example of a specialization that seems to be genetically potentiated is the tendency for *Bombus terricola* to become more than 90% flower constant on blue flowers within 50 visits when flowers have equal reward, as compared to the requirement of more than 250 visits to white flowers before becoming reliably constant (Heinrich *et al.*, 1977). A bee's floral choice may be biased by its sensory capabilities; for example, to an insect the colour red does not contrast against green foliage, and so red flowers may be visited less frequently than those of other colours. Hummingbirds, on the other hand, prefer red flowers.

10.1.2 Conditioning of perception

In conditioning of perception the cryptic or camouflaged prey type upon which an individual animal learns to feed is available before the animal begins to feed on it, but the animal is unable to perceive it. For example, individual captive jays (*Garrulus garrulus*) and common chaffinches (*Fringilla coelebs*) had difficulty finding stick caterpillars (Geometridae) on the floor of a cage mixed with twigs from trees on which the insects live (de Ruiter, 1952). However, after a bird found one caterpillar, often accidentally, it rapidly located others. It would seem that the birds did not distinguish between twigs and caterpillars until experience allowed them to do so.

Conditioning of perception refers only to situations in which the predator's ability to detect prey improves as a function of recent encounters with the prey type (Dawkins, 1971), and not for particular prey preferences, differences in palatability, ease of capture or handling time among the prey types, or avoidance of an unfamiliar food object. Although conditioning of perception should not be confused with conditioning of response, it is in fact somewhat difficult to separate the two conditioning types on the basis of behavioural observations alone. Both mechanisms may increase discovery rates, and both may result in search biased toward one food type when several are available in the habitat. The extent to which either mechanism is employed toward search for a particular target is likely to depend on the reward rates associated with alternate target types. Despite these similarities, there is one significant difference that has proved operationally useful in distinguishing between these two kinds of learning: conditioning of perception is most effective when the target is cryptic, whereas the efficacy of conditioning of response should not vary with the degree of crypticity of the target. If increases in discovery rates with experience are observed only under cryptic conditions, conditioning of perception is probably the mechanism involved.

Search image formation is the best example of conditioning of perception, whereby a predator selectively attends to one particular type of cryptic prey (Dawkins, 1971), and the prey for which a predator has no search image are effectively not encountered (i.e. the predator does not see them). It results in a temporary improvement in a predator's ability to detect a certain type of cryptic prey (Tinbergen, 1960).

Dawkins (1971) examined the ability of domestic chicks (*Gallus domesticus*) to detect rice grains on a background of stones that were glued to the substrate. Grains were dyed green or orange and attached to backgrounds of the same colour (cryptic condition) or opposite colour (conspicuous condition). The chicks required more time to find five cryptic grains than to find five conspicuous grains during the beginning of an experiment, but their ability to locate the cryptic grains improved by the end of each test period. To determine if chicks ignored the cryptic grains initially because they did not notice them, Dawkins compared the amount of time chicks spent in the head-down position when searching for cryptic or conspicuous prey. At the beginning of the tests chicks spent approximately 34 s in head-down position before pecking at a cryptic grain, and about 1.3 s before pecking a conspicuous grain. The results support the idea that chicks did not notice the grains initially. Interestingly, tests on subsequent days showed that chicks did not retain the search image for cryptic prey between test days. This is probably an adaptation to enable an animal to switch search images as resources change in profitability. Finally, Dawkins showed that experience in locating cryptic grains of one colour improved a chick's ability to locate cryptic grains of another colour, whereas finding conspicuous grains did not improve their success with cryptic grains. Search image formation is therefore generalizable.

Gendron (1986) investigated search image formation of Bobwhite quail (*Colinus virginianus*) in an arena covered with crushed green corncob. Coloured lard pellets, used for prey, were distributed in densities of either 4 or 9 per m^2 (112 or 252 prey). The quail were experienced foragers in this setting after several months of trials. Search image formation was substantiated by an increase in the probability of prey detection with increased number of prey captured in several immediately preceding sequences. Search image was reinforced whenever an appropriate prey type was captured, and it decayed between captures. Gendron's data also confirm Tinbergen's (1960) idea that search image should be stronger when reinforced more frequently, as when a high prey density condition (9 per m^2) is compared to a low prey density (4 per m^2) condition.

Earlier, Gendron and Staddon (1983) predicted that the probability of detecting an encountered prey item depends not only on a prey's crypticity but also on the predator's search rate (area searched per unit time). Their model states that as search rate increases, the probability of detecting an encountered prey item, P_d, increases according to:

$$P_d = (1-s/m]k)1/k \qquad k > \Phi$$

where
s = search rate
m = maximum possible search rate
k = measure of prey conspicuousness

Gendron (1986) recorded a decrease in the search speed of quail as prey were made more cryptic. At low speeds (when prey were most cryptic), the probability of detection was lower when the distance between the predator and the encountered prey was greater. As prey became more conspicuous, speed increased and the probability of detection in all of the series converged on a value of about 0.9. Thus, as contrast between prey and background decreased, locomotory rate decreased from 25 m/min to 12 m/min, and the probability of detecting a pellet decreased from 0.9 to 0.4. The overall result is a four-fold decrease in prey capture. Goss-Custard (1977) found something similar in redshanks, where the fast-walking shore bird may be able to detect large worms but overlook the small ones that occur among them. If large worms are rare, however, the bird may be forced to slow down so that it can detect small prey. Smith (1974a) also showed that thrushes search faster when prey are conspicuous. Hence, the mechanism of prey selection may involve consideration of both a choice between search speeds and a choice between different stimuli.

Recently Guilford and Dawkins (1987) reassessed the concept of search image and the evidence behind it, and concluded that most of the findings are either equally consistent with, or are better explained by, the simpler search rate hypothesis of Gendron and Staddon (1983). A predator simply learns to slow down when hunting for cryptic prey. Thus, the existence of search images remains an open question.

10.2 SAMPLING BETWEEN PATCHES

Although animals are endowed with a variety of patch orientation skills and assessment mechanisms, an important additional asset where patches differ in quality is the ability to sample among patches and to use this information to best advantage. One way for an animal to determine if the patch it is currently utilizing still has a higher quality or higher density of resources than some other patch in the habitat is to compare patches continually through sampling. However, the value of sampling to a forager must be viewed relative to the potential increase in rate of intake that will result from the sampling (Stephens, 1982). Patch sampling is most valuable when variance in patch quality is high, but then that is also the condition in which an animal is likely to recognize the differences in patch quality by direct sensory cues with little or no sampling. Nevertheless, animals should spend part of their time sampling even when the environment is stable, as a hedge against possible gains in nearby patches (Oster and Heinrich, 1976). Moreover, not all species can recognize patch quality using direct sensory cues, nor do all kinds of resources provide such cues. Finally, continuous sampling would be too costly in most situations, especially if patches were far apart or if patches were nearly equal in quality. Such activity would compete with time needed for feeding, ovipositing, or mating. Thus, we would expect the incidence of sampling between patches to be affected by the variance in patch quality, distance between patches, and perceptible sensory cues related to patch quality.

An experiment with houseflies shows how an animal can sample among patches and feed primarily in the most profitable patch without counting or calculating or learning (Fromm, 1988). Two patches 9 cm apart, each consisting of nine 0.5 μl sucrose drops, were arranged in an arena. The sucrose concentration in the two patches was either the same (0.125 M or 2.0 M) or different (0.125 M and 2.0 M). A fly was led to the centre of one of the patches and released when it began to ingest sucrose. As shown in Figure 10.2, flies searched and located drops in the first patch, left when some of the resources had been ingested, and moved to the other patch. When patches were of equal concentration the flies spent the same amount of time in each patch (Figure 10.3c and d). When the patches had different sucrose concentrations, the flies spent more time in the 2.0 M than in the 0.125 M patch (Figure 10.3a and b), regardless of which patch they started in. Why does a fly spend most of its time in the 2.0 M patch? Mainly because its local search pattern remains constricted longer after ingesting 2.0 M than after 0.125 M sucrose (see also Chapter 13.2.3), and thus time in the better patch accumulates over the period of the experiment (Figure 10.4). Thus the proximate mechanism for what might appear to be 'sampling between patches' is simply the consequence of differences in local search duration caused by the ingestion of different concentrations of sucrose.

In a graphic example of patch sampling, groups of six cichlid fish, *Aequidens curviceps*, adjusted their distribution and foraging time in an aquarium to match the profitability ratios of two patches of different densities of prey (Figure 10.5) (Godin and Keenleyside, 1984). When patches of prey were of equal density, the cichlids distributed

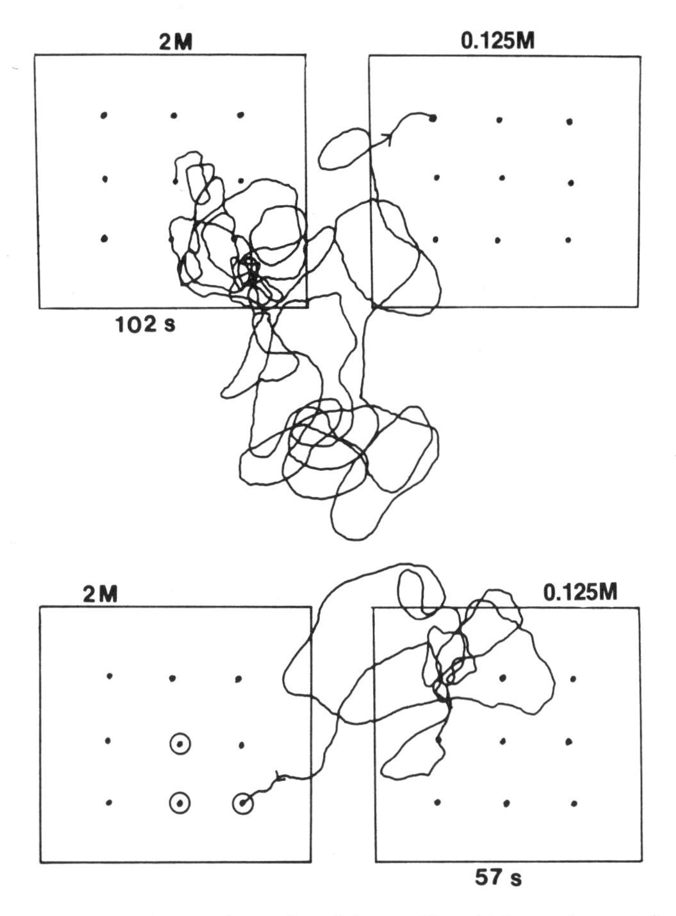

Figure 10.2 Typical search path of houseflies (*Musca domestica*) moving within and between two patches of sucrose drops. In the upper patches the fly began foraging in a 2.0 M patch and then after 102 s it moved to a 0.125 M patch. In the lower patches the fly returned to the 2.0 M patch after searching for 57 s in the 0.125 M patch. Drops were 3 cm apart. Dots with circles are ingested drops. (After Fromm, 1988.)

themselves approximately equally between the two patches. When the patches were of different densities (ratios of 2:1 or 5:1), the fish adjusted their distribution within 1 minute to match approximately the profitability ratio of the two patches: fewer fish were located in the poorer patch. The strategy used by the fish to assess patch profitability included sampling the available patches, although individual fish switched patches with varying frequency. Sampling had an associated cost, since high-frequency switchers had lower feeding rates on average than low- frequency switchers, but these fish

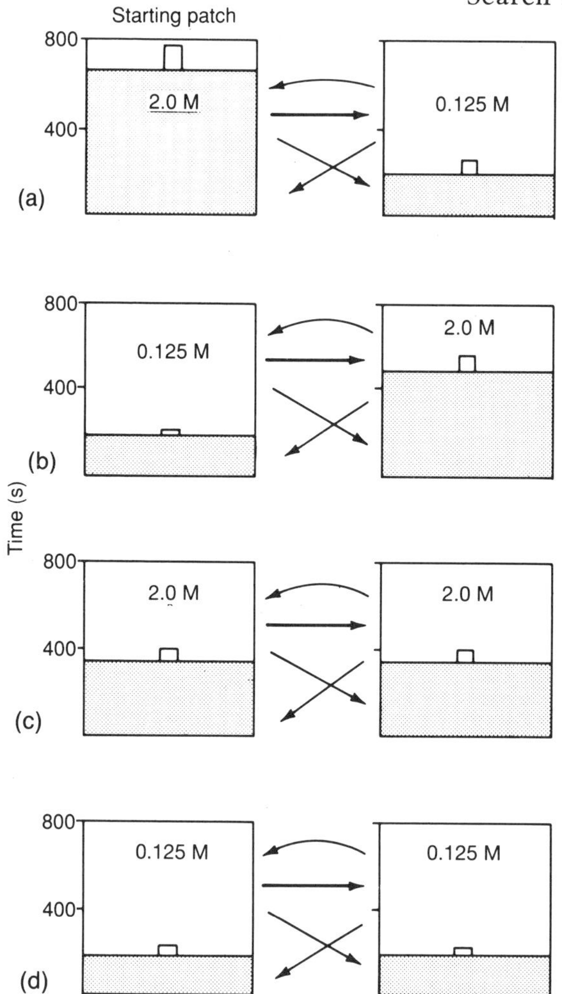

Figure 10.3 Mean time that houseflies spent searching in 2.0 M and 0.125 M patches. For each pair of diagrams, the fly began searching in the left patch. In (a) and (b) the drops differed in sucrose concentration between patches, whereas in (c) and (d) they were the same. Vertical bars are standard errors. (After Fromm, 1988.)

were able to feed in the best patch. Individual variation among fish in their competitive abilities for limited food resources, an interesting aspect of this study, suggests differences among individuals in perceptual or motor abilities or differences in ability to 'sense' which patch is the most profitable. The proximate mechanisms underlying sampling and decision-making in the cichlids are not known, but

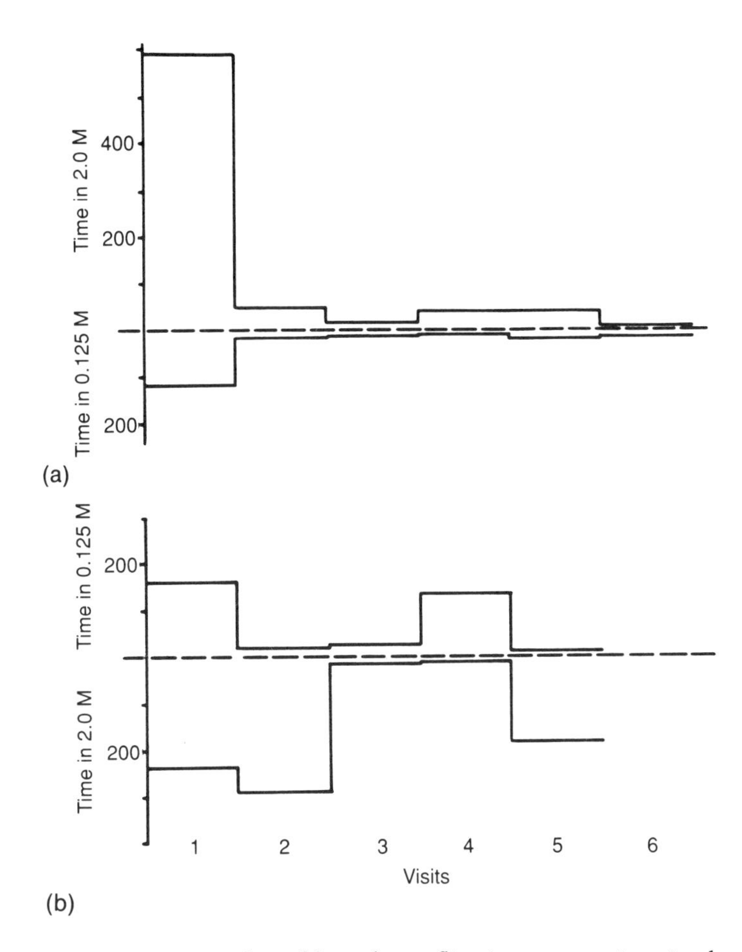

Figure 10.4 Two examples of how houseflies increment time in the more profitable patch of sucrose drops while 'sampling' between two patches (refer to Figure 10.3). (After Fromm, 1988.)

presumably are linked to perception of capture rate and the behaviour of other fish.

Krebs *et al.* (1978) devised a way to investigate the effect of patch heterogeneity on the decision to search in the best patch in a very controlled situation. They trained captive great tits to hop up and down on two perches at opposite ends of a large aviary; movements of the birds caused a computer-controlled disc containing mealworms to rotate. The system resulted in a different rate of prey presentation per hop and thus simulated two patches differing in quality. The researchers measured the number of hops required by birds sampling the two patches before concentrating their efforts on one of the

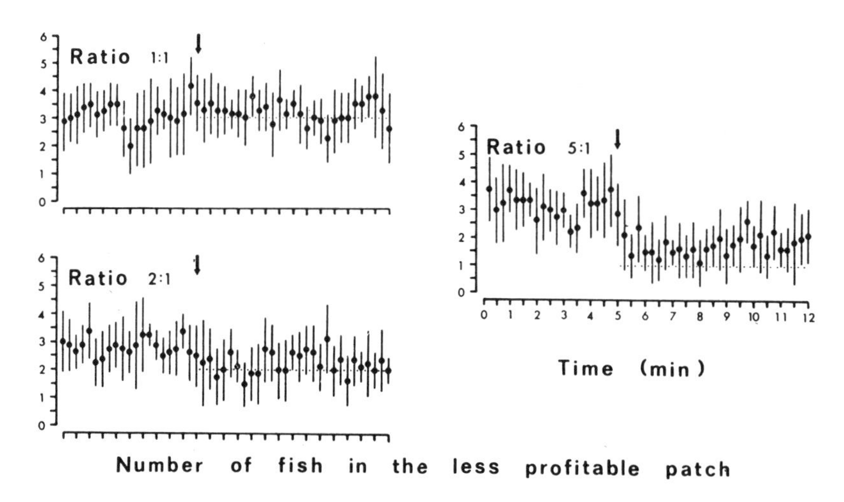

Figure 10.5 Mean number of cichlid fish (*Aequidens curviceps*) observed per 15 s in the less profitable patch, plotted against time elapsed since the start of the trial for each of the three patch profitability ratio experiments. Vertical bars are 95% confidence intervals. Arrows indicate the onset of the feeding period. Dotted line is the number of fish predicted by ideal free distribution theory. From Godin and Keenleyside, 1984. Copyright The Association for the Study of Animal Behaviour, reprinted by permission.

patches. The results showed that the number of hops the birds took to reach a decision decreased as the difference in rewards between the patches increased. Thus, it takes more hops (sampling) to come to a decision when patches are closer to equality than when they differ greatly in quality.

Among lepidopterous larvae, individuals establish a threshold leaf value, and choose to feed only on leaves above that threshold (Schultz, 1983). The threshold is lowered if no leaves of this quality can be found within a certain time period or after some number of leaves are sampled. For example, over a period of 5 hours several larvae of the moth *Heterocampa guttivitta* moved 970 cm, passing by at least 776 leaves, sampling (tasting) 98 leaves and accepting 30 leaves for sustained feeding (Schultz, 1983). This example is somewhat different from the others, although perhaps more typical, in that the animal samples patches as it moves along, and remains feeding at the best patches. This strategy is quite different from the fish, fly, and bird that sample among patches, 'decide' which one is best, and then feed at that best patch.

Cafeteria or free-choice feeding assay is a method used to determine preferences of herbivores for certain species of plants. Test animals select their food from a number of equally accessible plant species made available in approximately equal amounts. Larvae of the tobacco hornworm (*Manduca sexta*) have been tested extensively with regard to acceptable food plants (Jermy *et al.*, 1968; de Boer and Hanson, 1987). Circular disks of leaves are arranged on pins in a round dish, and a larva walks about tasting bits of the leaves. It eventually consumes the leaves it prefers, and the others remain. With wild mammalian ruminants, forage is supplied in containers or in pure-stand pastures. Smith and Hubbard (1954) and Smith (1958) used cafeteria feeding to determine plant preferences of mule deer (*Odocoileus hemionus*). The advantages of the method are that

1. plant species can be ranked according to animal preferences;
2. effects of unequal plant densities and their effect on preferences are controlled,
3. the animal can sample by moving about and spend more time at the most preferred station, and
4. it is possible to determine preferences for plant species not commonly eaten in the field.

A disadvantage of the method for large ungulates is the huge quantities of forage that must be collected by humans; this is not such a problem with caterpillar cafeteria.

10.3 SWITCHING BETWEEN PATCHES CONDITIONAL ON WHAT OTHER INDIVIDUALS ARE DOING

Fretwell and Lucas (1970) and Fretwell (1972) have suggested that where there are several habitats varying in resource value, we would expect individuals to be distributed such that each individual obtains the same fitness prospects; equality would be achieved by adjustments in competitor densities in the habitats. It is assumed that resource value decreases with increased competition, and so good habitats are filled first and contain more competitors than poorer habitats. Some of the poorest habitats may remain uncolonized, and would become occupied gradually if the population size increases. Davies (1982) uses an analogy of customers in a supermarket assessing the lengths of the lines at service counters. Each customer looks over the situation and then joins the shortest line, and given equal amounts of shopping items among customers and equal efficiency among clerks, the stable equilibrium is the same length for all of the lines. Fretwell and Lucas (1970) termed this kind of equilibrium spacing an 'ideal free distribution' (IFD), meaning that individuals are 'free' to enter or leave habitats, and selection favours various forms of competitor density assessment (e.g. conspecific cueing) and habitat selection so that all individuals experience equal gains. For individuals to behave optimally they must move to the appropriate site, and they must be able to monitor

1. the current resource patch quality and 'predict' the way it will change,
2. the current level of competitor density in a patch, and
3. the current level of competitor population in relation to the current general habitat quality.

The most clear-cut example of IFD is that of Harper (1982) who tossed pieces of bread to a colony of 33 free-living mallard ducks (*Anas platyrhynchos*) in a pond. The bread was thrown in the pond at two sites 20 m apart with site A receiving twice as many pieces as site B in some tests and the reverse in other tests. The ducks moved about, sorting themselves into the two sites to obtain food. The observed results are very close to the ratios of ducks predicted by the ideal free distribution model. When the two sites received equal number of bread pieces, the birds sorted out equally, with 16.5 ducks in each site (Figure 10.6a). When the bread was distributed in a 2:1 ratio, 11 ducks foraged in the least profitable patch and 22 in the most profitable patch (Figure 10.6b). This type of behavioural assessment also operates in generating equal profitabilities in different searching areas of feeding sticklebacks (*Gasterosteus oculeatus*) (Milinski, 1979) and in mating aggregations of toads (*Bufo bufo*) (Davies and Halliday, 1979). The proximate mechanisms promoting stable equilibria could include sampling the resource sites and choosing the one with the highest density of resources, or sampling the sites and remaining longest in the best site, or an individual could select a site according to the perceived number of individuals at the site, or all of the assessment 'criteria' could be employed.

In the yellow dung fly (*Scatophaga stercoraria*) both females and males search for dung of cattle, the females for oviposition sites and the males to locate females for mating (review: Parker, 1978). Both sexes are probably attracted by odour, and probably orient to the resource by flying upwind, as in other coprophagous flies and beetles, although visual cues are probably involved. The males arrive at fresh dung first, and females arrive shortly after. Males move repeatedly between the cow pad and the surrounding grass, with the sex ratio at any given time being approximately four to five males per female.

Since cow pads represent a patchily distributed resource, at the mating site there is keen competition between males for opportunities to mate with females. For example, males remain mounted on females during oviposition, and thus reduce the possibility of the female being mated by other males (note that eggs are fertilized by sperm from the last male she mates with). Males copulate with females long enough

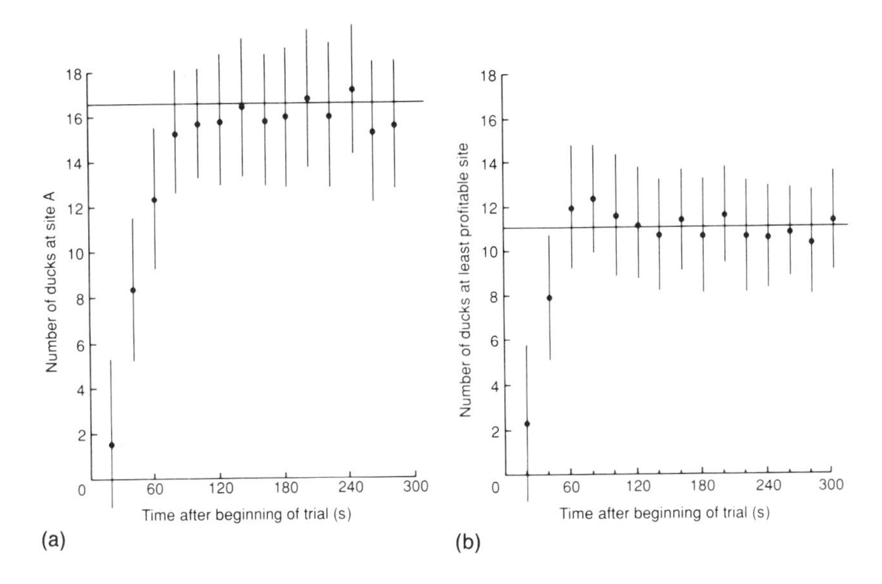

Figure 10.6 Distributions of ducks (*Anas platyrhynchos*) in two patches with equal numbers of tossed bread pieces as a function of time since the beginning of a trial. Horizontal lines indicate ideal free distribution prediction. From Harper, 1982. Copyright The Association for the Study of Animal Behaviour, reprinted by permission.

to ensure sperm precedence (35 minutes) but not so long as to decrease chances of locating and copulating with other females (review: Parker, 1978). Males that remain near a pad long enough for some of the other males to leave, benefit by increased yields. Thus, there can be an advantage to staying. However, this is disadvantageous if the arrival rate of females drops off or if the dung ages and becomes less suitable for oviposition.

Male dung flies usually arrange themselves upwind of the cow pad, and females usually land upwind of the pad and walk toward the pad, or land directly on the pad. In low male densities, the best tactic for a male seems to be waiting directly on the pad. However, in high male densities, because a male's fertilization rate is dependent upon number of competitors and arrival rates of females, the best tactic is to wait upwind of the pad. Parker showed that the number of males searching in grass, as compared to on a pad, decreases with the age of the pad and increases with male density at the pad. Thus the ideal free distribution is approached.

Hanski (1980) presents a point of view opposing that of Parker on movement patterns of *Scatophaga stercoraria*:

'Instead of advocating loose arguments about conceivable reasons for such and such 'strategies' more attention should be paid to informed modelling of the behaviour itself: the ultimate question, why the animals do what they do, should be easier to master after we have a good hypothesis for how they do it.'

While admitting to the strength and completeness of Parker's study, Hanski takes exception to the idea that the optimization model has been 'tested through direct observation'. This is so because 'functional' models are supposed to deal with evolutionary aspects of behaviour and so, strictly speaking, no tests on the evolution of the system have been carried out, for obvious practical reasons. While Parker shows that the proportion of males searching for females in the surrounding grass, instead of on the pad, changes with both the total male density and the age of the pad, Hanski proposes a simple Markov chain model to describe these same changes. Parker's data seem to fit the Markov model somewhat better than the optimization model, especially in predicting the proportion of males in the grass immediately after deposition of the pad and at low male density. However, Hanski's explanation does not deal satisfactorily with why the flies are more likely to move to the grass when the density on the pad increases.

Most studies, in addition to pointing out how various data sets generally support the IFD, also show that all individuals do not actually reap equal rewards. Dominant ducks obtain more food than subordinate ducks (Harper, 1982) and in other animals large males may obtain more matings than do smaller ones (Arak, 1983). Individuals with the highest competitive ability would be expected to occur in the best patches, or in patches where competitive differences are greatest; individuals with the lowest competitive ability would be found in the poorest patches, or in patches where competitive differences are least; individuals with intermediate phenotypes are ranked between these two extremes. An intriguing idea is that good competitors may simply be those that decide more quickly where to search and that switch less often; they do not need to be the largest, fastest, or the most dominant. In fact, Godin and Kleenleyside (1984) postulate that the observed variation in feeding rates by cichlids, *A. curviceps*, within patches of prey is related to individual differences in the ability to assess (learn) rapidly and accurately the profitability of the patches and in the ability to detect and attack prey first.

10.4 SAMPLING AND SPATIAL MEMORY

If animals have an opportunity to learn where the best patches are located in their environment, this information can improve their efficiency in relocating patches without wasting time visiting the poorer patches. In other words, the more an animal learns, the less it will have to rely upon sampling. There are not many examples of animals learning the positions of the best patches through sampling in nature, but there are a few hints that they may do so. For example, recall the artificial pinecone experiments in Chapter 8, the kestrels that organized their search routes so as to concentrate on profitable sites (p. 213), and several species of insects and primates that trapline along a learned route from one resource site to another (see Chapter 9.2). The following experimental results reveal insights into mechanisms that combine sampling and spatial memory.

10.4.1 Using the best patch

Several experiments with birds foraging in patches in aviaries or in outdoor enclosures show that they are remarkably sensitive to the density of prey in a patch, and that they can remember the spatial positions of profitable patches. For example, individual great tits (*Parus major*) combine learning with patch sampling to maximize their searching efforts in experiments with four food sites (Figure 10.7a) varying in density (1, 4, 8, or 16 mealworms per site) (Smith and Dawkins, 1971). The birds allocated their search time to match approximately the rewards of the four sites (Figure 10.7b). When the sites were exchanged (the site that previously had the highest density became the lowest and vice versa), the birds continued to visit the patch that had previously been the most profitable, but over several trials they began to concentrate on the new best site (Figure 10.7c), substantiating a learned assessment of the profitability of different spatial areas.

The evidence suggests that individuals associate environmental cues with different quality sites, such that the birds recognize the varying quality sites by the places where the sites are located. Learning to search in the correct places can occur quickly, as in red-winged blackbirds (*Agelaius phoenicus*), where two 30 minute training sessions were sufficient to condition them to search in the right place (Alcock, 1973). Only one reward was required to stimulate great tits to search preferentially in one of several types of potential food sites (Krebs *et al.*, 1972)! Smith and Sweatman (1974) performed experiments along similar lines with individual captive great tits. In one of their experiments, the size of prey items was varied among four areas containing equal numbers of hidden prey. The birds learned to search selectively in areas containing larger prey and, as a result, obtained a greater total mass of prey. There are very few examples of patch

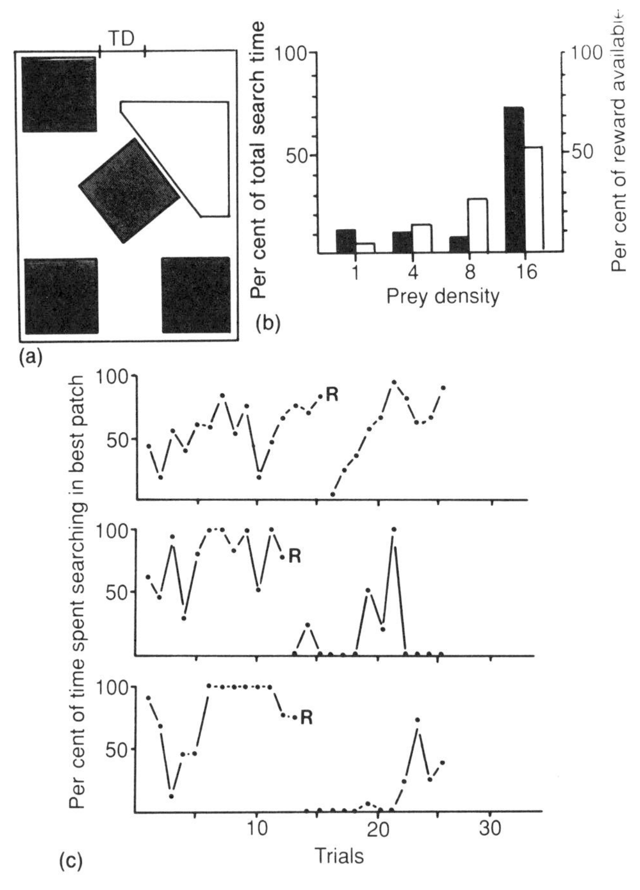

Figure 10.7 Great tits (*Parus major*) learn to forage in the most dense food patch (after Smith and Dawkins, 1971). (a) Diagram of the greenhouse (3.7 m × 4.6 m) where experiments were performed, showing 1.2 m × 1.2 m patch sites (shaded areas). TD designates trapdoor through which birds enter. (b) Proportion of time birds spent at feeding sites differing in prey density. Closed columns are search time, open columns are reward available. (c) Typical examples of percentage of time a bird spent feeding in the patch with the highest density (16 mealworms per site), followed by a switch (R) in which the best patch now contained the fewest number of prey.

differences related to prey size, but these data suggest that tits are sensitive not only to density but also to prey size. These results support an earlier hypothesis of Royama (1970) that great tits distribute their search effort in relation to spatial differences in the profitability of feeding sites.

Interestingly, in all of these experiments there were marked differences in the performances of individual birds. This led Partridge (1976) to suggest that profitability may be based in part on individual efficiencies, as well as on differences in patches of resources. The preferences of eight adult, hand-reared great tits were tested on four different types of feeding sites. The results showed that each bird preferred the feeding site that correlated with its own particular feeding efficiency on the various sites. If the birds were behaving in a maximally efficient manner in the context of this experiment, then they should each have gone exclusively to the feeding type on which they performed best. The birds did not do this, however, perhaps because in the wild there is a need to sample all potential sources of food from time to time owing to changing resource abundance and distribution. Thus, while spending most of the time at the most profitable source of food, they should also sample others. It might be that different types of feeding places in the wild are associated with different prey items, and that the birds must in the same way balance their intake of nutrients by not feeding exclusively on one type of food. The correlation between preference and profitability did not improve during the trials, and so the correlation seems not to derive from experience during the course of the experiments.

In addition to pointing out that the variance is more interesting than the mean, Partridge's (1976) results also influence our views on intraspecific competition. Any individual which could exploit a certain type of food more efficiently than other members of its own species would be at an advantage, because it could both feed at a higher rate and obtain a higher proportion of a temporarily abundant food source than could others. Successful individuals might therefore be more likely to survive in cold winters or in times of food shortage. On the other hand, differences among birds in ability to exploit different sorts of food might mean that the competition between members of the same species is less than is commonly supposed, in that individuals can specialize on the food which they exploit most efficiently. This would mean that competition for food would be strongest between birds with the same particular food specializations and least strong between birds with different specializations. The limits to the degree of variability and individual specialization found within a species would be set by many factors, including both the need to obtain a balanced diet for each individual of a species, and competition with other species. We would probably predict individual variability in feeding behaviour mainly in situations where the food supply is variable in space or time, as is the case generally with the foods eaten by great tits. If they could all be expected to encounter the same constant food supply in their life times, every individual bird should behave the same way.

10.4.2 Ecological relevance?

Do laboratory experiments, such as those discussed above, have ecological relevance? Smith and Sweatman (1974) made observations on the food searching behaviour of blue tits (*P. caeruleus*) and great tits (*P. major*) to determine:

1. if food is actually distributed in patches within the feeding area of a wild great tit,
2. if the searching behaviour of wild tits is selective in space, and
3. if this selectivity affects feeding efficiency.

Since previous field studies suggested that individual birds of a variety of species tended to make a number of successive flights to one foraging site, Smith and Sweatman attempted to quantify the distance and direction of foraging flights from nest boxes to determine if spatial selectivity related in some way to foraging times. The terrain around each nesting box was mapped by measuring the distances and directional bearing from the nest tree to other large trees near the nest. Approximately 20 trees within 50 m of the nest were marked with two or three labels indicating the distance and compass direction from the nest to a given tree. The results showed that the searching behaviour of wild tits feeding their young is indeed selective in space over short time intervals, but considerably less so over longer time intervals. The short-term selectivity was often quite extreme, with individual birds sometimes returning to within a few cm of a previous foraging site on the following flight. They were also more likely to return to a site if they found prey there quickly. By defining a feeding site as being the 'same' as the immediately preceding site if it was less than 5 m away within the same habitat type, or if it was within the same individual tree, it was concluded that most sites are only visited once, but that up to nine successive visits may be made to the same site by an individual bird. Thus, Smith and Sweatman concluded on the basis of laboratory and field studies that great tits have the ability to use prey size and prey density to make an assessment of the average profitability of a feeding area. These observations also suggest that the laboratory experiments discussed in this chapter may indeed provide reliable insights into the food searching of wild birds. The following case study on woodpeckers confirms this conclusion.

Case Study Woodpeckers searching for seeds in tree holes

A field study that questions the idea that birds count or even estimate capture rate was performed by Lima (1984) on the downy woodpecker

(*Picoides pubescens*) foraging in a small rural woodlot. The study utilized a natural population of birds, drawing them into the plot from surrounding areas. Patches in this case were thin 37 cm long saplings, each drilled with six groups of four holes. Lima placed a piece of sunflower seed in each hole and taped the holes shut. The birds visited the woodlot and learned to peck at the tape to examine the contents of a hole. Sixty patches were arranged in a rectangular area, each 4 m apart. In the experiments there were two kinds of patches: empty patches and patches with seeds in at least some of the holes. The number of seeds in non-empty patches in successive series of tests (termed 'environments') was initially 24, and then 12 and then 6. The problem for the birds was to determine, on the basis of sampling a patch, if the patch was sufficiently rich to continue feeding. A bird had to decide how many holes to sample without success before giving up the patch as empty. For example, in the first environment the site had patches with either no seeds or 24 seeds, and so we know that if the first hole opened did not contain a seed, then it would not be worth continuing to sample the patch. In the second environment, the site had patches with either no seeds or 12 seeds; in this case if, say, the first 11 seeds are found in the first 17 holes searched, we know that it would not be worth searching further to get the remaining seed. Thus, the general idea is similar to Gibb's (1962) survey, except that Lima was able to control the major variables. Lima predicted that as the intensity of the patch contents became more uncertain (i.e. 0/24 to 0/12 to 0/6), there would be a significant increase in the number of holes sampled without success before the woodpecker would leave a patch.

Figure 10.8 shows the number of holes opened in empty patches and patches with seeds. Before the first day of testing, the birds were exposed only to patches containing 24 seeds, and so on day 1 the woodpeckers 'expected' all patches to have 24 seeds. This is similar to the situation in the experiment of Krebs *et al.* (1974) where the birds should have expected 1 larva per 'cone'. During days 2 to 8 the woodpeckers learned that some patches were empty and that others contained 24 seeds, such that by days 9 to 11 they opened a mean of 1.7 holes in the empty patches. On day 12, the environment was changed so that patches contained either 0 or 12 seeds. By days 20 to 22 the birds were opening approximately 4.5 holes before leaving empty patches, but continued to open most of the holes in patches containing 12 seeds. On day 23 the environment was switched to patches containing either 0 or 6 seeds, and by days 29 to 31 there was a significant increase to 6.3 holes opened before leaving patches containing no seeds, but the birds continued to open nearly all holes in patches containing 6 seeds.

While the data qualitatively support Lima's prediction, the birds did not behave exactly optimally. For example, only one hole should have been opened in the first environment, whereas the birds opened more than one hole, and sampling was variable. There was also a significant increase in the number of holes sampled without success before a bird left an empty patch, as the identity of a patch (some seeds versus no seeds) became more uncertain with each successive environmental switch. To explore further the information used by foraging woodpeckers, Lima took advantage of the tendency of woodpeckers to begin searching at the bottom of the log and to

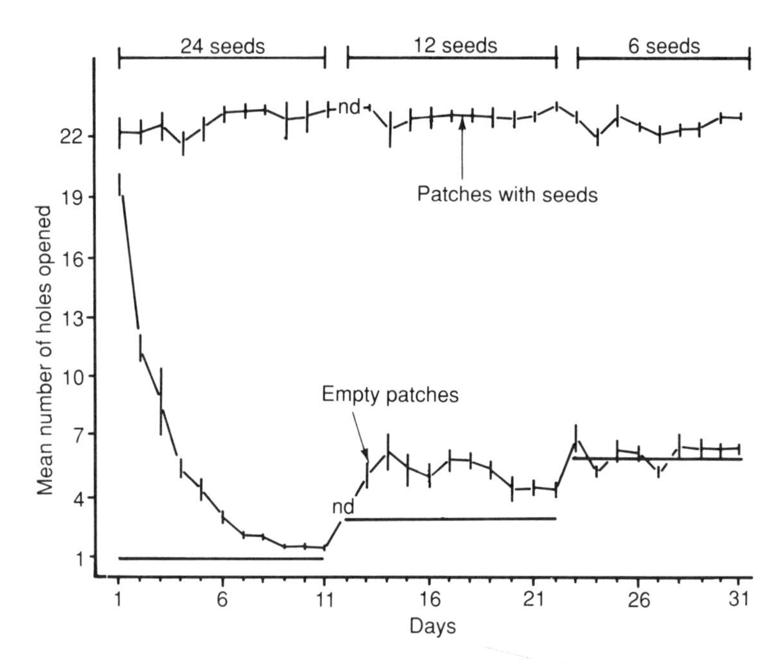

Figure 10.8 Mean number of holes opened in patches with seeds and in empty patches by downy woodpeckers (*Picoides pubescens*). Horizontal lines indicate the time period during which each 'environment', of 24, 12 or 6 seeds, was available. These lines also represent the predicted number of holes a bird should sample without success before giving up as empty. No data (nd) are given for day 12. Vertical lines are standard errors. From Lima, 1984. Copyright Ecological Society of America, reprinted by permission.

work toward the top. Thus, two additional kinds of patches were set out on day 22 (the last day of the 0 versus 12 seed environment). First, a few patches were presented with 12 seeds randomly distributed among the bottom five groups of holes (top holes empty). Of the five patches visited, 23.4 holes were opened per patch. This value was not significantly different from the number of holes opened in the patches containing 12 seeds randomly dispersed throughout the patch (23.3 holes opened), suggesting that the birds did not leave a patch after finding exactly 12 seeds, but opened some of the holes toward the top. Similar experiments in the 0 versus 6 seed environment also indicated that the birds had no idea as to how many seeds were in a patch. Finally, and even more telling, patches were presented that had only 2 seeds in the bottom group of holes, and the birds still opened 22.0 holes! Apparently they simply opened nearly every hole in any patch that contained seeds.

As it became more uncertain, with each environmental switch, as to whether a patch was empty or had some seeds, the birds responded by sampling significantly more holes without success before giving a patch up as empty. The strategy of opening every hole in a patch once a seed was discovered is simple and efficient for the 0 versus 24 seed environment, but less efficient for the 0 versus 6 or 0 versus 12 environment, and is a very poor strategy for the 0 versus 2 environment. Woodpeckers apparently do not or cannot count, or cannot remember their figures as well as they should in order to behave optimally. They can clearly distinguish between empty patches and those with some seeds, and this information apparently becomes the basis of a 'learning set', specifying that some patches have seeds and some do not. Lima suggests that perhaps this strategy is adaptive in adjusting to possible changes in the environment, since in these experiments the environment was changed twice. If the birds were aware of these changes they might have over-sampled to detect such changes. It should also be noted that the birds foraged in habitats outside the experimental plot during the experiments, and so the degree of satiation was not known, nor was the relative preference for sunflower seeds known as compared to other prey items in the area. The data indicate that woodpeckers do not use a simple fixed-time or number strategy to determine when to leave a patch, but that they 'recognize the distinction between empty patches and patches with seeds'. It is obvious that if Lima had switched environments even every three days, the woodpeckers could not have responded quickly enough to continue foraging efficiently. Thus, if the birds naturally adjust to changes in patch density, they can only do so within certain time constraints related to their learning-regulated assessment abilities.

0.4.3 Path analysis

Investigations of Zach and Falls (1976a,b,c) on foraging ovenbirds (*Seiurus aurocapillus*) add an important dimension to studies of sampling and learning, by presenting detailed path analyses underlying the proximate mechanisms of searching. Ovenbirds, which depend mainly on visual cues to detect prey as they walk on the ground searching for invertebrates in the leaf litter, foraged in a natural understory of leaf litter, dead twigs, stones, and mixed vegetation created in a 6 m × 6 m enclosure (Zach and Falls, 1976a). The area was subdivided with a grid of string to delineate 2.25 m^2 patches of mealworm pieces. Four patches, each with either 2, 4, 8, or 16 mealworms, were presented on two successive days. On the second day of testing the position of patches 2 and 16 were interchanged and patches 4 and 8 were interchanged. The investigators recorded data each day during four separate 1 hour observation periods.

By measuring path lengths in each of the four patches, Zach and Falls (1976a) showed that on the first day the level of search effort (path length) in patches was proportional to prey density. On the second day, considerable searching took place at the site where the density 16 patch had been (which was now density 2), but still more search effort was devoted to the 'new' density 16 patch. The birds also visited the higher density patches more often than the lower density patches. Consequently the search effort results are due to the combined tendencies of the birds to visit higher density patches more often, and to search longer in higher than in lower density patches.

Analysing the data set further, Zach and Falls (1976a) showed that the paths became less straight with increased mealworm density, and that the length of the search paths correlated negatively with straightness. Locomotory rate also increased with patch density, as in Smith's (1974b) thrushes and Gendron's (1986) quail that moved faster at the higher prey density or when hunting for more conspicuous prey. The sequence of visits to the four patches did not depart from randomness, thus failing to support either runs of visits to the denser patches or avoidance of patches already visited. Nor was there a sequence in visits from higher to lower density patches (e.g. density 16 to 8 to 4 to 2), which would indicate an optimal patch visitation mechanism. Finally, the actual number of prey eaten by individual birds differed markedly, as each continued to forage until satiated.

To investigate the search mechanism of ovenbirds further, Zach and Falls (1976b) observed birds over four observation periods in an arena containing one patch with 14 prey (treatment 1), 27 prey (treatment 2), or 54 prey (treatment 3). Analysis of the search paths revealed several interesting aspects of the ovenbirds' behaviour. First, the birds quickly learned the spatial position of the patch, as indicated by more local search in patch squares than in nonpatch squares. In fact, some birds only visited the patch once during the first observation period, and even then they were able to locate it quickly during the second observation period! Second, shorter search paths were required to locate the patch in periods after the first observation period, and the variance of mean path lengths decreased over observation periods. Third, more 'exploratory search' outside the patch was observed in treatment 1 than in treatments 2 and 3, suggesting that the amount of exploration decreased at higher prey densities. No differences in exploratory behaviour were observed between observation periods. Fourth, search paths prior to and after locating the patch were straighter then while in the patch, and search paths were less straight at the beginning of period one (searching for a patch) than between patches or after finding patches. Thus, initially the birds searched the area thoroughly until a patch was located; they then searched a patch thoroughly.

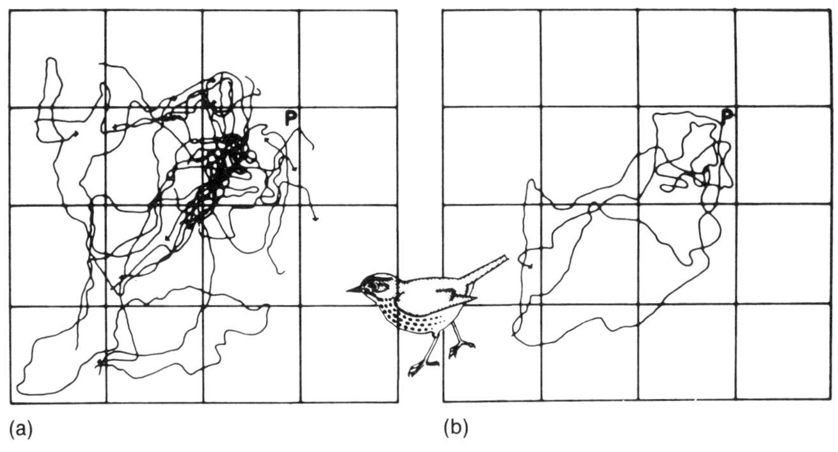

Figure 10.9 Typical example of the search of an ovenbird (*Seiurus aurocapillus*) on day 1 (a) in an arena with prey in the patch location marked P, and on day 2 (b) in an arena devoid of prey. From Zach and Falls, 1976b. Copyright National Research Council of Canada, reprinted by permission.

10.4.4 Local search released by spatial memory and expectation

In this and in other studies of birds (Zach and Falls, 1978; Smith, 1974a,b; Smith and Sweatman, 1974), locating and utilizing resources released local search behaviour. It is not clear, however, if local search can only be released by resource utilization, or if it can also be released by the expectation of resource utilization. The problem lies in distinguishing between local search caused by the actual finding of prey and the effect that learning has on searching. Can local search indeed occur as a result of expectation, based on earlier experience and information stored in spatial memory, rather than exclusively upon the immediate response to finding prey? To determine if the spatial memory of a profitable patch can release local search, Zach and Falls (1976b) exposed each ovenbird on the first day of testing to one of the patches described above (14, 27, or 54 prey), followed by a test run in which the arena was devoid of prey on day 2. On day 2, with no prey or patches, there was significantly more local search in the previous patch site than in other grid squares, even though there were no mealworms to collect (Figure 10.9). Thus, the patch site itself was learned, as shown by the way that the birds visited it directly and by the length of search path in that area. Interestingly, there was no effect of prey density on the intensity of searching in the empty patch site; the intensity was the same when the patch on the previous day contained 14, 27, or 54 mealworms.

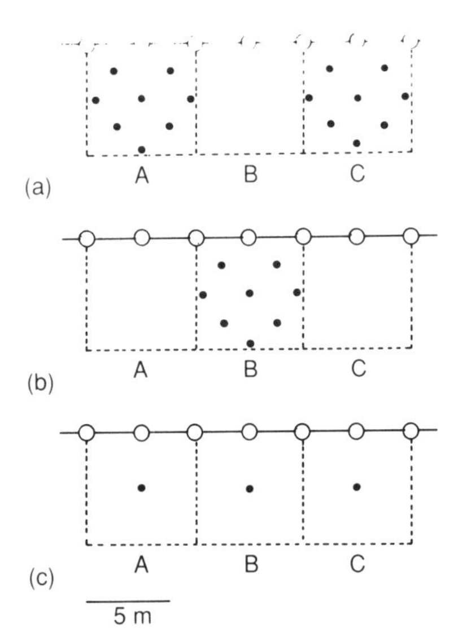

Figure 10.10 European badgers (*Meles meles*) learn to discriminate among patches varying in amount of food. Diagrams show arrangement of food in three adjoining patches (A, B and C) during training trials (a or b) and test trials (c). From Mellgren and Roper, 1986. Copyright The Association for the Study of Animal Behaviour, reprinted by permission.

The results of ovenbirds searching in an area where a patch previously had been located are important in pointing out what may be a major difference in patch assessment mechanisms among species. The search paths convincingly show that ovenbirds know in this situation that their prey exist in a spatially distinct area of the environment, and even when the patch is devoid of prey, they respect patch boundaries.

European badgers (*Meles meles*) discriminate among patches according to the amount of food in the patches (Mellgren and Roper, 1986). Badgers engage in local search when they find food in a patch, and then after a single trial they remember the location of profitable patches they previously visited. For example, Figure 10.10 shows tracks of badgers trained on food distributions (a) or (b) in Figure 10.11. Note the concentration of search in the patches previously endowed with food. In each case the badger was trained to one type of distribution in trials 1 to 6, and then in trial 7 all three patches had food. In trial 9 the distributions were switched so that badger 1 received distribution (b) and badger 2 received distribution (a).

Zach and Falls wrestle with two remaining questions. First, is the strategy of the ovenbird efficient? Second, why do ovenbirds bother

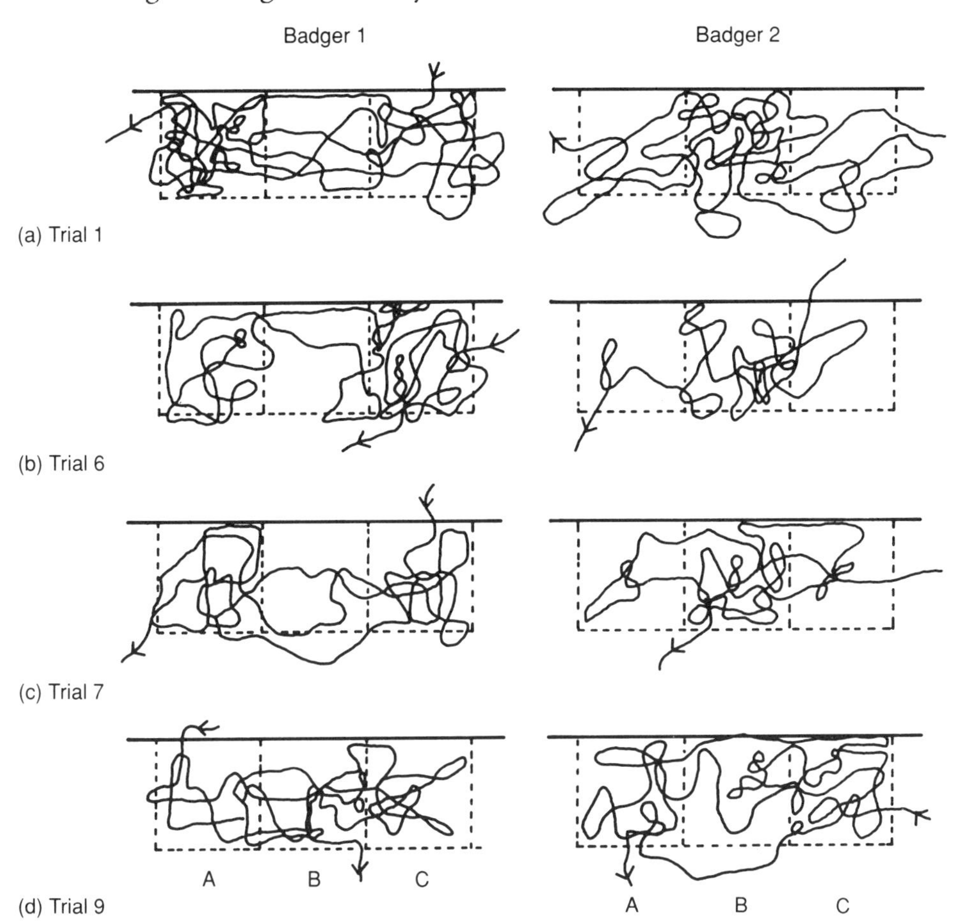

Figure 10.11 Search paths of two individual badgers (refer to Figure 10.10) in training trials (1 and 6), when all patches had food (7), and after the patch arrangement was switched so that badger #1 was exposed to the arrangement of food used to train badger #2 and vice versa (9). From Mellgren and Roper, 1986. Copyright The Association for the Study of Animal Behaviour, reprinted by permission.

spending any time at all searching in low density patches, especially after they have experienced the higher density ones?

10.4.5 Efficiency

The efficiency of the ovenbird strategy can be estimated by the relationship between the number of prey taken and the length of the

Table 10.1 Mean length of search path per prey found in patches of different densities (all periods and both days of testing combined) (from Zach and Falls, 1976a)

Prey density	*N*	*Mean length*(m)*
2	9	4.70 (0.72)
4	9	2.53 (0.26)
8	9	1.29 (0.05)
16	9	0.90 (0.05)

*Standard errors in parentheses.

search path (Table 10.1): more search per prey item was expended finding prey in lower density patches than in higher density patches. What would an optimal performance of ovenbirds be in this situation? If 'optimal' means that a foraging bird should almost always feed in the patch with the highest density, then the birds did not pass the test. They spent considerable time searching and feeding in low density patches. Their performances improved, however, from the first to the fourth one hour recording sessions. By session four of day 1 the observed and expected number of prey taken from the density 16 patch were not significantly different, suggesting that the ovenbirds had adjusted to the switch in prey densities (i.e. that they had learned to forage almost optimally). On day 2 the performances were again not optimal until the later sessions. Thus, although search effort in the four patches increased exponentially with prey density and a higher percentage of prey was found in the denser patches, foraging was not exactly optimal.

10.4.6 Why forage at all in low-density patches?

To understand why ovenbirds spend any time at all searching in low-density patches, especially after they have experienced ones with the higher density, consider the following explanations. First, the birds seem to sample all of the patches from time to time, perhaps as a hedge against changes in patch density over time or perhaps because of labile memory. Second, they may have taken prey from low-density patches while walking to or from resting spots; it is not easy to determine if a bird knows that it is actually searching in a patch as delineated by the experimenter.

10.5 MORE ON SPATIAL MEMORY

In all of the examples of sampling and learning discussed here, it would seem that the most important process involved is 'spatial memory'. By sampling different spatial areas the animals learn where to find food. Two related topics, radial arm maze learning and recovery of cached food, will be discussed briefly, because both rely upon the same spatial memory abilities that are needed in searching behaviour.

10.5.1 Radial arm maze learning

An apparatus that has been used frequently to study spatial memory is the radial arm maze. The maze consists of a central hub with several arms (typically eight or 12) radiating from the centre. The end of each arm is baited with a piece of food and an animal searches the arms to clear the maze of food. An error is defined as entering a previously visited arm. Rats are extremely good at this task, usually making less than one error in the first eight choices on an eight-arm maze (Olton and Samuelson, 1976). Although not all researchers are convinced (e.g. Markowska *et al.*, 1983), apparently explanations not based on memory, such as the use of odour trials or response biases, have been eliminated (review: Roitblat *et al.*, 1984).

Thus, an animal uses spatial memory to keep track of which arms have been visited and which have not. It is possible that animals encode this information in memory retrospectively if they remember the arms that have already been visited (past events), whereas the information is held in memory prospectively if they remember the arms that have yet to be visited (future events). Cook *et al.* (1985) showed that rats encode retrospectively early during a run and then prospectively later on. Hence, they probably use whatever code resulted in the shortest list of items to be held in memory.

10.5.2 Food caching and spatial memory

The caching of food in discrete dispersed locations, and its subsequent recovery, is characteristic of several avian and mammalian species. Although the caching and recovery phenomenon has been known for some time it was not until recently that several principles have become clear. Van der Wall (1982) found that cache recovery behaviour in Clark's nutcracker (*Nucifraga columbiana*) depended upon visual memory. Most caches were made in soil within 5 cm of conspicuous large objects, and then later on the nutcrackers relocated caches using the large objects as remembered visual cues. Soil microtopography and small (<2 cm diameter) objects may be used as cues to facilitate cache recovery, but they are not essential. Nutcrackers that had not been involved in caching seeds were nevertheless able to locate caches by using soil disturbances at cache sites as visual cues and by searching preferentially near objects where caches were concentrated. However, the success rates of non-seed-caching nutcrackers ranged from 8 to 12%, whereas those of seed-caching nutcrackers ranged from 52 to 78%. By limiting the sites where the birds could cache seeds, the role that preferences for certain caches could play in the accurate recovery of seeds was eliminated or reduced (Shettleworth and Krebs, 1982; Kamil and Balda, 1985). Food-storing birds are able to remember the spatial locations of large numbers of scattered caches for periods ranging from a few days to several months, and they use visual cues to do so (Sherry, 1985). The birds can perform a number of operations on the remembered set of storage sites, including the ability to recall which caches have been previously exploited and which caches have been lost to other animals. Cowie *et al.* (1981) provided direct evidence that seeds hoarded by wild marsh tits (*Parus palustris*) disappeared more rapidly than control seeds placed in identical sites 100 cm away, suggesting that the birds remember their exact locations. Seeds also tended to be stored and recovered in the same sequence.

10.6 SUMMARY AND CONCLUSIONS

Learning may influence either how an animal searches in a resource unit or the choices it makes among the available resources.

Conditioning of response occurs when the probability of an animal exhibiting a certain response increases with repeated presentation of stimuli. Sometimes animals learn which resource type is the most profitable and then remain constant to that type. Remaining constant to one particular type of resource is not an advantage in a changing resource environment, however; a higher average reward can be attained by employing a mixed strategy, such as majoring on one resource type while minoring on several others, which allows foragers to track changes in resource availability.

Conditioning of perception occurs when the probability of perceiving a stimulus increases as a result of repeated presentations of stimuli. In conditioning of perception the cryptic or camouflaged prey type upon which an individual animal learns to feed is available before the animal begins to feed on it, but the animal is unable to perceive it. The most common type of conditioning of perception is search image formation, in which a perceptual change results in a temporary improvement of a predator's ability to detect cryptic prey. As resources become more conspicuous, the probability of detection improves, and search speed increases.

In an ideal free distribution (IFD) individuals distribute themselves such that each obtains the same fitness prospects. Equality is achieved by adjustments in competitor densities in available habitats. The proximate mechanisms might include sampling the resource patches and choosing the one with the highest density of resources, sampling the patches and remaining longest in the best patch, or an individual could select a patch according to the perceived number of other individuals at the patch. All individuals seldom obtain exactly equal rewards; dominant individuals obtain more food than subordinates, and large males may obtain more matings than do smaller ones.

11
Exploratory behaviour

Exploratory behaviour serves to acquaint an animal with the topography of its range, and to familiarize it with potential food sources and escape routes. Exploratory behaviour is probably a more significant component of searching than is usually appreciated. For example, Balda (1980) observed nutcrackers (*Nucifraga caryocatactes*) first carrying out an exploratory survey of an area before searching around for a specific site in which to cache food. Since the birds proved to be highly efficient at locating caches several months later, the survey of the area apparently enabled the birds to learn local topography (Balda, 1980). The black bear (*Ursus americanus*) which is mainly a vegetarian feeding on berries and insects, occasionally heads off to a site as far as 16 km from its home range to visit a rubbish dump or a salmon stream (Jonkel and Cowan, 1971). Although bears might possibly be stimulated to make this foray because they detect odours from the distant resources, it is more likely that their search for these resources is guided by information previously gained through exploration. Of course both mechanisms could be involved.

Exploration can also be effective in learning topography which might be instrumental later in avoiding predation. The meadow vole (*Microtus pennsylvanicus*) (Ambrose, 1972) and the white-footed mouse (*Peromyscus leucopus*) (Metzgar, 1967), released in field sites that contain nest boxes but that were unfamiliar to the animals, were preyed upon by captive screech owls (*Otus asio*) more than individuals that were given several days to explore the new environment before owls were introduced. Among examples in which exploratory behaviour can minimize predation are resident mice becoming aware of danger more readily than transients, residents becoming familiar with terrain and being able to escape more quickly, and transients being more active and attracting the attention of predators (Metzgar, 1967).

Exploration by rats includes investigation by sniffing, head rotation, and physical contact, as well as locomotion within previously unencountered (novel) surroundings. Thus an animal's sense organs are brought to focus on or into contact with novel objects and situations. Of course, different species use different means to explore. Animals with keen olfactory abilities use olfaction, whereas animals with better visual abilities use vision in exploring. Racoons (*Procyon lotor*) use their forepaws (Welker, 1961), whereas cockroaches (*Periplaneta americana*) use their antennae (McCoy, 1984). The degree to which a given species explores may also be related to the complexity of the discrimination and manipulation used in its feeding (Glickman and Sroges, 1966). For example, stereotyped feeding responses might in many cases be associated with low exploratory tendencies.

Animals often explore their range or parts of their range immediately after feeding. They often exhibit more exploratory behaviour when satiated than when hungry (Chapman and Levy, 1957), suggesting that having met their immediate needs, they monitor other resources of possible use in the future. Exploratory behaviour is also affected by food or water deprivation in interesting ways (Fowler, 1965; Gross, 1968; Cowan, 1977). For example, exploration usually takes precedence over feeding or drinking when fasted or water-deprived rats are given access to food or water in unfamiliar situations, or where they are confronted with unfamiliar stimuli (Chance and Mead, 1955; Zimbardo and Montgomery, 1957). Several lines of evidence indicate that exploration is prolonged the greater the difference between the familiar and unfamiliar feeding situation (Fink and Patton, 1953), but it then declines as the length of deprivation is increased (Bolles, 1962).

Exploratory behaviour immediately after feeding increases the chances of racoons (*Procyon lotor*) locating fresh food sources (Bider *et al.*, 1968). Similarly, rats commonly explore their (familiar) surroundings after eating, suggesting that the causal factors stimulating exploration were inhibited by feeding, and that exploration was disinhibited after feeding (Barnett *et al.*, 1978). An alternative explanation is that rats may search as a result of food utilization, and as in the racoon, the release of local search might aid in localizing other food sources.

Some species gradually expand the size of the area explored in establishing a familiar area. This kind of behaviour has been observed in a field mouse (*Peromyscus cranitus*) (Brant and Kavanau, 1965), the grey field slug (*Agriolimax reticulatus*) (Newell, 1966), and the goldfish (*Crassius auratus*) (Kleerekoper *et al.*, 1974). Figure 11.1 shows the movements and behaviour of an individual slug on five consecutive nights in an experimental arena (68 cm × 91 cm) as it searched for vegetation and then roosted each night in a hole in the ground. When crawling, the slug secretes a clear, colourless mucous which forms a durable trail. Exploration, characterized by a low turning rate and infrequent crossing of the previous track, maximizes efficiency of

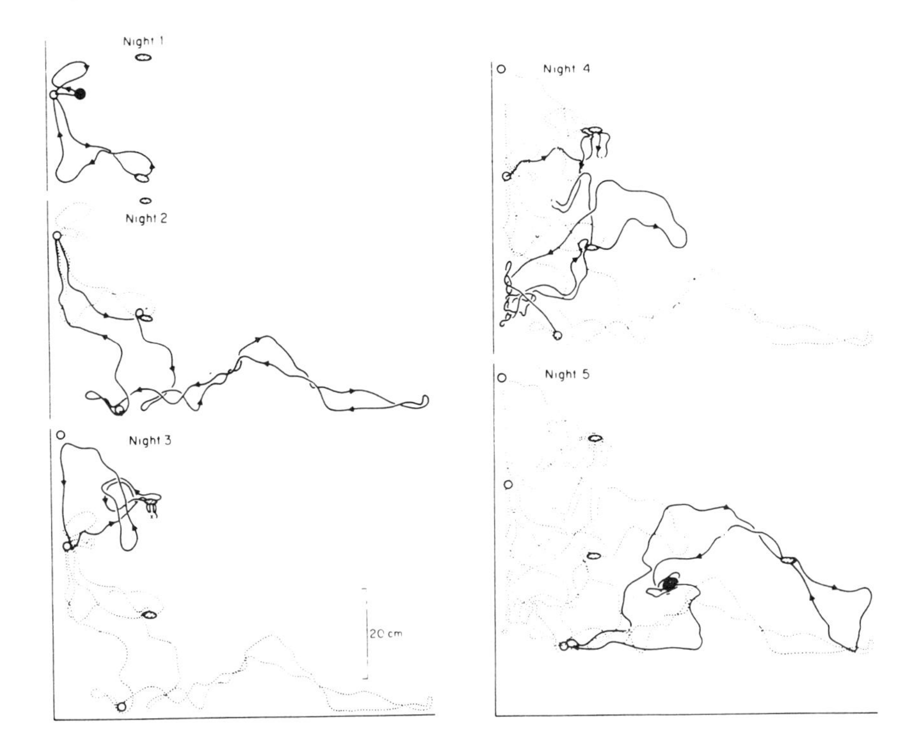

Figure 11.1 Exploratory behaviour and establishment of a familiar area by an individual grey field slug (*Agriolimax reticulatus*). The diagrams show the movements on five consecutive nights in an experimental arena (91 cm × 68 cm). The movements were recorded by time-lapse photography, one frame being exposed every 15 s using high-speed flash illumination. Solid line indicates the track of the animal on a given night; dotted lines indicate the tracks of the animal on previous nights. Diagonally hatched oblong shapes are the positions of slices of carrot provided for food. Open circles show the positions of holes in which roosting could occur. From Baker, 1978. Copyright by R.R. Baker, reprinted by permission.

search. The impression from Figure 11.1 is that the slug builds up a series of tracks in an area such that movement to a given point in the area (e.g. the roosting site), can be accomplished from any other site. The outward trip is generally quite straight, whereas the return trip meanders sufficiently for an animal to maintain contact with the outward trail, while still minimizing overlap with the trail laid down on the outward journey. Systematic components of exploratory behaviour in goldfish include a series of different locomotory patterns, each of which has a centre of activity that is spatially defined.

Tolman (1948), attempting to explain how rats learn a maze, was probably the first to formalize the concept that animals form cognitive maps of their environment. Animals could use such a map to find resources. An animal might learn the location of food, for example, during periods when it is neither food deprived nor consumming. Later when food deprived, this learned information might lead the animal to the food. Development of a cognitive map is thought to occur in the following way (O'Keefe and Nadel, 1978; Olton, 1979). First, there is an association between the information within the nervous system and the environment (i.e. a mapping of the environment onto some structural area in the brain), and second, the cognitive maps are causal agents in the animal's locomotory behaviour (i.e. it steers through its environment using the map). For example, when chimpanzees (*Pan troglotytes*) are placed in a novel environment they explore systematically, and each day expand outward to cover more area. The chimps appear to establish certain distinct landmarks which are used as references from which to conduct further exploration, suggesting the formation of a cognitive map (Menzel, 1978). Some species form versatile maps, as for example in a gobiid fish (*Bathygobius soporator*) which moves among scattered tide pools at low tide when they cannot see from one pool to another (Aronson, 1971). They are able to orient through the maze of pools very well, however, because at high tide they form a map of the area as they explore over the pools. For diverse examples, refer to Ellen and Thinus-Blanc (1987).

Exploration first facilitates construction of a cognitive map, and then leads to revisions of the map. New objects do not have representation in an already existing cognitive map, and so the map is updated as an animal notices new items or changes in its environment. Nadel and O'Keefe (1974) and O'Keefe and Nadel (1978) suggest that the hippocampus in the vertebrate brain is the site where cognitive maps are formed. Thus, hippocampal-damaged rats show no investigative (exploratory) behaviour to novelty, novelty being a quality of the immediate environment which does not have a representation within the cognitive map. In novel surroundings gerbils with hippocampal damage have increased locomotory activities, but decreased investigative behaviour (Glickman *et al.*, 1970). Many factors impact on exploratory behaviour and cognitive map building. For example, in a novel environment, socially-reared rats had higher initial rates of exploration than those reared in isolation, but the rate declined more quickly. This suggests that socially reared rats assimilate information more quickly about their environment (Einon and Morgan, 1977).

The proximate explanations for why an animal explores are not always clear, but most studies agree that a change in familiar surroundings stimulates exploration. Whereas exploration is a response to unfamiliar surroundings, patrolling is systematic movement through a familiar area. Patrolling usually refers to regular movements which some animals make through their familiar home range, although it would be possible for a patrolling animal to encounter cues that stimulate exploration. Many species of mammals scent mark (Johnson, 1973), as they move through new areas, and by doing so they establish regular routes of movement through their territory, and they tend to follow the same routes (Stoddart, 1980). This behaviour is not unlike the trail system established by the slug. Patrolling may also refer to regular, systematic, or cyclic search at a particular site or several sites. For example, male sweat bees systematically patrol over nests of emerging females (e.g. Michener *et al.*, 1979).

11.1 SUMMARY AND CONCLUSIONS

Exploratory behaviour serves to acquaint animals with the topography of its surroundings, and to familiarize it with potential food sources and escape routes. Changes in familiar surroundings often stimulate exploration.

In exploratory behaviour there is a mapping of the environment onto some structural area in the brain. These cognitive maps are causal agents for future searching behaviour as an animal steers through its environment using the map.

What an animal learns about during exploration might be instrumental later in locating resources and in avoiding predation or other stress sources. Exploratory behaviour immediately after feeding may increase the chances of an animal locating other nearby food sources.

12
Central place foraging

If a resource is an oviposition site, a nesting site, or a potential mate, an animal locating such a resource will most probably utilize it on the spot. Resources such as food or water, however, may be carried back to some fixed point such as a nest or a colony and consumed or stored there. Animals that carry resources back to a particular site are called central place foragers (CPFs), and they generally have a nest to which they bring resources. Central place foragers differ from the rest in that their activities include an outbound journey, a period of searching, and then a return journey. For example, birds carry food back to nestlings or capture prey while diving in the sea and carry it back to a terrestrial feeding site; bees transport nectar and pollen back to the hive.

To be sure, central place foraging combines various search mechanisms that have already been discussed in previous chapters, but it also has some unique features. First, because of travel costs, we would expect CPFs to search so as to locate the most profitable patches and to minimize revisits to already used patches, to regulate the load to be carried back relative to the distance travelled, and to assess distances travelled relative to predation risk. Second, efficient CPFs should be selective in establishing the central place in relation to the resources they require. For example, the male peacock butterfly (*Inachis io*) selects a protected spot opening out onto a wide open expanse as a good site for intercepting passing females (Figure 12.1). Third, because most CPFs are social species with potential for communication between individuals, we would expect that exchange of information would occur commonly. Fourth, all species that employ central place foraging have at least one common attribute, in that they can find their way back to the central place after foraging away from it.

12.1 DISTANCE TRAVELLED AND PATCH CHOICE AS RELATED TO METABOLIC COSTS

Orians and Pearson (1979) make several predictions about the foraging behaviour of animals relative to distance from the central

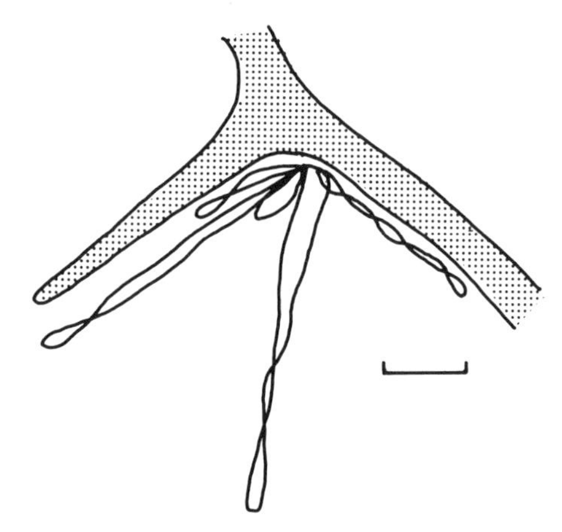

Figure 12.1 Track pattern of a male peacock butterfly (*Inachis io*) in relation to an area of high suitability for intercepting females. The male leaves its guarded corner and flies out to investigate any large insect that comes by; if unsuccessful in locating or copulating with a female, it returns to its perch. Shaded area indicates a tree line. The bar indicates a distance of 40 m (After Baker, 1972.)

place. First, as long as patch quality remains constant, the optimal load an animal carries should increase with increasing distance of the patch from the central place; otherwise the trip may not be profitable relative to losses incurred in energy used for travel. Second, time spent in a patch will increase with increasing distance of the patch from the central place; in other words, short trips should correlate with short patch times, and long trips should correlate with longer patch times. For a constant distance between the patch and the central place, the optimal load will increase slightly with patch quality, whereas time spent in the patch will decrease significantly. Third, since the predator weighs more on its return trip than on its outbound trip, it can increase its rate of net energy delivery to the central place by shortening the return trip relative to the outbound one. This latter hypothesis is borne out by observations of desert ants that have meandering outgoing paths and straight return paths (see Chapter 9.1).

The first and second hypotheses have been tested in breeding colonies of the white-fronted bee-eater (*Merops bullockoides*) in Kenya. When the distance of foraging from the nest was increased from 25 to 575 m there was an increase in search bout duration, time spent at the colony between bouts, and the mass of insects carried back to the nest (Hegner, 1982). Similar results have been reported by Kasuya (1982) for foundresses of the eusocial Japanese paper wasp (*Polistes chinensis antennalis*). The wasps find and carry water back to their nests for regulating nest temperature and for making pulp used in nest construction. There is a positive correlation between the time for water sucking and the time for the outward bound trip. Hence, as predicted, the load carried increased with increasing distance of the patch from the central place, and time in a patch increased with increasing distance of the patch from the central place.

Animals may also choose foraging sites on the basis of travel time, collecting time, or according to some unequal weighting of these two elements. Kramer and Nowell (1980), studying the Eastern chipmunk (*Tamias striatus*), set up two patches of seeds in which a rich patch with a short collecting time was 31 m from the burrow (minimal travel time of approximately 122 s), and a poor patch with a long collecting time was 9 m from the burrow (minimal travel time of approximately 50 s). The chipmunk detected the near patch and collected there several times. It then approached the more distant sites, but returned and collected at the near site. Over several subsequent trips it alternated between near and far patches, and then collected at the far site for several trips. It appeared, however, that the chipmunk had been taking fewer seeds from the near site than were available. When the far site was removed, the animal first searched in other areas and then concentrated on the near site, and obtained full loads from the near site. Thus the animal had been previously departing the near site with incomplete loads. It seems that chipmunks sample between patches and assess travel time, collection time, and the spatial distribution of seeds. Since chipmunks are territorial, the best strategy might be for them to bring in food from a distant patch first, since food in an individual's own territory is protected to some degree.

12.2 DISTANCE TRAVELLED AS A FUNCTION OF PREDATION RISKS

Animals deal with risk in several different ways, from simply allocating time to watching for predators (vigilance) to intrinsic mechanisms which indirectly reduce the chances of predation. Central place foragers exemplify some of these intrinsic mechanisms. For example, grey squirrels (*Sciurus carolinensis*) trade off energy intake rate against predation risk in the following way. Maximum energy efficiency is realized when they consume food immediately upon finding it on the ground, whereas carrying items to the safety of trees provides minimum exposure to predation, but incurs additional energy expenditure. As a compensation mechanism the squirrel's tendency to carry an item of food decreases with distance of food from cover (travel time) and increases with size of the item (handling time) (Lima *et al.*, 1985).

For many animals, it is much safer to be in the central place than to be at a foraging site, and under conditions of predatory risk they tend to search closer to home. For example, Martindale (1982) used stuffed specimens to simulate conspecific attacks on male nest-defending Gilia woodpeckers (*Melanerpes uropygialis*). Following the attacks they remained closer to their nest while foraging, foraged for shorter times per patch, and delivered smaller loads than would be predicted for delivery rate maximization. The effect was more pronounced for patches further from the nest, such that the male left distant patches sooner than near ones, and delivered smaller loads from further away. Because the two objectives, food delivery and nest defence, require different behaviour sets, an individual cannot maximize both nest defence and delivery rate. For monogamous species in which both parents raise the young, the situation is simplified somewhat because individuals need not forage independently. Instead, the mates can work as a team to pursue two objectives simultaneously.

12.3 GROUP EFFECTS

Species that have nests and that live in societies, such as ants and bees, or in social groups, such as bats and many birds, increase their foraging efficiency through two important mechanisms. First, nest-site selection can have an important impact on the foraging success of a colony, depending on the distances to food sources during the period of nesting. When there are several nests, their spatial organization may be important relative to foraging in nearby resource patches. Second, efficiency can be improved by communication between colony members concerning changing resource availability.

12.3.1 Geometrical food distribution hypothesis

If food resources are evenly distributed and stable, then optimal foragers will tend toward regular dispersion of individuals or small social units, rather than toward aggregation at a central point; if resources are mobile and unevenly distributed or clumped, the optimal strategy is aggregation of the foraging population at a central location, rather than toward dispersion (Horn, 1968). Thus, according to Horn, travel costs to and from food patches can be minimized by dispersed breeding when food is uniformly distributed and by colonial breeding in the geometric centre of all food patches when food is spatiotemporally clumped. In Figure 12.2, d equals the round-trip

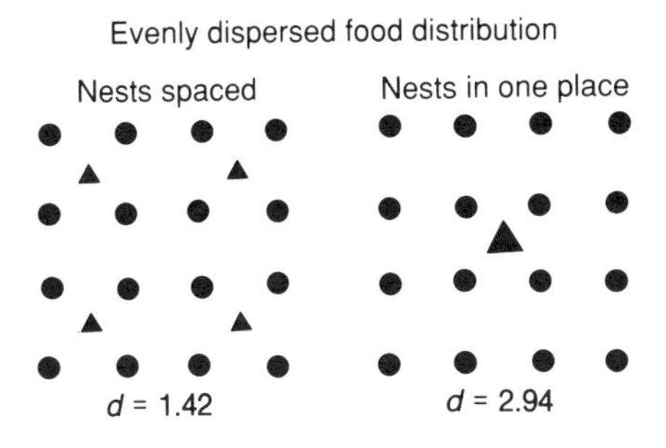

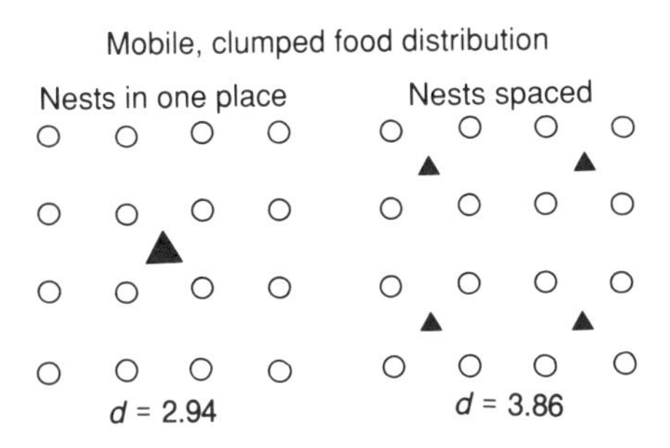

Figure 12.2 A geometrical model showing how colonial nesting of birds can increase the energetic efficiency of exploiting spatiotemporally clumped food resources. (After Horn, 1968, on colonial nesting of Brewer's blackbirds.) Solid circles represent continuously available food sources, each containing 25% of the food required by one nesting pair; open circles represent spatiotemporally clumped food sources that contain food one at a time, and at that time sufficient food for four nesting pairs. Small triangles signify single nests; large triangles signify colonies containing four nests. d equals the average distance required to travel to obtain food, based on the assumption that each adult visits the nearest available food source on each trip.

distance from the nest to resource sites (weighted by the probability of successfully locating food). The optimal situation is the one requiring an individual to travel the shortest distance in order to locate food successfully.

If maximization of rate of energy delivery to a central place is the sole criterion influencing selection of a central place, then the optimal location is one that minimizes travelling time, that is, the site that lies at the 'centre of gravity' of the food distribution (Horn, 1968). Wittenberger (1981) discusses the distributions of food sources for several social vertebrates and concludes that indirect evidence supports the idea that some but not all colonial vertebrates exploit spatiotemporally clumped food resources, as predicted by Horn's (1968) geometrical model. However, the distribution of food resources exploited by colonial animals is often difficult to measure. Many types of food are inherently difficult to sample, and the various food types captured by a species may differ considerably both in distribution and importance.

Aerial insects provide the staple food for many colonial birds, such as swallows, swifts, and for many bats (Lack, 1968; Dalquest and Walton, 1970; Bradbury, 1977), and the insects are often locally concentrated as a result of wind currents and updrafts. For example, Brewer's blackbird (*Euphagus cyanocephalus*) is a colonial marsh-breeding species that generally exploits adult damselflies, dragonflies, and other insects that emerge periodically from aquatic larval stages (Orians and Horn, 1969). Damselflies emerge during the morning and dragonflies emerge at night (Orians, 1980) along shorelines and then they disperse into the surrounding uplands. Colonial nesting of great blue herons (*Ardea herodias*) is also associated with spatiotemporal clumping of food resources (Krebs, 1974). Individual herons breeding in colonies primarily hunt invertebrates and fish along the margins of shallow ponds. That prey are spatiotemporally clumped is suggested by the way that herons from breeding colonies feed in different locations each day, and that the number feeding at any given location varies from hour to hour. In contrast, herons that breed in widely spaced trees and defend large territories forage along irrigation ditches in pastures where they capture mice and other more evenly distributed prey.

Not all colonial birds exploit spatiotemporally clumped food resources. Piñon jays (*Gymnorhinus cyanocephala*), for example, breed in colonies near abundant sources of piñon pine cones (Balda and Bateman, 1971). The cones are spatially clumped, but the best foraging locations are fairly constant on a short-term basis. Cone and nut crops fluctuate greatly between years in any given area, however, causing species that feed on them to breed in different localities each year.

Wittenberger (1981) outlines several possible problems with Horn's (1968) geometrical model. First, it does not easily explain the formation of very large colonies. Resources in the area of large colonies are soon depleted, forcing individuals to commute longer and longer distances for food. From an energetic standpoint individuals might be able to forage more efficiently by segregating into several smaller colonies, each centered in its own foraging arena. Second, coloniality does not minimize travel distances to spatiotemporally clumped food resources unless the colony is centrally located. Colonies are usually situated at sites that impede access to predators, and such sites are often not conveniently available in a central location. Consequently, many colonies are really not located in the centre of their foraging areas. It is not yet clear how acentric a colony site can be before the advantages of shorter average commuting distances are lost, but many colony sites probably do not fall within that limit. Finally, as shown in Figure 12.2, dispersed individuals must commute further than clumped individuals when food is spatiotemporally clumped because their nests lie near the edge of an isolated foraging area. However, if the foraging area is not isolated, each pair could nest in the centre of its own food distribution and still remain dispersed. Dispersed individuals would then have the same average commuting distance as clumped individuals. The model therefore works only if the foraging area surrounding each colony is isolated. This may be the case for blackbird and seabird colonies, but probably not for many other colonial birds.

12.3.2 Information centre hypothesis

Ward (1965), Ward and Zahavi (1973), and Zahavi (1971) have proposed that central gathering places of group-foraging animals may function as information centres, enabling individuals who lack knowledge of the locations of patchily distributed but abundant food concentrations to profit from the searching abilities of other individuals in the social group. The information centre hypothesis suggests that sharing of information within a social group can be advantageous for tracking variable resources. It is particularly interesting for cases in which neither predator avoidance nor thermoregulation provides a likely explanation for aggregation.

Individuals unsuccessful in finding food must either become informed of food source locations through active signals exchanged with successful foragers, or they must recognize successful individuals on the basis of their appearance or behaviour, and then follow them. If, as is sometimes the case, individuals are genetically related, such sharing of information increases the fitness of the individual who gives the information, because the receiver of the information who benefits is likely to be related genetically.

Acquiring information from other colony members is most beneficial when the locations of good foraging areas are relatively unpredictable. Food should also be sufficiently concentrated so that several individuals can exploit good foraging sites profitably, and food should persist in good areas long enough for returning individuals to still find food there, but not so long that most individuals would know about it without obtaining information.

Among insects, recruitment strategies based on visual, chemical, and auditory communication have been extensively investigated. Although most examples derive from the social insects such as ants and bees, presocial insects also have recruitment mechanisms. For example, in the Eastern tent caterpillar (*Malacosoma americanum*) recruitment is elective and contingent upon individual assessment of the quality of potential feeding sites, and assessment includes the quality of leaves and the degree of crowding (Fitzgerald and Peterson, 1983). Sibling groups of 50 to 300 larvae live in a silk tent on branches of rosaceous trees. The tent serves as a staging area from which the leaf-feeding larvae launch group marches in search of food during each of three to four daily activity periods. Larvae lay down strands of silk and trail pheromones as they disperse. Unlike unsuccessful foragers, successful foragers reinforce the pheromone trail as they return to the tent after feeding. Reinforced trails can then be followed preferentially later, and thus act as guides to feeding areas and enable larvae to relocate productive feeding sites during the next activity period.

The recruitment strategy of honeybees has been widely studied, especially after Frisch (1967) unravelled the honeybee dance language, and seems to exemplify the information centre model.

Case Study Honey bee dance communication

The following is a brief account from Frisch (1967), Michener (1974), and Seeley (1985) of how honeybee foragers relate information to other bees in the hive by dancing. The two extreme forms of the dance are the round dance when resources are found close to the hive and the tail-wagging dance when resources are distant.

When a successful forager returns and enters the hive she promptly offers her crop contents (unless she has collected only pollen) to 'storage bees' waiting about on the combs. If the resource was located near the hive, the forager may perform a round dance, running in a small circle, reversing and going in the opposite direction after every turn or two (Figure 12.3a). Other bees that are attracted and that contact the dancer may be recruited to the food source about which she is dancing. Recruited foragers take honey into their crop as fuel needed for the trip, and then leave the hive. The recruited bees are able to find the resource based on the odour of the nearby source obtained by antennating the dancer and from tasting some of her nectar. The more vigorous and long lasting the dance, the more bees are recruited to the food source.

The tail-wagging dance, which is used when distances greater that approximately 80 m are being indicated, consists of a short straight run with the bee turning to one side and returning by a semicircle to the starting point, followed by repetition with the bee making the second semicircle on

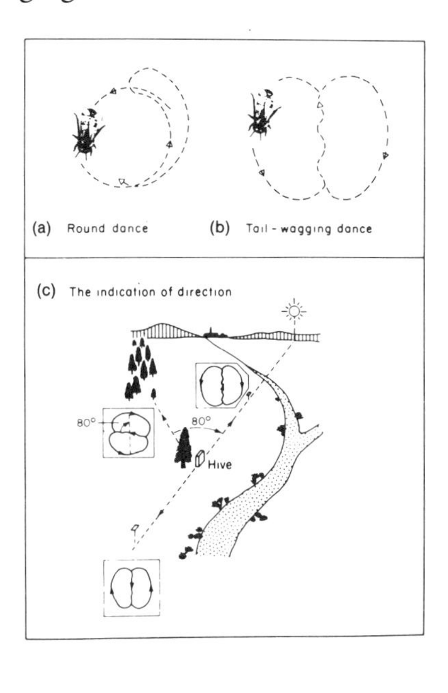

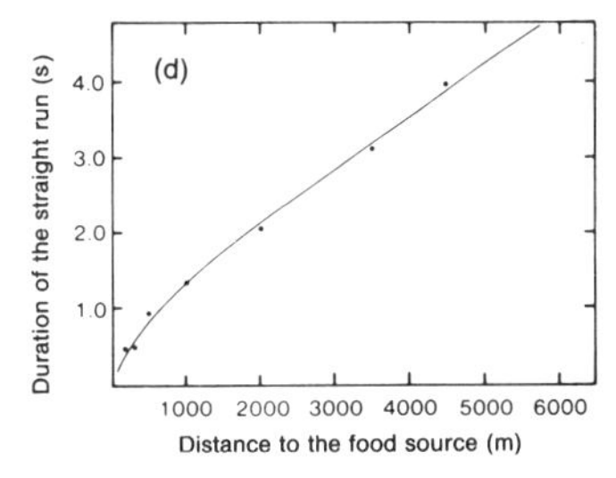

Figure 12.3 Honeybee dances used in communicating distance and direction to food sources. (a) Round dance used to announce resources discovered less then 80 m from the nest. (b) Waggle dance, consisting of a figure-eight configuration; the abdomen is waggled during the straight portion of the run. (c) In the waggle dance, direction to the food source is communicated by maintaining the same angle of the food source relative to the sun while dancing on the vertical comb relative to gravity. Thus, a dance with the straight portion 80° to the left of the upward direction would indicate food located 80° to the left of the direction to the sun. From Baker, 1978. Based on observations of Frisch 1967. Copyright 1978 by R.R. Baker, reprinted by permission. (d) Distance to the food source is communicated by the duration of the straight run (s). From Seeley, 1985. Based on data of Frisch 1967. Copyright 1985 Princeton University Press, reprinted by permission.

the opposite side of the straight run (Figure 12.3b). The straight run is emphasized by vigorous wagging of the abdomen from side to side and usually by a buzzing sound.

Distance to the food source is reported by the number of circuits (half of the typical tail-wagging dance) danced by a bee per unit time. As shown in Figure 12.3d, duration of the straight part of the dance is closely correlated with distance to the food source. It is therefore the feature believed by Frisch (1967) to be the most probable distance cue for the recruits. This view is supported by the attention given by recruits to the straight run: they often lose contact with the dancer on the return part of the circuit, go to the place where the next straight run will begin, and touch the dancer again with their antennae while following the run.

Dances are performed in the dark on vertical surfaces in the nest. Here the dancer converts the angle with respect to the sun to an angle with respect to gravity. A straight-up run means 'fly toward the sun'; down means 'fly away from' or '180° from the sun'; and 30° to the left of straight up means 'fly 30° to the left of the direction toward the sun' (Figure 12.3c). The recruits leave the nest and fly in the direction with respect to the sun indicated by the dancer.

Bees seemingly know in advance what they are going to do. Thus, bees enter a hive in different directions depending on the direction to the location of the forthcoming dance and whether they will dance near the entrance (as for a nearby food source) or in the interior (as for a distant one). They also ingest an amount of honey that is neither too little nor too much for the upcoming trip, as though the bee planned its trip in advance based on information learned by monitoring dances, and took on a fuel supply accordingly.

The minimal concentration of sugar required to elicit dancing varies with the internal conditions of the nest, availability of other sources, and taste. Lindauer (1949), studying variation in the concentration threshold over a 3-month period, found that in early summer a high concentration (up to 2.0 M) was required to cause dancing, whereas in midsummer when nectar in the field was scarce, a sugar concentration as low as 0.12 M caused dances.

Honeybee colonies track rich food source patches over time. Visscher and Seeley (1982) mapped out day-by-day the forage patches worked by a wild colony located in a forest. A 4-day sequence, shown in Figure 12.4, reveals that the overall pattern of a colony's foraging is a 'rapidly changing mosaic' of food source patches, and that consecutive days never share the same pattern of labour allocation to patches. They also found that for any given day a colony's foragers are distributed over a small number of patches, but over a large area (50 to 10 000 m) from the hive. One is left with the conclusion that a honeybee colony monitors a vast area around the hive, and somehow pools the reconnaissance information of foragers to focus a colony's work force on a few patches. In other words, the colony seems to act like an information centre (Heinrich, 1978; Visscher and Seeley, 1982; Seeley, 1985).

The strategy of a colony seems to rely on the following mechanisms:

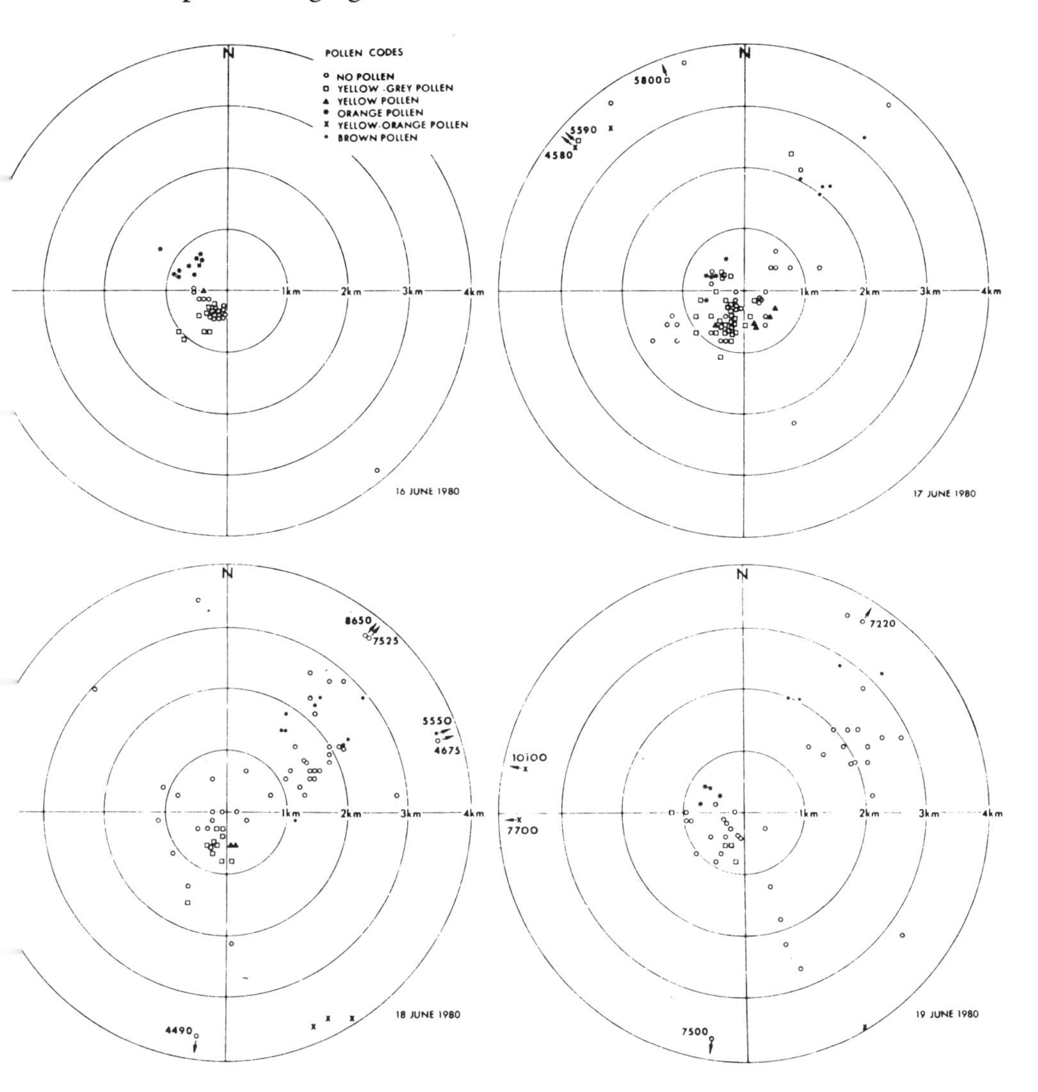

Figure 12.4 Maps of foraging locations inferred from honeybee recruitment dances over a period of 4 days. Different symbols code the colour of pollen, if any, borne by a dancing bee. Locations beyond the edges of the maps (4 km) are not shown. From Visscher and Seeley, 1982. Copyright 1982 Ecological Society of America, reprinted by permission.

1. workers locate new patches by searching and presenting recruitment dances, and
2. foragers are allocated only to patches of relatively high profitability.

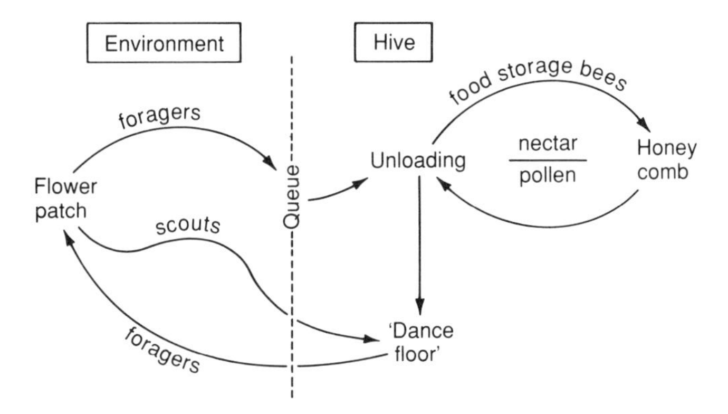

Figure 12.5 Summary of interactions between bees outside and inside of the hive. Scouts seek new nectar and pollen resources, and then foragers learn from the dances of scouts where food sources are located. Foragers fly to the reported patches and return to the hive with nectar or pollen. Food storage bees preferentially unload foragers carrying higher quality sources before they unload those carrying lower quality sources. As the queue of bees waiting to be unloaded increases, fewer bees can be unloaded, and hence fewer bees dance; the overall effect is to reduce foraging activity of the entire hive if stores are becoming filled, and to direct foragers (through dances) to the higher quality patches of flowers.

The net effect is to promote abandonment of poor patches and foraging at better patches. It is readily apparent that the key puzzle here is 'the mechanisms by which an individual bee can know the quality of her patch relative to the other patches being worked by her colony's foragers' (Visscher and Seeley, 1982).

The honeybees, as a colony, not only assess information about the available food in the environment, but recruitment is also geared to the nutritional status of the hive; more foragers are allocated when food stores are low than when the stores are substantial. How do they accomplish this feat? Food storage bees unload incoming foragers, transport the nectar or pollen to the honeycomb stortage sites, and then return to repeat the process (Figure 12.5). Once foragers are relieved of their resources, they dance; but, if they have to wait a long time to be unloaded, they do not subsequently dance. Thus, if the food store is quite full, or if many foragers are returning with food, the queue of foragers to be unloaded backs up and fewer bees dance, and hence fewer bees are recruited. In other words, if foragers are unloaded by food storage bees quite quickly, the colony need must be great, but if they are forced to wait to be unloaded, the colony need must be low.

Food storage bees also tend to unload preferentially those bees that return with the highest quality food before unloading bees with lower quality food. This behaviour tends to favour unloading bees from the better sites first, and releases them to dance. One other factor is that foragers grade their own responses (dance or not to dance) to food sources according to the distance to the food patch, quality of the nectar or pollen, and the weather. Thus, there is both a distance and a food quality effect, such that recruitment to a site decreases if it is moved further away *or* if the sucrose concentration is decreased. The change in recruitment to a given site occurs within minutes.

The colony switches recruitment from one resource patch to another because bees that are dancing and drawing recruits are those that are being unloaded most quickly. Thus the attention of the colony switches from poor to better resource patches over the course of a day or several days. Seeley (1985) lists three properties of a nectar-yielding patch that may be important to a forager's assessment process (Figure 12.6):

1. the time required to fly from the patch (travel time, T_t) to the hive,
2. the time required to load up with nectar while in a patch (patch time, T_p), and
3. the sugar concentration of the nectar in the patch.

Gross energy (GE) obtained can be estimated as:

[loaded nectar volume] × [sugar concentration] × [4 calories per mg sugar].

As shown in Figure 12.6, the slope of line AB represents the gross rate of energy intake from a patch. In order to estimate the profitability of her current patch, a bee would have to 'calculate' the relative energy yield per trip to the patch and the relative duration (T_t + T_p) of her foraging trips. Given that all workers load equal volumes of nectar, the relative energy yield per trip could be evaluated simply by sugar concentration, which all bees can estimate. To perceive the relative duration of a foraging trip the bee could simply use the one-way distance to the patch, which we know can be translated into dance circuits by honeybees. The only hitch would be factors such as wind currents that make distance disproportionate from time, but even this rough estimate would serve the required purpose.

Aside from honey bees, humans best illustrate the use of signals in information centre strategy. Human language reduces foraging overlap even for individuals who forage alone, as long as they share information about areas recently foraged or plans for future foraging (Winterhalder, 1981b). Smith (1981) has attempted to calculate whether the size of contemporary Inuit hunting groups is adjusted so as to maximize the net energy return per individual participating. The Inuits (Canadian Eskimos) were studied in and around the village of Inukjuak, with a population of 600 along the eastern coast of Hudson Bay in Canada. Groups seeking Beluga whales have a mean of 10.3 hunters (range 5 to 16), and for winter caribou a mean of 4.0 (range 1 to 7). The efficiency declines linearly from small to large groups for Beluga, and efficiency peaks at about 6 or 7 hunters for

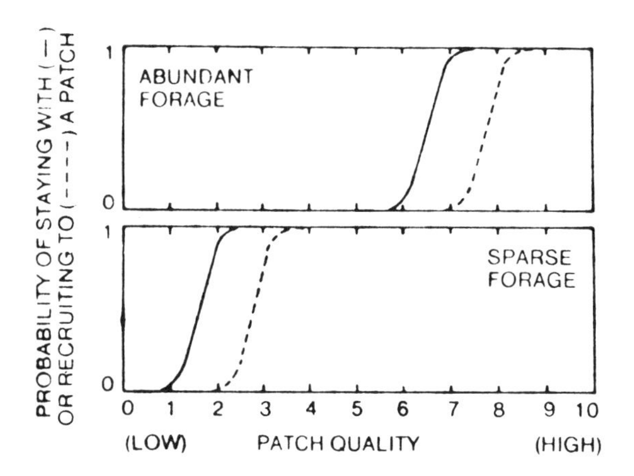

Figure 12.6 A model of how foragers may integrate information about patch quality and forage abundance in deciding how to respond to a patch of flowers. A forager can make an 'assessment' of the overall quality of a patch of flowers and can adjust the thresholds of patch quality for accepting and recruiting to a patch in relation to the abdundance of forage. By raising these thresholds when forage is abundant, a forager can focus efforts on only the highest quality patches, and by lowering them when forage is sparce, a forager broadens the spectrum of acceptable patches. From Seeley, 1986. Copyright Springer-Verlag, reprinted by permission.

caribou. The data do not overwelmingly support the idea that group size is tightly optimized to maximize individual energetic efficiency, but then again, many factors could influence group size, including differences in individual abilities and experiences.

Birds may have an opportunity to learn where food is located by following other colony members. Such opportunities should be especially prevalent when colony members feed together, as in most colonial birds (Lack, 1968; Newton, 1972; Ward and Zahavi, 1973). Some exceptions are herons, hawks, and vultures, which breed or roost in colonies and forage alone, but that might still benefit from learning better places to search for food (Ward and Zahavi, 1973).

Identifying a likely individual or flock to follow from a colony is not as difficult as it might seem. Several potential cues have been suggested, although it is not certain if any are actually used. For example, individuals returning from good feeding areas should have full crops or large amounts of food in their bills, both of which might be monitored by neighbouring individuals. Consistently successful foragers should have better fed chicks, and in winter they should maintain heavier body weights. Perhaps other individuals observe these indicators. Proof that individuals actually follow other colony members to food is not easily obtained. Birds depart winter roosts in consecutive waves, and in some species individuals often lag behind each wave, vacillating between leaving the roost and returning to it. Ward and Zahavi (1973) suggest that these stragglers are birds that left poor foraging areas the previous evening and are trying to decide which flock to follow.

Birds in breeding colonies seem to arrive and depart in a disorganized manner, and are seldom organized into flocks. Nevertheless, direct observations show that individuals do sometimes follow others from colonies of several species (Horn, 1968; Walsberg, 1977). For example, common terns (*Sterna hirundo*) tend to leave in approximately the same direction as birds leaving just before them when departing from the colony to forage (Waltz, 1987). Relatively unsuccessful foragers are more likely to follow others when departing from the colony than are relatively successful ones.

One problem with the information centre hypothesis, at least for nonbreeding situations, is that it fails to explain why individuals or groups might leave superior foraging areas and return to a central roosting or nesting place. In groups such as bird flocks, by returning to the colony an individual would very likely recruit competitors the following morning. In cases where the resource is present for only a short time period this would not be a problem, since the site can provide food for more birds than can possibly get to it. A good strategy might be to roost near the food source until it is exhausted and then visit the colony to seek information about new feeding grounds. Of course, if every individual adopted such a policy, colonies would provide little information for anyone, and this advantage of sociality would be lost. One possible reason why successful foragers return to the colony and share information is that they expect others to reciprocate in the future. However, reciprocal altruism requires the ability to discriminate against cheaters, and nothing prevents cheaters from returning to a colony on days when they fail to find good foraging areas and remaining away on days when they do find good foraging areas. For breeding colonies the information centre hypothesis is more tenable because individuals must return regularly to feed offspring and to defend nests or territories. In species such as honeybees, information sharing is embedded in stereotyped behaviour, and so cheating is minimized.

Information sharing lends itself to cheating, since an individual does not have to reciprocate in providing information to others. In fact, an individual could simply become a parasite on naive information sharers. A number of investigators have dealt with this issue, leading to the development of the producers and scroungers concept (review: Barnard, 1984). The idea is simply that some individuals find resources (the producers) and other individuals (the scroungers) take advantage of the producers' yield in various ways. In studies of foraging by house sparrows (*Passer domesticus*) Barnard and Sibly (1981) sorted out some of the ways that scroungers may operate. Area copying was the most common form of interaction among the sparrows and involved a scrounger responding to some cue from a producer (not necessarily a successful find) by moving across to search in the immediate area around the producer. In following behaviour one bird moved along behind another without exhibiting any searching behaviour itself. The scrounger usually maintained a distance of not more than 30 to 35 cm between itself and the recipient, and the scrounger frequently snatched prey from the bill of the producer, or, if prey was dropped, from the substratum. Snatching was the least frequent form of interaction in the house sparrow and usually consisted of the scrounger taking prey directly from the producer's bill.

Barnard and Sibly (1981) conclude that scroungers do better when there are fewer producers. They do badly, however, when there are a lot of producers, perhaps because scroungers are in some way outcompeted for the small food supply when they are heavily outnumbered by producers. Producers had a mean capture rate almost 2.5 times higher than scroungers in groups of one type only, and appeared to achieve this by searching food patches more thoroughly. Like other predators which exploit patchy food supplies, the sparrows showed area-restricted searching while foraging on an experimental grid (Barnard, 1978). By localizing their searching effort in an area where they have just been successful, they maximize their chances of finding more food. In fact, scroungers search less intensively than do producers (index of path straightness 0.31 for producers, 0.54 for scroungers). The experimental evidence suggests that searchers and copiers operate on a 'more permanent and less flexible basis than a motivational one', but it is not possible to determine whether the basis is genetic, developmental, or a response to individual capability.

Giraldeau and Lefebvre (1987) demonstrated that pigeons (*Columba livia*) who shared food discoveries of others (scroungers) did not learn the food-finding technique used by discoverers (producers). On the other hand, individually-caged birds that were prevented from scrounging, quickly learned the food-finding technique from a tutor. Thus, scrounging may interfere with cultural transmission of food-finding behaviour.

12.4 SUMMARY AND CONCLUSIONS

Central place foragers (CPFs) carry resources back to some fixed point, such as a nest or a colony, and consume or store it there. CPFs are unique in that they must be able to orient or navigate to find their way back to the central place after foraging away from it.

Most species that are CPFs have nests or colonies and live in societies or in social groups. They increase their foraging efficiency by selecting nest sites in relation to the distances to food sources, and many use communication between colony members to track changing resource availability.

Among the general rules that apply to central place foraging, the following are most important:

1. If patch quality is constant, the load an animal carries should increase with increasing distance of the patch from the central place, otherwise the trip may not be profitable relative to the energy used in travel; likewise, time in a patch should increase with increasing distance of the patch from the central place.
2. Since the forager weighs more on its return trip than on its outbound trip, it can increase its rate of net energy delivery to the central place by shortening the return trip relative to the outbound one.
3. The central place is usually safer than foraging areas. Thus, under risk of predation, nesting CPFs may remain closer to the nest while foraging, forage for shorter times per patch, and deliver smaller loads than would be predicted for delivery rate maximization. The tendency of an animal to carry an item of food decreases with distance of food from cover (travel time) and increases with size of the item (handling time).

PART FOUR
Sources of variability

Most animals live in dynamic and uncertain environments characterized by temporal variability and spatial heterogeneity

Gass and Montgomerie, 1981, summarizing Southwood's 1977 address to the British Ecological Society.

As with any type of behaviour, searching differs among individuals, even among individuals of the same species. In this section we examine the sources of variability affecting search behaviour and search success. These sources derive from abiotic and biotic conditions of the external environment, physiological factors of the internal environment, genetic factors, maternal effects, and factors related to early experiences of juveniles. All of these factors combine to shape the search behaviour of an individual during its lifetime, and then natural selection operating on the phenotypes of all competing individuals would be expected to influence search behaviour of future generations.

In Part 3, various search mechanisms were described and analysed. Here we find out how these search mechanisms are modulated or adjusted so as to respond efficiently to changes in the internal and external environment. Thus, if resources are scarce or if an animal is starving, it may search in such a way as to expose itself to predation to a greater extent than if resources are abundant or if it is satiated.

13
External environment

Resources vary temporally and spatially in availability; patches of resources also differ in the same way within a habitat; and even habitats change within and between seasons. In addition, external environmental factors, including abiotic factors such as shifts in temperature or biotic factors such as influences of other individuals, may affect searching requirements or abilities. These sources of variability add to the risks and problems of animals searching for resources and have given rise to many interesting adaptations.

External environmental factors can alter searching behaviour rather directly by acting upon environment-sensitive physiological processes. For example, ectotherms move more slowly at lower temperatures. It is also possible for an animal to perceive changes in environmental factors, usually through the nervous system, and based on that information it may alter its behaviour. Such alterations in searching behaviour may be monotonically related to changes in an environmental factor, or the animal could switch its searching tactic or the resource it seeks as a result of some environmental change. Extreme changes in environmental variables, too large for changes in searching tactics to cope with, lead to drastic responses such as migration, hibernation, or dormancy.

13.1 ABIOTIC FACTORS

Changes in abiotic variables such as temperature, humidity, and solar radiation can affect both resource availability and forager searching efficiency. The important point is that basic physiological principles account for variability in searching behaviour in response to abiotic factors.

1. All endotherms and many ectotherms operate most efficiently at relatively high constant body temperatures.
2. Many ectotherms depend on environmental sources of heat, often warming up by capturing solar radiation; they may also use muscular thermogenesis to attain internal temperatures needed for movement.

3. Endotherms, while able to regulate their body temperatures, use more energy doing so at lower than at higher temperatures, and may have to supplement with heat gained from solar radiation.

13.1.1 Direct effects of environmental factors

Environmental conditions may directly affect searching behaviour of ectotherms by reducing their ability to move or at least by decreasing their rate of movement. These factors include weather conditions, such as wind velocity and temperature, humidity, light levels, and the amount of oxygen in the medium.

Weather conditions may indirectly, but drastically influence the oviposition patterns of butterflies by inhibiting flight when the weather is overcast, windy, or rainy. However, oogenesis may continue during such periods in butterflies such as the monarch (*Danaus plexippus*) as long as daylength exceeds 12 hours and daytime temperature is greater than 20°C (Barker and Herman, 1976). Then, because searching for host plants is affected by the number of mature eggs in the oviducts, as shown for the small white butterfly (*Pieris rapae*), bursts of egg deposition occur when favourable flight conditions follow unfavourable conditions (either on an hourly or on a daily basis) (Gossard and Jones, 1977). Such bursts of egg deposition may lead to egg clustering in species that ordinarily lay eggs individually or in small groups.

Searching behaviour in two fish species, the roach (*Rutilus rutilus*) and the Eurasian perch (*Perca fluviatilis*) is strongly influenced by temperature (Persson, 1986). The swimming speed of roach was observed to be nearly constant between 12 and 15°C, and then increased exponentially from 15 to 21°C, whereas the swimming speed remained relatively constant in perch from 12 to 21°C. This is reflected in prey capture rate: prey capture rate in perch is fairly similar between 15 and 21°C, whereas prey capture by roach was lower than perch at 15°C and higher than perch at 21°C. Although the two species overlapped in their vertical distribution (i.e. both were restricted mainly to the upper (epilimnion) and middle (metalimnion) layers of the lakes), a substantial segregation was evident, with perch penetrating deeper than roach in each lake. These distributions correlate well with their foraging efficiencies at different temperatures.

It is well established that the reactive distance in fish increases proportionally with prey size (e.g. Vinyard and O'Brien, 1976; Kettle and O'Brien, 1978), but visibility is also an important factor. In natural settings the diel vertical migration of zooplankton places them

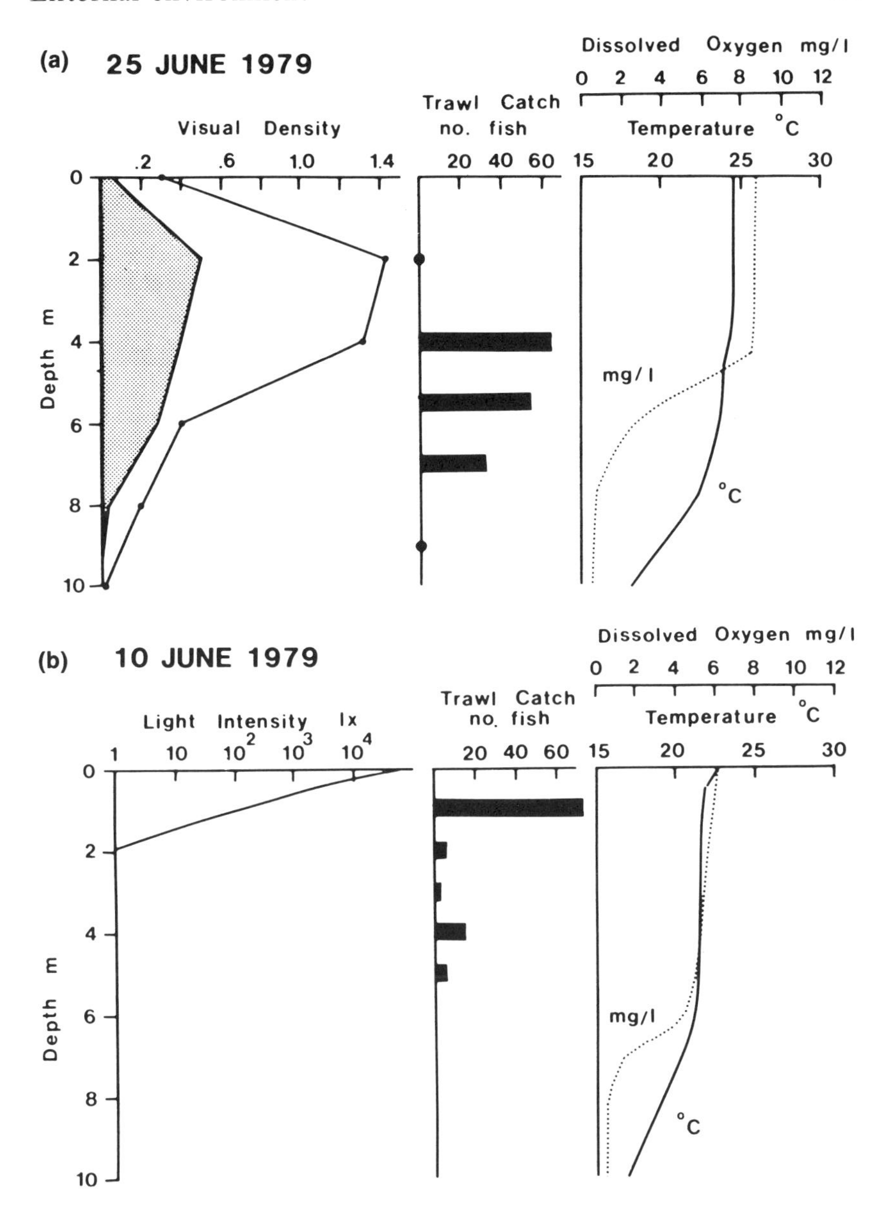

Figure 13.1 Vertical distribution of density of prey (shaded portion refers to *Daphnia* spp.; unshaded portion refers to total zooplankton), density of white crappie (*Pomoxis annularis*) (expressed as the number of fish caught per trawl run), dissolved oxygen (dotted line), and temperature (solid line) for (a) 25 June 1979, and (b) 10 June 1979. From O'Brien and Wright, 1985. Copyright E. Schweizerbart'sche Verlagsbuchhandlung, reprinted by permission.

in the poorly lighted deep areas of lakes during the day and greatly reduces their visibility to predators. Comparing the trawl catch of white crappie (*Pomoxis annularis*) in a small lake on a day when turbidity in the water was relatively low (Figure 13.1a) with a day when it was much higher (Figure 13.1b), it is evident that the fish migrate vertically to zones where they can see zooplankton (O'Brien and Wright, 1985). As might be expected, the size of prey and the amount of turbidity interact in determining reactive distance, such that the reactive distance decreases as a function of turbidity. Thus, the fish can see larger prey more readily than smaller prey at any given turbidity level. As with roach and perch, water temperature has an effect on foraging in crappie, with more rapid prey capture at higher than at lower temperatures (Evans and O'Brien, 1986). It would appear that white crappie compromise by stationing themselves in those depths where zooplankton can be visualized but where the temperatures are low enough to reduce energy costs and high enough to allow for efficient prey capture. However, as shown in Figure 13.1, the lower level to which crappie could move was always determined by the minimum oxygen concentration in the lake. Thus, the crappie balances light level, turbidity, water temperature, and available oxygen in order to find the optimal level for feeding.

Individual components of searching behaviour may differ in their sensitivity to temperature changes. For example, as *Pieris rapae* larvae were starved from 0 to 48 hours their locomotory rate increased rapidly at 25°C as a result of starvation, but to a lesser extent at 14°C (refer to Figure 14.2) (Jones, 1976). Temperature effects on head-waving tendency and directionality were much less dramatic. Rapid locomotion at higher temperatures may be adaptive because larvae may need to find a host plant faster at higher temperatures than at lower temperatures owing to a greater depletion of energy and water reserves.

Hunting in predatory birds such as kestrels is seriously impeded or reduced by winds greater than 12 m/s or less than 4 m/s, a constraint that may be related to increased energetic demands above and below the optimum flight speed of about 8 m/s (Rijnsdorp *et al.*, 1981). Foraging of ringed (*Charadrius hiaticula*) and grey plovers (*Pluvialis squatarola*) is also adversely affected by various environmental conditions, and also when the birds switch from diurnal to nocturnal conditions (Pienkowski, 1983). Pecking rates on dark nights were significantly lower than in daylight, whereas waiting times were longest at night, and the mean distance moved to take prey was longer in daylight than at night. Thus, the rate of capturing prey is depressed at night, even though in some situations there is an increase in prey activity and availability.

Unless conditions are unfavourable, lions (*Panthera leo*) hunt mainly by night, most likely because they are able to stalk with greater chance of success under the cover of darkness. Schaller (1972) suggests that lions 'may be well aware of the advantage that darkness gives them'. They frequently watch prey in late afternoon but wait until dusk to begin hunting. Similarly, a burst of activity often follows the disappearance of a bright moon. Schaller noted that one night the Masai pride heard zebra calling about 1.5 km away. The lions did not respond, but when a dark cloud obscured the moon they walked rapidly toward and stalked the zebra. However, lions may also hunt whenever the opportunity arises. Those in the plains, where cover is sparse, generally hunt at night, but one group which centred its activity around a reedbed caught wildebeest there in daytime. Woodlands lions usually hunted at night when prey were dispersed but also often during the day, when they were near the rivers.

13.1.2 Indirect effects through energetic or hygrothermal budgets

Environmental factors can influence searching behaviour indirectly as animals respond to these factors according to their needs. An energetic budget depends on inputs, with food obtained and utilized providing one source and extra energy obtained from basking or resting on warm substrates providing the other source. Energy output is the amount expended for work and for body maintenance (high in endotherms, low in ectotherms). When an animal searches for food, it must maintain a positive energetic budget by gaining more than it uses. The smallest birds and mammals, which have the most serious energetic problems, maintain positive energetic budgets by either

1. daily strategies characterized by feeding activity spread over day and night, as in shrews (Crowcroft, 1954), or
2. a period of torpor to save metabolic energy during the part of the day or night when feeding is either dangerous or inefficient, as in hummingbirds, bats and some small rodents (Kayser and Heusner, 1967).

In some cases searching behaviour is organized into bouts, reflecting physiological and/or morphological constraints such as size of crop or rate of digestion and assimilation.

The amount of time that endotherms spend foraging and the periodicity of foraging trips may be directly related to thermal conditions, which are important determinants of metabolic demands. For example, the percentage of time spent searching for food

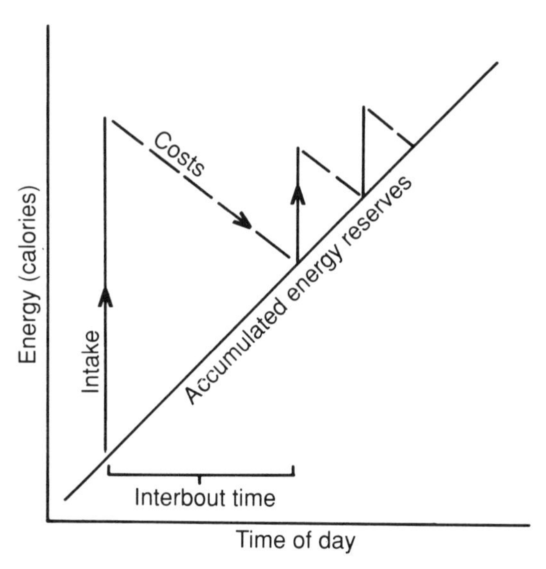

Figure 13.2 Summary of the apparent relationships between energy expenditures and intake for laboratory-tested hummingbirds. (after Wolf and Hainsworth, 1977). Dashed line represents rate of energy expenditure. The figure indicates that the length of the time interval between feeding bouts is regulated both by the amount ingested at the beginning of the interval and the rate of energy expenditure over the interval. Solid line indicates that, on average, feeding bouts also are timed to ensure a gradual accumulation of excess energy reserves throughout the day.

increases proportionally with a decrease in body weight of several species of small English birds during cold weather in midwinter (Gibb, 1954). Hummingbirds in an alpine habitat have a bimodal feeding distribution, with peaks in the morning and in the evening. To maintain their energy balance, individual birds have to eat more during relatively cold periods or when direct solar input is unavailable (Gass and Montgomerie, 1981). Foraging of white-crowned sparrows (*Zonotrichia leucophrys*) decreases as sunrise approaches (Morton, 1967), and chickadees and titmice reduce their foraging rates as the sun's rays begin to strike the treetops (Morse, 1970).

Hummingbirds apparently are able to adjust their energy intake relative to energy expenditure (Wolf and Hainsworth, 1977), allowing them to maintain positive energy budgets under variable conditions (Figure 13.2). This is confounded, however, by differences in the availability or quality of food; for example, if rufous hummingbirds maintained at 10°C in the laboratory on 50% sucrose are then switched to 25% sucrose, they respond by increasing meal frequency (more foraging trips) (Gass, 1978). In a comparative study on several species of microtine rodents, Daan and Slopsema (1978) showed that the period of the short-term feeding rhythm also seems to be adjusted to the metabolic demands of the species, but the voles adjust their meal *size* rather than meal *frequency* to temperature-specific energy requirements: feeding bouts are longer at low than at high ambient temperatures. The relation to stomach size and time to utilize food in the stomach is simply reflected in the time between feeding.

Energetic costs of foraging in bumblebees (*Bombus terricola*) increase as the temperature drops, because a bee must maintain a high thoracic temperature in order to continue flying (Heinrich, 1979a). Since different species of flowers produce different amounts of nectar, it is possible to calculate how many visits to each plant species are required to maintain a positive energy budget at different temperatures. Certain flowers such as wild cherry blossoms and lambkill (*Kaalmia angustifolia*), yielding low energy rewards, are only utilized at relatively high air temperatures.

Activity of ectotherms often correlates with the way in which thermal effects act relative to their size. For insects of a similar shape and coloration, their actual dimensions (determining surface to volume ratio), establish the rate of heating in sunlight and thereby affect the timing and patterns of activity. This, in turn, affects their foraging efficiency. For example, weather conditions affect foraging in female sphecid wasps (*Cerceris arenaria*), and hence the number of eggs laid in provisioned cells (Willmer, 1985). Female wasps fly in search of prey as frequently as possible, but they also have to allocate time to warming up prior to take-off and for cooling down after flights. They also must allocate time for burrow construction to serve larval inhabitants. An insect that is large and dark can warm up more effectively early in the day, because it can gain heat faster than a small and light coloured one, but the large insect may overheat in full sun when smaller insects can still be active (Willmer, 1983). Even a two-fold weight range in female wasps leads to a marked difference in searching patterns and prey capture rates. In fact, the largest females store at least twice as many weevils in their burrows as do smaller ones, because they can start flying earlier in the morning, perform more and shorter successful trips, and can fly more regularly on cool days. More prey in a nest may also contribute to greater hygrothermal stability, as the cached weevils gradually lose moisture and hydrate the larval cells.

Another complex way in which temperature impinges on searching behaviour and resource selectivity is by altering the water balance of the animal. Higher temperatures may be associated with higher desiccation rates, and desiccation in herbivores often stimulates searching for host plants because these are frequently an important source of water. For example, in the red cotton bug (*Dysdercus koenigii*), either high temperatures or water deprivation decreases the time required to locate host plants, apparently by increasing the probability of directed orientation once an insect enters a zone sufficiently close to localize a plant (Saxena, 1969). The effect is two-fold: the number of insects in the zone near plants increases as dehydration progresses, and dehydrated insects are less discriminating in selecting among potential host plant species. As soon as the bugs become water-satiated, they also become more selective and the enhanced orientational response to the succulent leaves of any host type disappears. Responses of vertebrates to changing water availability are discussed in Section 13.4.2, regarding migration.

13.2 BIOTIC ENVIRONMENT

Most species use motor patterns in searching that coincide with the typical distribution patterns of the resources they require, or else their motor patterns satisfy some requirement for finding resources in a certain distribution. For example, the whitefly parasitoid (*Encarsia formosa*), whose hosts are randomly distributed on leaves of several plant species, forages more or less randomly on leaf surfaces (Van Lenteren *et al.*, 1976), whereas the aphid parasitoid (*Diaretiella rapae*), whose hosts are distributed in clumps on plants of a particular family, forages nonrandomly between plants (Read *et al.*, 1970), between leaves (Hafez, 1961), and on leaves (Ayal, 1987).

Few types of resources ever occur repeatedly in the exact same arrangement, density, or temporal sequence, and so within a certain range of plasticity animals must be able to cope with temporal and spatial changes in the distribution of resources. The spatial distribution of resources can affect either search behaviour or success in finding resources, or both.

13.2.1 Effects of resource density on resource accrual

It makes intuitive sense that the number of resources an animal finds would be proportional to the number available, and although there are some exceptions, in most cases that is exactly what happens. For example, in most instances predators attack more prey as the number of available prey increases (review: Taylor, 1976); and, parasitoids generally oviposit in more hosts when hosts occur in dense groups (review: Hassell, 1978). For predators, Holling (1959) referred to this

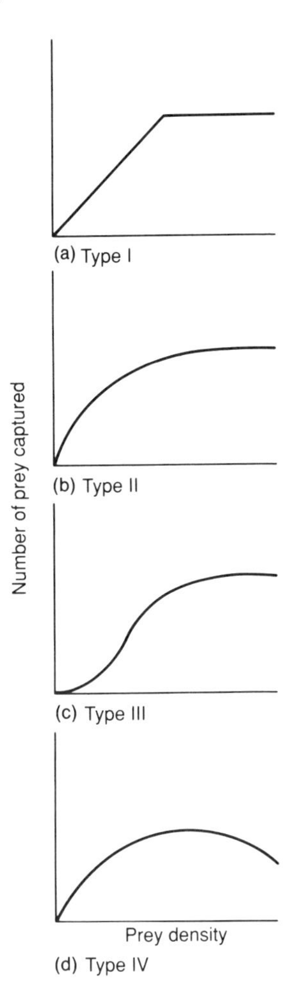

Figure 13.3 Four types of functional response curves. (After Holling, 1959.)

phenomenon as a *functional response*, of which there are four types (Figure 13.3).

Type I and II responses reflect either abrupt or gradual satiation processes, and are thought to occur in one of two ways. The rate of feeding declines continuously or abruptly as a function of the filling of the gut, going to zero when the gut becomes full. In other cases, quelling, killing, subduing, capturing, eating, and cleaning, which are possible components of handling, reduce the time available for searching. As prey become more abundant, predators initially improve their feeding ability on them, but then handling time soon takes up a sizable portion of the available time, producing the function plotted in Figure 13.3b. For example, handling time in the lady beetle (*Coccinella septempunctata*) increases with the number of aphids eaten (Cock, 1977).

The sigmoid type III response curve is typical of many invertebrate parasites and vertebrate predators. The shape of the curve may result from one or more of at least four possible changes in the forager's behaviour as influenced by alterations in prey density. The forager may

1. learn to find prey more readily (e.g. forms a search image) at some critical prey encounter rate,
2. alter the rate of search at some critical prey encounter rate,
3. spend less time engaging in non-foraging activities at higher prey encounter rates, or
4. emigrate from the resource patch more readily at low host densities.

The type III functional response of apple maggot flies (*Rhagoletis pomonella*) (Roitberg *et al.*, 1982) is principally due to emigration. At low host fruit density most flies leave a tree before discovering any fruit. However, if a fruit is located and 'success motivated' search initiated, the chances of other fruit being encountered and exploited greatly increase. The functional response of the 'finder only' individuals (unfilled circles in Figure 13.4) is characteristically a type II response, and lends support to the suggestion that the type III response of 'all flies' (filled circles) is primarily due to emigration by a large proportion of flies before they encounter hosts at low densities.

The type IV responses occur when capture rate actually decreases at the highest densities of resources. This response is exemplified by the defense of sawflies (*Neodiprion swainei* and *N. pratti banksianae*) when attacked by their heteropteran (penatomid) predator (*Podisus modestus*) (Tostowaryk, 1971). At densities of two to five larvae per branch the sawflies tend to disperse; at a density of 10 they form a loose aggregation; and at densities of 20 to 40 they form tight aggregations. When attacked, both sawfly species struggle vigorously and secrete a sticky material from their mouthparts. The result is that at the highest densities of sawfly larvae the capture rate of the predator decreases because of the struggling and the sticky secretion which can effectively immobilize them.

13.2.2 Effects of resource density or availability on searching tactic

Discontinuous variability, characterized by individuals adhering to one or more searching tactic, can be conditional upon environmental

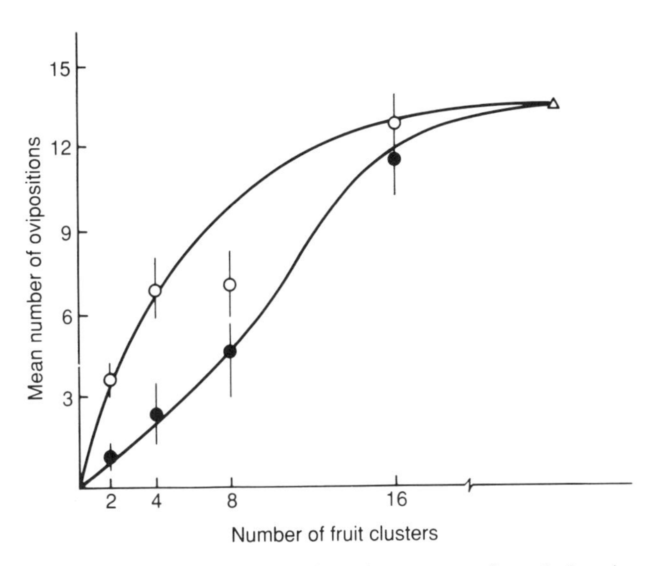

Figure 13.4 Functional response of apple maggot flies (*Rhagoletis pomonella*) to different host-fruit cluster densities. Solid circles represent all flies, open circles represent 'finders only'. Laboratory-derived data on the maximum number of ovipositions are indicated by a triangle. Vertical lines are standard errors. (After Roitberg *et al.*, 1982.)

situations. In the following examples, based on combinations of abiotic and biotic environmental variables, each system is discussed and then an attempt is made to uncover and compare the proximate mechanisms by which environmental cues are perceived and utilized in determining the switchpoint.

(a) *Mate finding*

Male speckled wood butterflies (*Pararge aegeria*) have two tactics for locating females: perching and patrolling. Which tactic will be played depends on the availability of sunspots in the environment (Davies, 1978): active patrolling is appropriate in the canopy where females are scarce, while perching is a better tactic in sunspots where females are abundant. Butterflies spend the night in the canopy, 5 to 15 m off the ground, and in the morning when they begin to fly about, males descend to the ground level where the sun casts pools of sunlight onto the woodland floor. Males tend to remain in sunny spots, rather than in shade, because they require solar radiation to remain active. Individual males often spend the entire day on one sunspot, shifting with the spot as the sun moves across the sky, although not all males are able to locate and occupy sunspots at any given time.

In the experimental transect observed by Davies, 5.2 males were found in the canopy and 8.6 males occupied sunspots. Sunspot males remained in a small area, flying out to inspect passing objects. If a passing object was another male, the sunspot holder defended its area. If a passing object was a female, the male pursued and attempted courtship and copulation. Alternatively, canopy males using the patrolling tactic continually moved about, flying up and down over stretches as long as 30 m. That the tactic employed is indeed a function of environmental factors is suggested by the finding that all males patrolled early in the morning before sunspots were abundant. Later in the day when sunspots became available, some males were able to find and occupy sunspots. On cloudy days when sunspots were totally unavailable, all males engaged in patrolling both in the canopy and at ground level. To determine if sunspots were important resources, Davies removed territorial males from their sunspots, and found that other males occupied the vacant positions within a few minutes. Finally, Davies showed that a male holding a sunspot has a mating advantage over a canopy male, in that they encounter more females than do canopy males. The key element in this example seems to be that any individual male can use both tactics, and which tactic it uses is simply a function of sunspot availability. When sunspots are totally unavailable, all males patrol, because this tactic is the most effective way to locate females in the absence of sunspots.

Behavioural plasticity is also characterized by adult males of a scorpionfly (*Hylobittacus apicalis*) which mimic female behaviour, and thereby enhance their copulatory success and probably their own survival (review: Thornhill, 1984). In its most common mate-finding tactic, a male scorpionfly finds a prey item (usually a fly) and feeds briefly. It then holds the prey in its hind legs and hangs by its forelegs from a leaf or twig, and releases sex pheromone to 'call' females. A female, attracted by the pheromone over distances up to 13 m, lands beside the male and evaluates the prey by feeding on it during courtship. Females refuse to mate with males offering bad tasting or small prey. After copulation the female and male struggle to retain the prey, and if the female wins, the male has to search for another prey; if the male wins (64% of the time), it may reuse the prey.

The plasticity in the searching strategy is as follows. If a male first encounters a prey, it will probably adhere to the common behaviour described above (searching tactic). If a male first encounters another male with a prey or a copulating pair with prey, it may attempt to steal the prey (pirate tactic), or it may fly to a calling male and assume the wing postures and abdominal movements of a female (imposter tactic). Although males always offer the prey to real females, they do so to imposters in 67% of cases. Imposters successfully fly off with the prey item in 22% of cases, suggesting that this tactic is reasonably profitable. Forced attempts to steal prey are only successful in 14% of cases, a seemingly marginal technique with respect to profitability alone.

When factors in addition to finding prey are considered, however, the benefits of pirate and imposter tactics become more obvious. In hunting for prey, males following the searching tactic expend considerable time and energy, whereas imposters and pirates that rob other males of their prey reduce their flight time. This is important for two reasons. First, since search time for prey comprises about 50% of the period between copulations, males that steal prey rather than catching their own reduce intercopulatory time by 42%. The reduction in searching time would thereby allow imposters to copulate more frequently. Secondly, pirates and imposters also reduce the associated risks of searching, because they move shorter distances than do males using the primary tactic. During the six years of Thornhill's study the most important predator of the scorpionflies was found to be web-building spiders, whose webs capture the insects flying through the vegetation. Predation by spiders is significantly male-biased, suggesting that *H. apicalis* that steal prey would have reduced risks of predation. Thus the tactic followed by a given male depends largely upon the availability of other males and of prey, and on the sex ratios and sizes of the population which would be expected to fluctuate in time and space.

(*b*) *Foraging mode*

Adult spotted flycatchers (*Muscicapa striata*) use two very different methods of foraging (Davies, 1977). On cold days, and in the early morning and evening of warm days, when the abundance of flying prey was low, flycatchers foraged high in the tree canopy, hovering in the foliage and sometimes hopping about on the terminal twigs of trees. They seldom remained for long in the same place, but continually moved in their search for prey.

During the middle part of warm days, when large numbers of flying prey were available, flycatchers fed from perches 0.5 to 2 m from the ground. They used a 'sit-and-wait' searching tactic, flying out from a favourite perch to capture prey in mid-air, and then often returning again to the same perch. The birds switched from actively searching for prey in the canopy to the sit-and-wait strategy near the ground when the abundance of large Diptera reached a density of

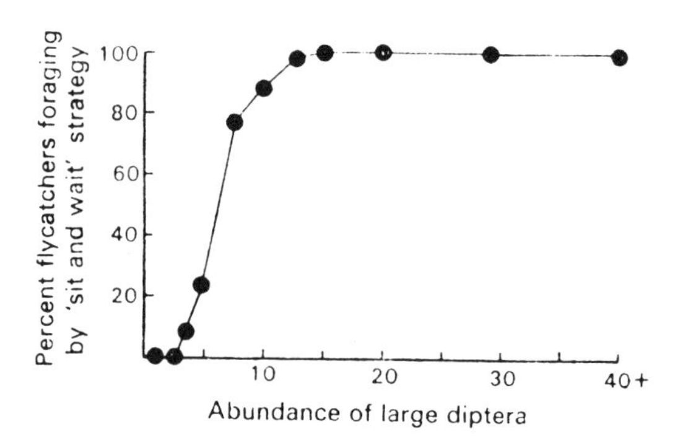

Figure 13.5 Adult flycatchers (*Muscicapa striata*) switch from a searching tactic in the tree canopy to a 'sit and wait' tactic near the ground as the abundance of large Diptera increases. From Davies, 1977, copyright The Association for the Study of Animal Behaviour, reprinted by permission.

approximately ten per two-hour catch in a Malaise trap (Figure 13.5). Davies (1977) tested the hypothesis that at this abundance of large prey the sit-and-wait tactic becomes more profitable than the canopy searching tactic in terms of energy gain. When feeding in the canopy, aphids comprised 46% of the diet, followed by 30% small Diptera and 15% large Diptera. At high prey abundance, when the birds were foraging close to the ground with the sit-and-wait tactic, they took mainly large diptera (40%), small Diptera (26%) and Coleoptera (13%).

The flycatchers incorporated more large Diptera into their diet as the abundance of these insects increased, whereas the abundance of small Diptera and aphids did not influence their contribution to the adult diet. This is consistent with the view that the birds preferred large Diptera, and only when the abundance of this preferred prey was low, did they switch to small Diptera and aphids.

For the following reasons, Davies concluded that energy intake is the criterion by which the flycatcher solves the problems of where to feed and what prey to select.

1. The switch from feeding in the tree canopy on small prey to exploiting large prey near the ground probably occurred when the latter tactic became the more profitable one in terms of energy intake per unit time.
2. Changes in diet were influenced by the absolute abundance of the preferred prey (large Diptera) and not by that of the alternative prey (small Diptera and aphids). The preferred prey were those that, when abundant, gave the greatest energy intake.
3. From the large prey that were available the adults concentrated on

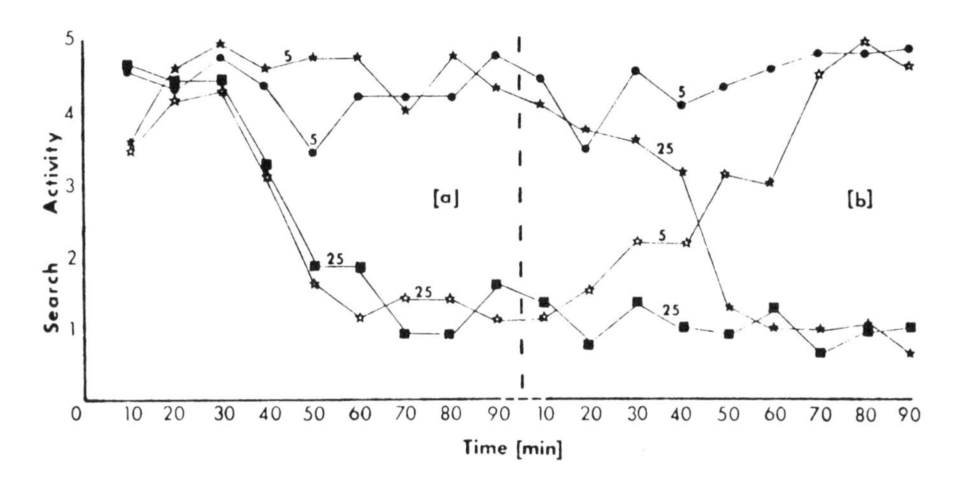

Figure 13.6 Search activity of centipedes (*Scolopendra polymorpha*) as a function of increasing time exposed to two prey densities (5 or 25) during two 90-minute periods (a and b). Closed dots and squares are controls in which prey density remained constant in both time periods. Closed stars are centipedes exposed to a prey density of 5 during the first period and 25 in the second; open stars are those exposed to a prey density of 25 during the first period and 5 in the second. From Formanowicz and Bradley, 1987. Copyright The Association for the Study of Animal Behaviour, reprinted by permission.

those that yielded the maximum energy return. They ignored prey that were difficult to handle (bees and very large Diptera) or difficult to capture (Muscidae and butterflies).

4. When hunting for large prey, the behaviour of the flycatcher in leaving perches and its waiting time were consistent with the view that they were searching for patches that yielded the most prey per unit time.

With a change from low to high prey density, centipedes (*Scolopendra polymorpha*) switch from active searching to a more ambush-like or sit-and-wait tactic (Figure 13.6a) (Formanowicz and Bradley, 1987). The utilization of more than one search tactic may be adaptive for the centipede because searching actively at low prey densities increases the probability of encountering prey, while adopting a more ambush-like tactic at high prey densities may reduce energy usage in searching activities. In another experiment, the lag-time associated with switching between search tactics was determined, and found to be between 30 and 40 minutes after a switch in prey density. Formanowicz and Bradley (1987) rule out capture rate, inter-capture interval,

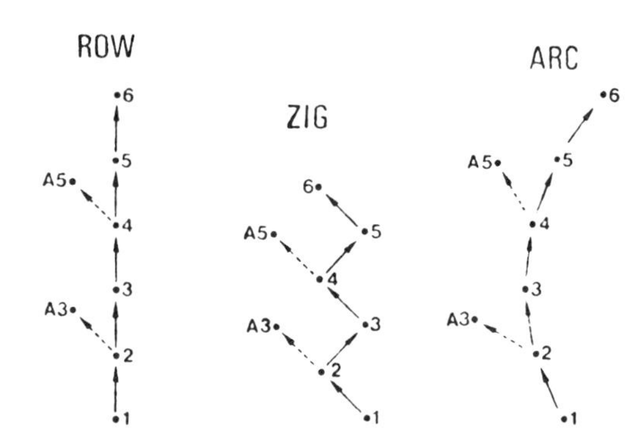

Figure 13.7 Patterns of sucrose drops arranged in configurations of ROW, ZIG and ARC used in experiments with *Drosophila melanogaster*. (After Tortorici *et al.*, 1987.) Drops labelled A refer to alternative drops outside each pattern.

and satiation as mechanisms by which centipedes might sample or assess prey density. They suggest that the centipedes may have used an encounter rate mechanism in sampling or assessing prey density and making subsequent 'decisions' about the type of search tactic to use.

13.2.3 Effects of resource dispersion and texture on searching behaviour

In addition to density, dispersion and texture are two important aspects of resource distribution. Dispersion has to do with the spatial configuration of resources relative to each other, whereas texture refers to the spatial configuration of resources relative to other resources or nonresource objects.

A simple experiment with *Drosophila melanogaster* shows how perception of the spatial arrangement of resources can restrict the area searched and the geometry of the local search pattern of a fly as it works through a patch of sucrose drops (Tortorici *et al.*, 1986). Flies were tested on three patterns of sucrose drops (Figure 13.7): a straight line (ROW), a zigzag (ZIG), and a semicircle of of drops (ARC). The patterns were designed to provide spatial diversity, but with similar linear dimensions and equal distances between drops. Alternative sucrose drops were placed at drop positions 3 and 5 in each array to determine if a fly, after locating one or several drops in a given pattern, is progressively less likely to locate alternative drops that are the same distance from drops in the pattern but are outside the pattern (drops A3 and A5 in Figure 13.7).

The results showed that the number of times the alternative drop was found at position 3 did not differ significantly from a 50:50 ratio in any pattern, whereas at position 5, the alternative drop was located significantly fewer times than a 50:50 ratio. Thus, *D. melanogaster* has insufficient information concerning the spatial configuration of resources at drop 2 in each pattern, and it forages with an equal chance of encountering the next drop in the pattern or the alternative drop at position 3. By drop 4, the foraging behaviour of the fly has become 'constrained' by the spatial configuration of drops in the ROW, ZIG, or ARC, and it follows the sequence of resources to the exclusion of another resource which is in the immediate vicinity.

An alternative explanation to gaining information about the spatial configuration of the drop pattern is that a fly deciphers the temporal sequence of the resources. This explanation seems unlikely, however, since the interdrop time intervals for an individual fly are highly variable. The benefit gained by modification of search orientation relative to the spatial configuration of resources is short-term constraint of search orientation parameters to a resource arrangement that is predictable for a short time period. It is possible that *D. melanogaster*, and perhaps other organisms, exploit resources in a food patch by processing proprioceptive information obtained from locomotory movements, producing short-term retention of the spatial patterning of resources.

Similar experiments have been performed on houseflies and parasitoid wasps. Individual houseflies (*Musca domestica*) foraged in patches of sucrose drops arranged in a line (ROW) or hexagon (HEX) (refer to Figure 8.2) (Fromm and Bell, 1987). In both arrays the flies located a median of four drops before leaving the patch. Area coverage of completed search paths of *M. domestica* in HEX and ROW arrays is markedly different, and even though few differences were detected in interdrop measures (e.g. locomotory rate, turning rate), the pattern of resource distribution restricted the flies to a linear orientation pattern in the ROW and to a circular orientation pattern in the HEX. Thus, even though the tactics of the flies do not change relative to the pattern of drops, *finding the drops leads to drastic alterations in the shape of the area covered*. Rather than supporting the idea that a fly 'assesses' the resource pattern, 'integrates' this information, and changes its tactics accordingly, these findings suggest substantial plasticity in the searching tactic of the fly, and that finding the resources in a certain pattern constrains its orientation to that pattern.

Laing (1937) arranged eggs of the cereal aphid (*Sitotroga cerealella*) in two rows of 12 eggs (Figure 13.8). A female parasitoid wasp (*Trichogramma evanescens*) was placed on the paper containing the eggs of its host, and observed until 24 contacts with eggs were completed.

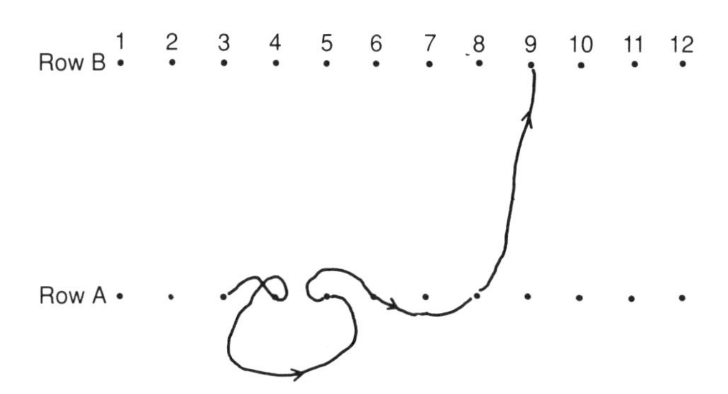

Figure 13.8 Arrangement of eggs of cereal aphids (*Sitotroga cerealella*) in experiments on the parasitoid wasp (*Trichogramma evanescens*) searching for hosts in which to oviposit. (After Laing, 1937).

Laing was particularly interested in whether the wasp would be more likely to locate an egg next to one contacted than an egg at some other position. If a wasp walked in sequence from one egg to the next in a row, within the inter-egg radius (10 mm), it was counted as a near-neighbour move. Thus in a sequence such as 3A, 4A, 5A, 6A, 8A, 9B (see Figure 13.8), sequences 3A to 4A and 5A to 6A were counted as near-neighbour hits, whereas 4A to 5A, 6A to 8A and 8A to 9B were not. The data revealed that 53.3% of moves were made directly from one egg in a row to another after oviposition, 46.6% after examination, and 25.4% after contact. Laing concluded that the turning movements released by oviposition or examination are responsible for 'an increase in the chance of meeting the neighbouring host'.

From the results from all three studies we can conclude that resource dispersion influences locomotory search patterns, mainly because locomotory searching patterns become constrained by the distribution pattern of resources. Apparently, search tactics are sufficiently plastic to allow efficient search on various types of patterns, and the looping local search patterns of arthropods seem adaptive for locating adjacent resources.

13.2.4 Resource dispersion and texture affect success of resource finding

The effect of resource distribution on search success is sometimes different from the effect it has on search behaviour. Insect herbivores provide good examples of how resource dispersion and texture affect searching success.

(a) *Texture*

Animals can respond in various ways to resource texture, depending on their search strategy and their perceptual mechanisms. For example, a specialist herbivore responding to host-specific cues might be confused or even repelled by nearby nonhost plant species (Atsatt and O'Dowd, 1976), whereas a polyphagous herbivore, orienting to generalized plant characteristics (foliage colour, 'green-plant volatiles', or humid zones), may perceive diverse plant mixtures merely as dense stands of food. Mixing closely related plants together may actually increase the attractiveness of a site to an oligophagous species if the set of plant species all release similar attractive volatiles. Risch (1980) showed, for example, that polyphagous chrysomelid beetles reached higher densities in plots where corn, beans, and squash were mixed than in single species stands.

The perceptual mechanisms by which plants are located can also be important. For example, visual searchers may locate host plants according to their silhouettes against a homogeneous background or their basic shapes, whereas herbivores that primarily use odours may find it easier to locate clumped plants than sparse plants (review: Stanton, 1983). Species that recognize host plants on the basis of leaf shape, as in visually- scanning pipevine swallowtail butterflies (*Battus philenor*), suffer reduced oviposition on host plants amid uncut vegetation (Rausher, 1981). When host plants are found by olfaction, odours of surrounding vegetation may mask host plant odours or may produce repellent volatiles. For example, fewer eggs of the cabbage-root fly (*Delia brassicae*) are laid on their brassica hosts when they are intercropped with clover than in monocultures (Hawkes and Coaker, 1979). Flight chamber experiments revealed that the same number of cabbage root flies *arrived* at trays of biculture and monoculture patterns (Hawkes and Coaker, 1979). Although clover did not disrupt host-finding, it did increase activity upon arrival so that flies spent more time flying and less time ovipositing.

(b) *Dispersion*

The dispersion pattern of host plants severely impacts the frequency with which lepidopteran larvae locate their host plants. Larvae of the nymphalid butterfly (*Melitaea harrisii*), which feed on *Aster*, must find additional hosts on which to complete their development after defoliating their initial host plant. Dethier (1959) estimated that 80% of these larvae die while searching for another plant in a field with 2.86 food plants per m^2!

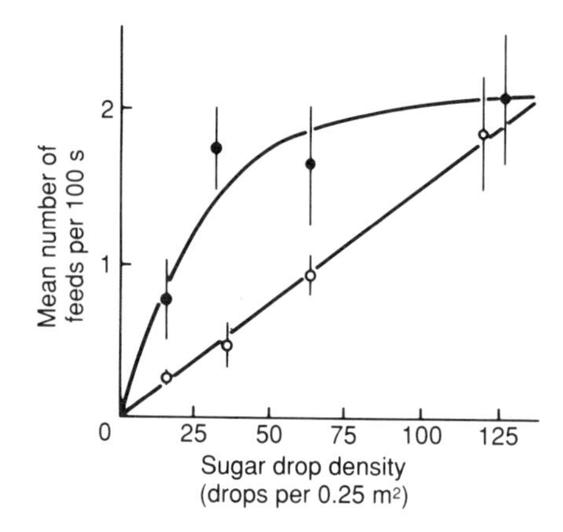

Figure 13.9 Relationships between drop density and the mean number of sucrose drops ingested by houseflies (*Musca domestica*). Regular distributions are indicated by open circles, clumped distributions are indicated by solid circles. (After Murdie and Hassell, 1973.)

Cain *et al.* (1985) investigated the effects of plant dispersion on host-plant finding by *Pieris rapae* larvae searching for collards (*Brassica oderacea*), where the collards were either uniformly dispersed or grown in clumps. Field-collected larvae were released at random locations within each collard plot where plot size and density over a stand were held constant, and the time required for larvae to locate a plant was recorded. Success in locating a plant was significantly affected by each of the variables tested, including plant distribution pattern, age of larvae, and nutritional state. The results suggest that plant dispersion affected the frequency with which larvae located collards only when the larvae were younger instars or unstarved; these individuals are the poorest searchers, because they are less mobile, and are able to locate collards spaced at regular intervals more readily than collards that are clumped. No effect of plant dispersion was found with the highly mobile fifth instar or starved larvae. Similarly, Murdie and Hassell (1973) found that the number of sucrose drops located by houseflies increased with increasing drop density. However, the relationship was linear when drops were arranged in an even distribution, and curvilinear when they were arranged in clumps (Figure 13.9). The search tactic of flies is adapted for searching for aggregated food particles, giving rise to the fast rising curvilinear function in clumped distributions; at extremely high resource densities no differences were noted in searching success.

13.2.5 Temporal distributions

The pattern of searching often reflects the interaction between the resource requirements of an animal and temporal changes in availability of that resource in its environment (review: Daan, 1981). As a consequence, much like the selection pressure on plants relative to spatial patterns of community structure, activity patterns of prey are often molded by the activity of predators.

One of the most informative studies of interactive prey and predator foraging rhythms centres around the European kestrel, *Falco tinnunculus*, and the vole (*Microtus arvalis*) it preys upon (Daan and Slopsema, 1978; Rijnsdorp *et al.*, 1981). The foraging activity of the voles reflects the consequence of selection pressure applied by the kestrel, and the foraging habits of kestrels reflect the temporal availability of voles.

Daan and Slopsema (1978) collected voles in reclaimed land in The Netherlands and transferred them to recording cages with running wheels, food, and nestbox. Their results show that wheel running occurs only during the dark phase of the photocycle, whereas feeding bouts occur through both dark and light periods. Feeding bouts during the day were at approximately two-hour intervals, as shown previously by Lehmann (1976) with *M. agrestris*, and in other rodents and insectivores (review: Aschoff, 1962). The pattern of the short-term feeding rhythm in the daytime and temporally dispersed feeding and locomotory activity during the night is not dependent on the light/dark cycle, as the same pattern holds with continuous darkness. The rhythmic pattern was much more synchronized when examined with groups of three voles as compared to individuals, suggesting that voles may be able to synchronize their feeding bouts with those of other group members. The short-term feeding rhythm cannot be explained solely on the basis of repetitive changes in satiation related to intake and digestion of food, but more upon the habitual daily pattern established when food is available.

Daan and Slopsema (1978) also investigated foraging rhythms of voles in the field. Voles live in burrows and occasionally emerge to the surface for feeding. Trapping during the day showed cyclicity in capture frequency, with peaks at 08:00, 10:30, 13:00 and 15:00. The cyclicity appears to be a population phenomenon, although the data are constructed from short-term trapability of individual voles. The two-hour rhythm suggested by trap data is in agreement with laboratory results, and probably represents the temporal distribution of surface feeding excursions alternating with subterranean digestive periods. These observations indicated to the researchers that predation on voles by diurnal predators such as kestrels may be restricted to the short bouts of feeding activity occurring once every two hours, and that perhaps predation has been responsible for shaping the foraging cycles of the voles. Risk to an individual vole from kestrels is actually two times higher during the vole population inactivity period as compared to activity periods.

Foraging of *F. tinnunculus* was observed in the same habitat and time frame as for the vole study (Rijnsdorp *et al.*, 1981). The highest rates of vole capture by kestrels occurred at the same time as peak foraging activity of voles, and the lowest rates during periods when voles were inactive. Similarly, Mikkola (1970) showed that pigmy owls synchronized their foraging schedules to the activity of bank voles (*Clethrionomys glareolus*). Many of the rodents captured by kestrels are not consumed immediately, but are cached, especially during the middle of the day, and then retrieved at times when food is scarce, mainly during the late portions of the day. The principle of caching and retrieving allows the kestrels to optimize the times of hunting with respect to both weather conditions and prey availability. It also enables them to uncouple food searching from its consummatory act, eating, and to optimize these separately according to time of day.

Enright (1975) has emphasized the role that circadian oscillators may play in adjusting an animal's daily sequence of behaviour to its own experiences in a periodic environment, rather than to those of its ancestors. Enright's hypothesis is that biological clocks in higher animals primarily find adaptive significance in the repetition of behaviour patterns from day to day. Daily similarities in the time course of environmental change would put a premium on such repetition, provided that modifications in response to experience are continuously incorporated in the circadian programme. The spatiotemporal pattern of consecutive day-to-day foraging flights of the kestrel is quite consistent (Rijnsdorp *et al.*, 1981) (Figure 13.10); as Enright (1975) has pointed out, having found food at a certain time and place makes returning to that place at the same time of day an appropriate strategy. An increased tendency to flight-hunt in response to prey capture 24 hours ago contributes further to the consistency. Since the immediate effect of prey capture is a depression in hunting tendency, the increase 24 hours later is not simply another expression of the same day-to-day correlation, but a clear adaptive adjustment. The area chosen most often is the one where prey has been captured 24 hours earlier, rather than any other time lag. Thus, both elements of Enright's hypothesis are present in hunting behaviour of kestrels:

1. there is a correlation in the temporal pattern of hunting behaviour on consecutive days, and

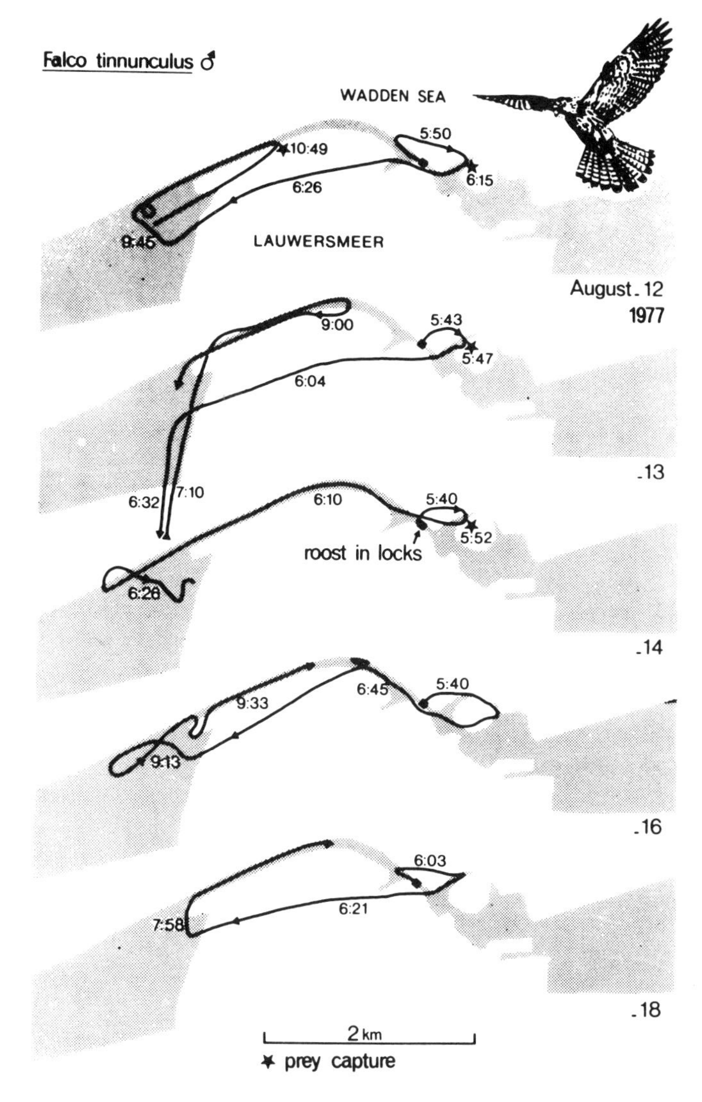

Figure 13.10 Foraging flights of a European kestrel (*Falco tinnunculus*) on five mornings. From Rijnsdorp *et al*., 1981. Copyright Springer-Verlag, reprinted by permission.

2. this pattern tends to be modified by experience such that prey capture is followed by increased hunting frequency 24 hours later.

One experiment by Rijnsdorp *et al*. (1981) with a freeliving kestrel provides at least modest support for the idea that the choice of foraging area is partly determined by daily habits, but is modified by the experience of capturing prey 24 hours previously. They offered laboratory mice to a kestrel named Kiki each day between 08:00 and 10:30, by releasing them in the vicinity of the bird's perch. Kiki usually caught the prey immediately. Prior to the timed feeding regime, Kiki rarely visited the area where the mice were released. It typically visited a favourite hunting area in the morning around 10:00, and again in the late afternoon. The habit of visiting this area in the morning was associated with a tendency to flight-hunt during the first morning peak in vole availability. The influence of the mice fed at the perch site was considerable. Kiki not only came back to the perch every morning but also stayed there for almost the entire day following capture of the prey. During two days when no mice were provided, Kiki returned and spent 77% of her time from 08:00 until 10:30 in the perch area as compared with only 45% of her time during the remainder of the day. Apparently experience can modify daily programs generated from a stored set of routines.

Temporal interaction between searcher and resource also typifies the relationship between nectivores and flowering plants. Gilbert (1975) monitored diurnal patterns of nectar production and pollen release in two sympatric species of *Anguria* (Cucurbitaceae) visited by the same species of *Heliconius* butterflies in Trinidad. Pollen is shed and nectar is produced by *A. umbrosa* early in the morning and by *A. triphylla* in the afternoon. The butterflies adjust their foraging by visiting *A. umbrosa* in the morning and *A. triphylla* in the afternoon, thus reducing uncertainty for the butterflies and ensuring against periods without resources. A similar situation occurs in hummingbirds in Costa Rica that temporally partition their visits to flowers according to whether the flowers produce nectar in the morning or in the afternoon. In many communities different species of plants bloom at different times over the flowering season (e.g. Gentry, 1974), thus reducing the uncertainty of pollinators locating required resources. Plants probably did not evolve these patterns for the benefit of pollinators, but rather to reduce competition for pollinators required for fertilization (Levin and Anderson, 1970). For example, in two wildflower species that share hummingbird pollinators there is a short period of overlap in blooming time, and both species have reduced seed set during the overlap period as compared to the nonoverlap periods (Waser, 1978).

Zielinski (1988) attempted experimentally to encourage mink (*Mustela vison*) and weasels (*M. erminea* and *M. nivalis*) to synchronize their foraging activity with prey availability. Animals were exercised on running wheels to increase their energetic costs, and thereby making it economical to search during the opposite phase of the dark/light cycle that they preferred. Of seven animals tested, six showed moderate switching to the opposite phase of the light/dark cycle, but only two (mink) approached the 180° redistribution of activity that would be expected for an animal to be totally efficient under these circumstances. Adaptive constraints other than energetic demands no doubt account for the inability of some species to switch foraging schedules.

Honeybees can learn to visit a feeding dish at the time of day when reward was present or highest the day before (Wahl, 1932; Beling, 1935). This shift is adaptive, since it would be most efficient to visit flowers when the rewards are highest. Kleber (1935) presented open poppy flowers (*Papaver rhoeas*) to marked honeybees. Those visiting after 10:00 did not subsequently visit before 10:00. They relied on time memory instead. Poppies generally are visited before 10:00, after which pollen availability normally wanes. They relied on time memory rather than on information from other bees in the colony. Bees also correctly selected between two scents, such as thyme and geraniol, when the reward schedule was alternated in a daily sequence every 45 minutes (Koltermann, 1971). That birds can tell time accurately is shown by cases where caged starlings showed anticipatory key pecking just before the one-hour mark when they had been were rewarded for key pecking at hourly intervals (Adler, 1964).

13.2.6 Resource quality

Resource quality can affect search duration, searching speed, capture rate and run/stop ratios. The effect of resource quality perception often acts to increase the efficiency of searching behaviour as outlined in the following section.

Resource quality may directly affect the course of time-dependent shifts from local search to ranging, such that an animal remains in a rich patch longer than in a poor patch. Search duration is related to the size of the prey consumed in coccinellid beetles, and in fact is dependent on the size of the most recently consumed prey: significantly longer periods of local search occurred when beetles were fed first on a small then on a large aphid, as compared to feeding first on a large then on a small aphid (Nakamuta, 1985). Dethier (1957) showed that the time spent by the blowfly (*Phormia regina*) searching near a drop of ingested sucrose solution increased with sucrose concentration (Figure 13.11). Similar data have been recorded for the housefly (*Musca domestica*) (Mourier, 1964) and the fruitfly (*Drosophila melanogaster*) (Bell and Tortorici, 1987). Houseflies remained in a patch of 2.0 M drops longer than in a 0.125 M patch, because

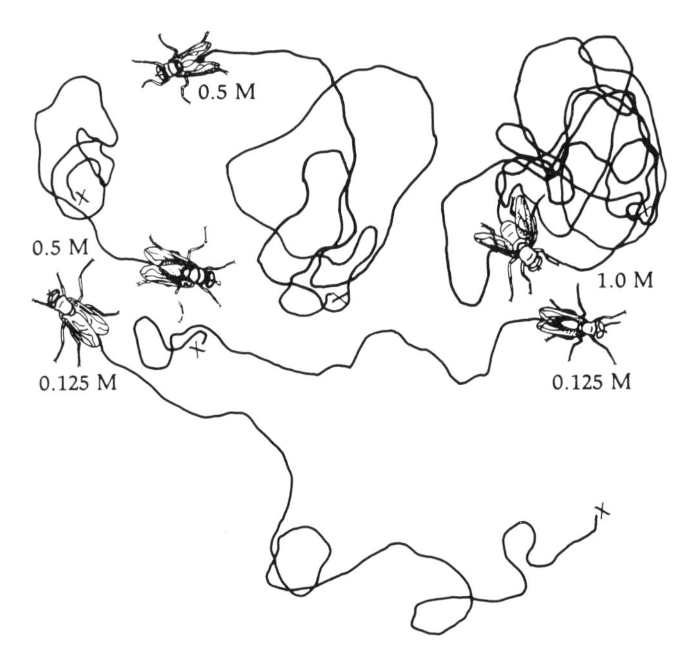

Figure 13.11 Pathways of blowflies (*Phormia regina*) fed different concentrations of sucrose. From Dethier, 1957. Copyright American Association for the Advancement of Science, reprinted by permission.

the duration of ingestion-dependent local search and handling times is longer after ingestion of each 2.0 M drop than after each 0.125 M drop (Fromm, 1988). Glen (1974) points out the effect of resource quality on search pattern in the bug *Blepharidopterus angulatus*: after feeding on a first instar aphid it switched to rapid locomotion and moved to other areas quite quickly, whereas after feeding on a fourth instar aphid, 20 hours were required before locomotory rate increased again.

Several studies have shown that nectivores leave a plant if they encounter low or no yield in the first flower visited. The frequency with which the bumblebee (*Bombus appositus*) stays or leaves after sampling the first flower visited in a patch depends on the amount of nectar in the first flower. In manipulation experiments, when either 0.5 or 1.5 μl was found in the first flower, Hodges (1981) found that the forager stayed to visit another flower significantly more often than when no nectar was found. The probability of staying was 0.9 after finding 1.5 μl as compared to 0.67 after finding 0.5 μl.

An interesting effect of resource quality is the way it alters the search cycle in fish. As shown in Figure 13.12, in both the white crappie (*Pomoxis annularis*) and the Arctic grayling (*Thymallus arcticus*)

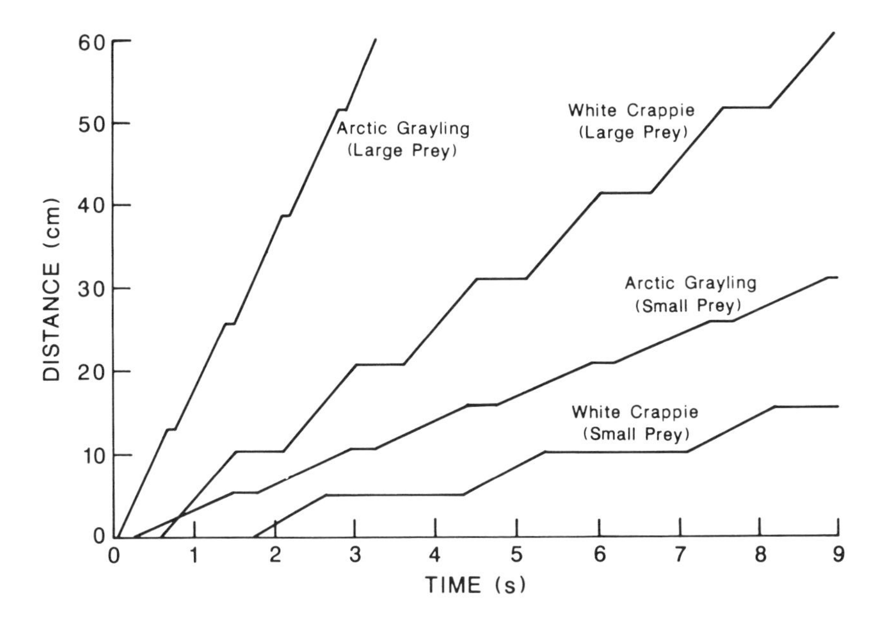

Figure 13.12 Alteration of the search cycle by resource quality (prey size) in the arctic grayling (*Thymallus arcticus*) and the white crappie (*Pomoxis annularis*). (After O'Brien *et al.*, 1989.)

the duration of stops, during which they scan for prey, increases from large to small prey availability (O'Brien *et al.*, 1989). The change is particularly dramatic in crappie where pause time increases from approximately 0.6 s to nearly 2 s. Obviously the fish require less time to fixate on the more apparent prey.

3.3 INFLUENCES OF OTHER INDIVIDUALS

The presence of other individuals can affect searching behaviour, resulting in either negative or positive effects on searching success. When groups can capture more resources per individual per unit time than can a lone individual, the influence is positive. When groups can forage more efficiently and watch for predators, capture larger prey, or share information, the effect is also positive. However, if groups make foraging more difficult, by scaring away prey or through direct competition for available resources, the effect is negative.

3.3.1 Positive effects

Positive group effects can be manifested either through gains in foraging efficiency or through gains in predator surveillance

Table 13.1 Effects of groups on searching efficiency

Positive group effects	*Negative effects of grouping*
Increased encounter rate	Direct interference
Increased capture success	Defence of territory at the expense of foraging
Increased prey size taken	Prey decrease owing to behavioural responses to predators
Reduction in foraging area overlap	
Passive information sharing	Prey frightened away
Active information sharing	Emigration
Better risk aversion	

efficiency. While grouping can increase searching efficiency for several different reasons (Table 13.1), most studies show that potential gains reach an upper limit at a certain population size.

(a) *Searching efficiency*

Flock formation in birds is usually beneficial or least neutral to the prey-catching of individual members, including the individual who locates the fish (the 'first finder') (Gotmark *et al.*, 1986). For example, the fishing success of individual black-headed gulls (*Larus ridibundus*) increases with flock size up to at least eight birds. Part of the reason is that a school of fish is more vulnerable when attacked at once by several gulls.

A good example of many individuals working together to subdue a resource that no single individual could possibly challenge is the invasion of a tree by thousands of bark beetles (review: Birch, 1984). The aggregation and boring activity of a large number of beetles over a short time period may be sufficient to overcome the defences and kill a healthy tree. Single beetles are ineffective because boring beetles can be 'flushed' out by sap flow in a healthy tree. Many species of bark beetles carry pathogenic bacteria and fungi that weaken a tree and reduce its ability to counterattack. Moreover, a mass attack simply makes it difficult for a weakened tree to successfully 'flush' out all of the invading beetles.

Schaller (1972), in his studies of lions on the Serengeti, showed that these predators benefit from stalking prey cooperatively (Figure 13.13). When several lionesses spot potential quarry they characteristically fan out and approach in a broad front, sometimes spread over 200 m of terrain. This fanning action may be well coordinated in that

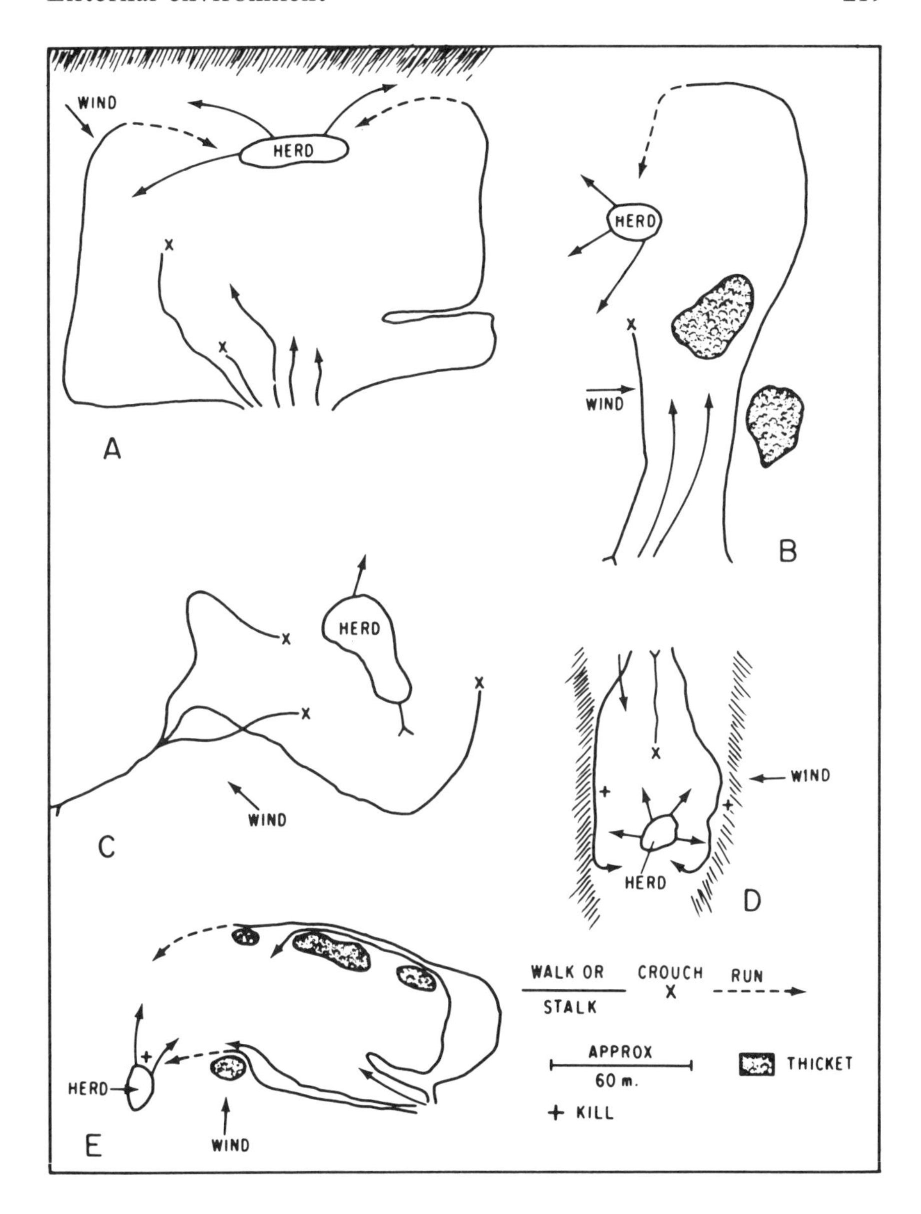

Figure 13.13 Five examples of routes taken by lionesses when stalking prey cooperately. From Schaller, 1972. Copyright University of Chicago Press, reprinted by permission.

those at the flanks walk rapidly in their chosen direction while those in the centre halt or advance slowly. By being widely spread out, lionesses increase their chances of coming into contact with prey which may either scatter, with one or more animals running at a hidden lioness, or veer to one side and inadvertently meet one of the hunters at the flanks.

When cooperating with each other it seemed as though lionesses oriented toward a common goal and that they coordinated their actions, with each animal patterning its behaviour according to what the others were doing. If only one quarry was involved, all individuals participated in catching it, but when several targets were available, cooperation declined as each tried to capture the nearest prey. In hunts where cooporation was clearly evident, the kinds of behaviour observed were relatively simple. However, on some occasions lionesses encircled prey, sometimes by detouring far to one side. The other lions waited during the flanking movements as if in anticipation of prey fleeing toward them. Figure 13.13 illustrates the movement patterns of lionesses in such hunts. In A, B, and D the encirclement of a herd succeeded, and one lioness caused prey to run toward the waiting ones, in C the herd moved off before the stalk was completed, and in E one lioness rushed before two others were fully in position. During such hunts lions integrate their actions solely by observing each other's posture and movement; no sounds are used nor are facial expressions employed which, at any rate, would not be useful at night. 'Encircling' implies that lions are aware of the consequences of their actions in relation both to other group members and to the prey.

Caraco and Wolf (1975), using Schaller's (1972) data on lions, found no specific foraging advantage to increasing group size if prey were small (e.g. Thomson's gazelles), but there is an advantage of foraging in a group of up to three individuals when the prey are large (e.g. zebra). Interestingly, Franks (1986) found that groups of army ants (*Eciton burchelli*) can carry food items so large that if they were fragmented the original members of the group would be unable to carry all of the fragments.

(b) *Predator surveillance efficiency*

Responses of individual starlings (*Sturnus vulgaris*) to a moving model of a flying hawk suggest that flocking can be a means of increasing the effectiveness of surveillance (Powell, 1974). In these experiments the starlings were housed in a large flight cage with a foraging area on the floor. A model of Cooper's hawk (*Accipiter cooperi*) was suspended from above, such that it could be moved across the top of the pen. The percentage of time spent foraging (per bird) increased with group size from 53% for single birds to 88% for birds in groups of 10, whereas frequency of surveillant observations decreased from 0.39 per s for singles to 0.19 per s for individuals in a group of 10.

Positive effects of grouping can be dependent upon several interacting factors including resource density, texture and distance from cover. Barnard (1980) studied feeding and scanning behaviour of flocks of house sparrows (*P. domesticus*) in two different habitats. In cattlesheds, where they fed on barley seed from bedding straw laid down for cattle, the birds were nearly completely sheltered from aerial predators, whereas in open fields, where they fed on seed taken from post-harvest debris or newly-sown winter barley seed, they were exposed to patrolling European kestrels (*Falco tinnunculus*) and sparrowhawks (*Accipiter nisus*). Barnard's point is that if food availability and distance from cover are not taken into account, the data suggest that flock size correlates positively with individual feeding rate and negatively with the rate at which individuals scan; if food availability and distance from cover are considered, however, the correlation with flock size disappears in shed flocks but remains intact in those feeding on open fields. The allocation of time to feeding and scanning by birds in the sheds was almost completely governed by food availability, whereas in the field flock size affected time allocation independently of food availability and other environmental factors.

13.3.2 Negative effects

Negative effects suffered through interactions with other individuals include an array of direct and indirect influences whereby potential prey are frightened or reduced in availability, or where the presence of other individuals necessitates defence or leads to competition (Table 13.1).

a) *Changes in prey behaviour*

A subtle way that one individual can affect the searching efficiency of another individual is by changing the behaviour of prey in a given area, making it more difficult for a second predator to capture them. This may involve changes in flocking behaviour, greater alertness, or reduction in activities that increase risk to predation (e.g. advertising, courting, and feeding). If the response is mostly a change in position, the prey may still be available to a second predator. If the change is increased alertness, however, it is likely that any predator hunting in the same spot soon afterwards would have reduced capture success. The following experiment illustrates how the presence of a predator can change prey behaviour, making them more difficult to capture (Charnov *et al.*, 1976). Mayflies of the genus *Baetidae* were used as prey, and a Kokanee salmon (*Oncorhynchus nerka*) as a predator in a 50-gallon aquarium with a dark slate bottom. Since the edges were dark metal, the four corners were dark and the risk of attack was lower for a mayfly in a corner. In fact, the mayflies remained in the corners when the salmon was present, and were more widely dispersed when it was not in the tank. The mayflies moved to the corners when a fish was placed in the tank, even when the fish was satiated and not actively hunting. In similar experiments with a substrate of gravel, leaves, and sticks, the movement of the mayflies was down or under, rather than to a corner, but the effect on availability to the predator was the same. Juvenile Atlantic salmon (*Salmo salar*) also changed their foraging tactic after exposure to a model trout predator (Metcalfe *et al.*, 1987). After seeing the predator, rather than orienting toward and capturing a food particle, the salmon waited until a passing food particle had reached its closest point before striking at it. Their behaviour would seem to reduce their conspicuousness and thus the possibility of being preyed upon. Foraging characterized by predator avoidance lasted for approximately two hours.

(b) *Increased territorial defence*

Red-backed salamanders (*Plethodon cinereus*) establishing territories in laboratory chambers, spent more time defending territories (displays and biting) at the expense of foraging as the degree of competitive threat increased. Thus less foraging occurred when a competitor or its pheromones were present as compared to when no competitors and no pheromones were present (Jaeger *et al.*, 1983). Another response to competitive threat was to shift from a specialized diet on the more profitable prey type to an indiscriminate diet, even though prey densities and the encounter rates of residents with each prey type did not change. The presence of unfamiliar pheromones led to a 50% decrease in the rates of net energy gain by residents, about 80% of which was due to the time taken away from foraging and 20% due to change in diet. Thus, changes in foraging time and diet both reflected the costs of territorial defence.

(c) *Interference*

Because visual foraging requires a relatively large search area per bird, ringed (*Charadrius hiaticula*) and grey plovers (*Pluvialis squatarola*) avoid concentrations of other birds. They especially avoid tactilely foraging bar-tailed godwits (*Limosa lapponica*) and dunlins (*Calidris alpina*), which feed near the tide edge where prey availability is generally highest (Pienkowski, 1983). A similar kind of avoidance was demonstrated by Morse (1980b) in the foraging of a small bumblebee species, *Bombus ternarius*, in the presence or absence of a larger species, *B. terricola*. Individual flower heads of goldenrod (*Solidago juncea*) are borne along several branches, each of which may be a few centimetres long. When foraging alone, *B. ternarius* visit flower heads on the proximal, medial, and distal parts of branches with similar frequency if the inflorescences are at their flowering peak. The larger *B. terricola* visit distal heads with a lower frequency than other parts. *B. ternarius* workers foraged progressively more distally on inflorescences as Morse added more and more *B. terricola* to cages in which the smaller species foraged. As with the plovers, overt interactions were not involved in the avoidance behaviour.

More aggressive interference occurs in oystercatchers (*Haematopus ostralegus*) feeding on mussels. Such interference may be the direct or indirect result of aggressive encounters over food which are common in this species and which increase in frequency as bird density rises (Vines, 1980). Ens and Goss-Custard (1984), observing oystercatchers wintering on the Exe Estuary in South Devon, England, showed that birds with high dominance scores almost invariably won over birds with lower scores. The dominance score was calculated as the percentage of interactions with any other bird which the focal animal won. The intake rates of six of the birds were significantly reduced at high bird densities, such that subdominant individuals actually avoided beds with high bird densities. Interference may have been due more to mussels being stolen than to foraging time lost, since the number of stealing attempts increased with dominance, especially at high bird densities. At the same time the frequency with which an individual was attacked decreased with dominance, although all animals were attacked more often at high bird densities. The overall result was indeed a net gain for dominants and net loss for subdominants, particularly at higher bird densities. In Forster's terns hunting for small fish in salt flats on San Francisco Bay, California, the rate of predation per bird declines as the number of birds in the area increases (Salt and Willard, 1971). In this case, in contrast to the aggressiveness of the oystercatchers, more time is spent avoiding collisions and less time is spent on searching.

Hassell (1978) points out several arthropod examples of reduced search efficiency as density of predators or parasitoids increases. For example, female predators may lay fewer eggs as predator density increases, as demonstrated by Evans (1976) for a predaceous bug (*Anthocoris confusus*) feeding on aphids. Many parasitoids have evolved threat displays and aggressive behaviour which they exhibit toward other females in the vicinity. For example, Spradbery (1970) found that female *Rhyssa persuasoria*, an ichneumonid parasitoid of wood wasp larvae, will threaten and if need be, drive an intruding female from the same area of tree trunk, and may maintain the territory for several days. Spradbery (1969) also showed that the ovipositing *R. persuasoria* ignores the very close presence of its cleptoparasitoid, *Pseudorhyssa sternata*, indicating that the aggression is intraspecific. Huffaker and Matsumoto (1982) demonstrated the effects of interference on parasitization. In *Venturia canescens* single wasps and groups of ten wasps were allowed to search for and parasitize hosts for 24 hours. Regardless of host density, the numbers of ovipositions per adult parasitoid are substantially reduced in groups of ten. This reduction derives mainly from interference.

Hassell (1978) stresses that the other side of the coin is that mutual interference leading to dispersal can serve to distribute predators more evenly among resource patches, and thereby improve searching efficiency. This can happen because prey are heterogeneously distributed, and in particular they occur in discrete units of the habitat. The end result is that populations of either predator/prey or parasitoid/host are stabilized. Aggregation leads to some host or prey patches being undiscovered, and interference-mediated dispersal prevents overkill and reduced searching efficiency.

13.4 ALTERNATIVES TO SEARCHING

Adversity in one form or another may stimulate more intensive search as an animal attempts to move away from stress sources, copes with changes in the abiotic environment, or attempts to locate rapidly declining resources. When the immediate responses of an animal are insufficient to solve the problem within a required time limit, the remaining options include hibernation or migration to a more suitable habitat. Such responses may be facultative (untimed, unpredictable) or obligative (highly predictable, generally heritable).

13.4.1 Hibernation

Hibernation, which generally occurs in response to winter cold, summer heat, or drought, is a period of dormancy and inactivity during which metabolic processes are slowed (review: Lyman *et al.*, 1982). Becoming dormant relates to search behaviour not only because the phenomenon is an alternative to continued searching for resources under adverse conditions, but also because in order to survive dormancy, animals must search for:

1. adequate food in preparation for periods of reduced metabolic rates,
2. an appropriate site to inhabit during the inactive period, and
3. food resources needed for recovery and emergence.

As we might expect, food hoarding occurs in many species of animals as a prelude to dormancy or for periods when food is unavailable. In the woodchuck (*Marmota monax*) hoarding usually occurs in late summer, well before the time to become dormant (Davis, 1967). Barry (1976) showed that temperature and photoperiod affected hoarding responses of several species of deer mice (*Peromyscus*). Locating an adequate hibernation site if it is not constructed by the animal also requires searching time. Limited numbers of available hibernation sites may actually control population size in some animals such as the Arctic ground squirrel (*Spermophilus undulatus*) (Carl, 1971) and the yellow-bellied marmot (*Marmota flaviventris*) (Andersen *et al.*, 1976).

13.4.2 Migration

Migration is a drastic process whereby animals leave one habitat and locate at another. Although migration offers a solution to certain types of adversity it is not without its problems: exposure to different conditions while on route, and encountering the unexpected when the destination is reached. For example, migrating animals have no way of knowing if their flyway or overwintering habitats have been destroyed or suffered climatic deterioration (e.g. Winstanley *et al.*, 1974), or what they will find during or at the end of their journey.

An important distinction is whether an animal moves between habitats because of immediate, unpredictable changes in the present resource unit, such as increased population density or deterioration of food supply (*facultative migration*), or moves at a certain time each year, regardless of conditions, and usually does so according to genetic information mediated by some external cue or endogenous clock (*obligative migration*).

(a) *Facultative migration*

Taylor and Taylor (1977) have made the point that movement is a fundamental biological response to adversity, by which organisms have a chance of locating a more favourable site. This is exactly what animals do in facultative migration. A decrease in the availability of resources is a major factor in facultative migration between habitats. In fact, Southwood (1962), among others, has pointed out that a predictable decrease in resource suitability is a major selective pressure on emigration, such that part of the adaptation of a species to a temporary habitat becomes involved in the evolution of obligatory migration. An example of facultative migration in response to reduction in food sources is the snowy owl (*Nyctea scandiaca*) in Canada. It preys on rodents, particularly lemmings, but periodically the owls move south into New England when the lemming population declines and the owl population increases. According to Baker (1978) the timing of these migration episodes suggests that owls respond to shortage of food by moving south. Low quality food or shortage of food also stimulates precocious flight in various species of insects (review: Dingle, 1980). Red kangaroos (*Megaleia rufa*) in central Australia, and wildebeest (*Connochaetus taurinus*) in the Serengeti region, move toward permanent sources of water at river sites in the brush during the dry season or during periods of draught, and they disperse over a wider area when water is generally available (Talbot and Talbot, 1963; Herbert, 1972). Other examples are discussed in Section 13.3 with regard to negative effects of other individuals on foraging success.

(b) *Obligative migration*

Much of the seasonal return migration of animals is adapted to predictable changes in the environment, particularly to changes in climate and availability of food. Obligatory migration represents the product of selection for endogenous mechanisms that trigger migration at the appropriate time.

In some species an endogenous rhythm, which ultimately controls the timing of migration, is set or rephased by exogenous events. For example, the onset of the post-nuptial moult in adult finches is in part a function of the date on which breeding ends (Newton, 1972). In juveniles of the bullfinch (*Pyrrhula pyrrhula*) and the chaffinch (*Fringilla coelebs*) the onset and duration of the post-juvenile moult is a function of the day of hatching (Dolnik and Blyumental, 1967; Newton, 1972). Young that are born late in the season also moult later, but they do so more rapidly and at an earlier age than do the young that are born early. Since the timing of the post-nuptial moult and of the autumn migration is also under endogenous control, the phasing of endogenous events in autumn is influenced by the photoperiod experienced 6 months earlier (Dolnik and Gaurilov, 1972). Thus the sequence of physiological changes that in autumn affect the timing and duration of moulting are part of an endogenous program that in at least some species has a free-running circannual periodicity.

Migrating animals seem to know when they have arrived at their destination. In some cases they home in on features of their destination that they recognize, and in other cases they land and then switch to local search for the final part of the journey. Physiological 'timers' may determine when some species stop migrating. For example, if the velocity of flight (calculated from release and recapture experiments) is multiplied by the duration over which caged birds showed 'migratory restlessness', the resulting time value yields an accurate estimate of the distance that warblers, *Phylloscopus*, normally fly along their migration route (Gwinner, 1977). This means, of course, that interruptions in flight along the way, if not compensated for in some way, could lead to precocious termination of migration in unsuitable habitats.

13.5 SUMMARY AND CONCLUSIONS

Biotic and abiotic environmental factors directly affect searching requirements, abilities, and efficiency.

Abiotic environmental conditions, such as changes in temperature or humidity, may directly affect searching behaviour of ectotherms by reducing their ability to move or at least decreasing their rate of movement, and of endotherms, because more energy is used in searching at lower than at higher temperatures. Changes in environmental temperature may directly affect searching behaviour. The effect may be mediated by alterations in locomotory rate or rate of prey capture, or influences on resource-finding behaviour and resource selectivity by altering the water balance of an animal. Factors such as poor illumination may affect reactive distance, and thereby influence foraging efficiency. In turn, animals such as fish may migrate vertically to zones where they can see their prey. Alternatively, poor illumination may benefit some carnivorous predators that hunt under the cover of darkness.

Biotic environmental conditions contribute to variability through variation in temporal and spatial availability of resources and patches of resources: resource accrual is generally proportional to resource density, a relationship referred to as a functional response; the spatial configuration of resources relative to each other (dispersion) affects searching by constraining the motor pattern to the spatial configuration of resources; resource quality affects search duration, searching speed, capture rate, and run/stop ratios, such that an animal tends to remain in a rich patch longer than in a poor patch; the presence of other individuals influences searching behaviour, resulting in either negative or positive effects on searching success.

The search tactic an individual employs may be conditional upon the environmental situation. Thus, an individual predator may switch from actively searching for prey to an ambush tactic when the abundance of prey reaches a certain density. Adopting an ambush tactic at high prey densities may reduce energy usage.

Under conditions of resource decline an animal may search more intensively. When an animal cannot locate the resources it requires through searching behaviour, it may execute an alternative to searching, such as migrating to a more suitable habitat or hibernating.

14
Internal environment

The internal state of an animal affects its searching behaviour, especially in the 'choices' it makes as to when to search and for what to search. Variability in the internal state reflects differences between individuals or changes within one individual over time in nutrient or energetic budgets, hormonal titres, and feedback from reproductive organs and other systems affecting behavioural thresholds. Thus, there may be differences within or between individuals in tendency to search, the search tactic employed, search intensity, and the sliding scale of acceptability of resources.

14.1 DEPRIVATION: TIME-DEPENDENT EFFECTS ON SEARCHING

Papaj and Rausher (1983) define 'time-dependent responsiveness' as an increase in the probability of an observable response to resource stimuli with time elapsed since performance of a consummatory act such as feeding, ovipositing or mating. According to their view, the change in responsiveness with time is independent of prior exposure to the resource stimuli or intervening performance of activities other than resource seeking, and presumably reflects some internal physiological variable that changes monotonically through time. The change in responsiveness is reversible upon performance of a consummatory act.

In all studies of searching behaviour in which the variable of deprivation has been assessed, deprived individuals search in a different manner from undeprived individuals. Deprivation is manifested in various ways, such as increasing reactive distance, reducing specificity of resource preference whereby animals become less choosy over time, or altering the motor pattern before or after an animal arrives in a resource patch. As more and more information becomes available on the effects of deprivation on searching behaviour, at least two important conclusions emerge:

1. Unless the experimenter knows or at least can estimate the time

since last feeding, oviposition, or other activity, enormous variability can be expected in the search behaviour and/or choices made by the animal.
2. As summarized by Zach and Smith (1981), one reason why most field situations are too complex to allow a meaningful prediction of optimal foraging performance is that 'wild predators cannot be forced to forage by deprivation, [and so] . . . one is never really sure what a predator is actually doing'. We do not know if an animal is being efficient because it is designed that way or because it is starving or has too many young to find food for.

14.1.1 Effects on the decision to search

There is an increased likelihood that an animal will search for a resource from which it has been deprived because deprivation increases the probability of responding to resource-related cues. For example, desert locusts (*Schistocerca gregaria*) that are progressively starved are increasingly likely to fly upwind toward grass odours in a wind tunnel (Haskell *et al.*, 1962). This upwind response decreases after contact with food in proportion to the length of time the animal is exposed to a food source (Moorhouse, 1971). In carnivores such as various species of small cats, while most aspects of prey capture seem to be virtually independent of hunger, initiation of search is highly correlated with hunger levels (Leyhausen, 1973). The same is true for lions (*Panthera leo*) (Elliott *et al.*, 1977) and spotted hyaena (*Crocuta crocuta*) (Kruuk, 1972). Similarly, isolation from males increases the phonotactic response of female field crickets (*Gryllus* spp.) to male calls (Cade, 1979).

Hunger or lack of available food can stimulate animals to move their ambush sites. For example, when ant-lion larvae are starved they move to another site and construct another pit, rather than increase the size of their current pit to enhance capture rate of prey (Griffiths, 1980). Turnbull (1964) showed that a web-building spider (*Achaearanea tepidariorum*) selects a site for its web after random movements. If the site turns out to be an unprofitable location, an assessment that the spider can make only by testing with its web, it moves to another place. Turnbull found that spiders would decamp if no prey were captured within three days. The typical feeding behaviour of the spotted fly-catcher (*Muscicapa striata*) involves flying to a feeding perch, sitting and waiting while scanning for prey, flying out to capture prey, and then either returning to the same perch or flying off to another one (Davies, 1977). Thus, the bird has to decide when to leave a feeding perch, and whether or not to return to the same perch. The key factor influencing its decision to return to the perch is the time spent waiting for a capture. If the capture was made quickly, this would indicate that the bird was in a patch of prey, and it would be advantageous to return to the same perch because the next prey would probably arrive soon also. In these examples there is no change in responsiveness to stimuli, but rather an increase in the probability that an animal will change its behaviour from ambush mode to actively searching for another ambush site.

Diel rhythmicity in tendency to search can interact with stimulation from hunger. Holling (1966) showed that in a praying mantid (*Hierodula crassa*), both gut content and time of day determined capture rate. If gut contents were held constant, more flies were captured per unit time in the morning than in the afternoon. Similarly, in adult coccinellid beetles (*Coccinella septempunctata*) foraging activity remained cyclic, with more activity during the day than in the night during 10 days of starvation. That diel rhythms predominate over hunger was shown by a much lower consumption of prey when offered during the night than when offered during the day, even though the beetles were starved (Nakamuta, 1987). Diel rhythmicity also dominates over hunger in African wild dogs (*Lycaon pictus*), and in African lions (*Panthera leo*) that generally hunt around dusk and dawn. If their hunting is unsuccessful, they wait and rest until the next morning or dusk period. When hungry because of conditions of low prey densities, lions will not increase their period of foraging activity, even when they have not found sufficient prey to feed their cubs (Schaller, 1972).

14.1.2 Effects on searching motor patterns

Feeding or ovipositing releases a bout of local search in many species. Search duration, i.e. the time for local search to readjust to ranging, increases with period of deprivation. As resources are utilized, however, there is a decrease in the time required for the transition from one pattern to the other.

A distinct effect of starvation was observed in digitized search pathways of *Drosophila melanogaster* after the flies fed on a drop of sucrose solution (Bell *et al.*, 1985). As flies were starved from three to 48 hours, the time required for local search to readjust to ranging increased, largely because the high turning rate remained in effect longer in starved than in unstarved flies (Figure 14.1). The study by Carter and Dixon (1982) on larvae of the lady beetle (*Coccinella septempunctata*) is probably the most extensive with respect to the effect of deprivation on various components of searching behaviour in

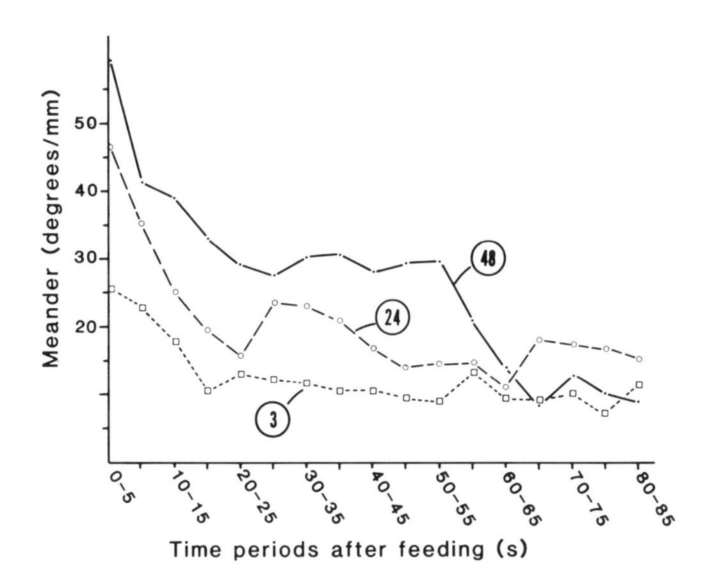

Figure 14.1 Mean rates of meander (degrees per mm) for *Drosophila melanogaster* fed after being starved for 3, 24 or 48 hours. (After Bell *et al.*, 1985.)

insects. Their results show that deprivation retards the decay of locomotory rate and turning rate back to control levels after feeding, so that in effect the larvae search longer when starved for 25 hours than when starved for 5 hours (refer to Figure 7.2). Deprivation translates into more time spent searching in a patch with the associated greater predation of aphids and more time spent handling each aphid through 12 captures. Searching time between encounters with aphids was relatively constant over a sequence of captures, except that in insects starved for 25 hours the first inter-catch interval was longer than the ensuing intervals. Larvae of the flower fly (*Metasyrphus corollae*) captured aphids more quickly when starved than when unstarved, and as in the coccinellids, the larvae spent much longer extracting fluids from aphids when starved than when unstarved (Leir and Barlow, 1982).

Starvation had no effect on ranging (i.e. pre-ingestive search) in coccinellid larvae, but only manifested an effect once the larva had ingested an aphid (Carter and Dixon, 1982). Before locating a sucrose drop *D. melanogaster* also engaged in ranging, and the effect of starvation on this activity was to decrease locomotory rate as starvation progressed; this is complicated by the fact that the percentage of time spent moving also changed as a function of starvation (Table 14.1). Thus, the amount of energy (EEV in the table) expended while ranging initially increased and then later decreased over 48 hours of starvation. Similarly, in predatory mites (*Amblyseius largoensis*) searching for prey, locomotory rate increased as they were starved, but then they later slowed down and became inactive (Sandness and McMurtry, 1972).

Thus, in many arthropods under conditions of starvation, and where food is unavailable, the locomotory rate may increase, but the percentage of time spent moving eventually decreases (Table 14.1) (e.g. Green, 1964; Bell *et al.*, 1985). This strategy after long-term deprivation would seem to prevent useless energy expenditures in the absence of information about resource availability. The chances of survival would depend on finding resources during very short quick movements, or on gambling on the fortuitous arrival of prey, or on chance resource replenishment. In those species where individuals seemingly continue to increase their locomotory rate and also move nearly continuously, as in larvae of the green lacewing (*Chrysopa carnea*) (Bond, 1980) and larvae of the small white butterfly (*Pieris rapae*) (Jones, 1977), perhaps this strategy at least promotes dispersal from areas of poor prey availability.

To differentiate between the effects of encounter rate and hunger on duration of local search, Carter and Dixon (1982) varied hunger from 23 to 63% of satiation, and varied encounter rate by manipulating the inter-catch interval from 7 to 21 minutes. As shown in Table 14.2, hunger has a greater influence than encounter rate in determining the duration of local search in a patch. In other words, the larvae do not 'assess' the patch using encounter rate as a diagnostic measure of patch worth. Hungry larvae also have longer handling times than do fed larvae, in part because they spend more time searching the substrate with their mouthparts to ensure that each prey is totally consumed. However, handling time did not decrease through the catch sequence as larvae captured and ingested aphids, and presumably as hunger decreased. Perhaps they measure hunger by the contents of their mid-gut and not their crop, resulting in a delay in their response to filling up. Carter and Dixon (1982) conclude that although many of the components of foraging behaviour of invertebrates have been shown to be hunger-dependent, this is an area largely ignored in the development of foraging theories.

'Instead, the desire to establish general strategic rules has necessarily sacrificed the consideration of detailed behavioural mechanisms and

Table 14.1 Energy expenditure values (EEV)* for ranging and local search of starved *Drosophila melanogaster* (from Bell *et al.*, 1985)

	Period of starvation (h)					
	0	*3*	*12*	*24*	*36*	*48*
Ranging						
Percentage of time moving	4.4	19.6	81.6	34.0	35.5	0.0
Locomotory rate (mm/0.2 s)	3.8	3.4	3.3	2.3	1.4	0.0
EEV	17	67	266	79	48	0.0
Local search†						
Percentage of time moving	90.4	44.5	66.4	64.3	60.5	71.2
Locomotory rate (mm/0.2 s)	3.9	2.2	1.7	1.6	1.7	1.5
EEV	352	97	112	105	100	101

*EEV = percentage of time moving × locomotory rate × 100.
†0 to 5 s after feeding on 0.2μl 0.25 M sucrose.

Table 14.2 Duration of local searching behaviour of *Coccinella septempunctata* larvae which experienced different encounter rates and levels of hunger (from Carter and Dixon, 1982)

*Encounter rate**	*Per cent satiation*	*Duration of local search (s ± SE)*
7	63	10.2 (1.5)
21	63	9.0 (1.5)
21	23	18.5 (2.0)

*Constant inter-catch interval in minutes.

the constraints to which specific predators and parasitoids are subject, in favour of generalized assumptions.'

Jones (1977) examined the effects of starvation on several species of caterpillars and was able to correlate their behaviour with distribution of their host plants. Fifth instar larvae of the small white butterfly (*Pieris rapae*) were starved and then allowed to search for host plants. Eventually the smaller larvae (50–120 mg) died, whereas larger individuals (150–220 mg) pupated. Before death or pupation, starvation led to increased locomotory rate, decreased head waving (scanning), and increased directionality (straighter movement) (Figure 14.2). Reactive distance also decreased, such that larvae were more responsive to nearby plants. All of these behavioural shifts could be reversed by allowing caterpillars to contact a crucifer plant. In contrast, there was no change in speed, head waving or directionality as larvae of the diamond-back moth (*Plutella maculipennis*) were starved; in fact, they always acted rather like fed *P. rapae*, in that they moved slowly, and turned and head waved often. Larvae of the alfalfa looper (*Plusia californica*) slowed down slightly as they were starved, but no change in directionality or head waving was observed. Both *P. rapae* and *P. maculipennis* feed on crucifers, but *P. rapae* is larger than *P. maculipennis*, and would probably be more likely to defoliate a plant and have to locate a new host plant than would *P. maculipennis*. Thus, the widely searching mode of *P. rapae* as it starves is probably an adaptation for its need to find other, perhaps distant, plants. *P. californica* occurs commonly on alfalfa, but it is polyphagous, and so its resource distribution is more homogeneous and its search pattern is appropriate for that kind of distribution. Thus, the differences in responses to starvation seem to be related to resource distribution.

As Curio (1976) points out, some components of searching motor patterns should not be affected by hunger. Predators such as cats, mantids, and salticid spiders must walk stealthfully regardless of their hunger level, or else their prey will be alerted and escape (Drees, 1952; Holling, 1966; Leyhausen, 1973). Hence, speed of capture remains constant regardless of hunger level.

14.1.3 Changes in reactive distance

The reactive distance at which the mantid (*Hierodula crassa*) first responds to a housefly increases with hunger (Holling, 1963, 1966). The elongate shape of the reactive field (Figure 14.3) relates to morphological properties of binocular vision. In addition, the amount of time spent pursuing the prey increases and 'digestive pauses' after each capture (a component of handling time) decrease with increased hunger. Holling (1966) was able to show that as food in the gut decreased, the average maximum reactive distance increased. The gut capacity is approximately 1 g, and it declines exponentially over time. Thus, the mantid could be estimating the rate of food intake by measuring the amount of food in its gut. In fact, among animals in general, this may be the most common proximate mechanism for assessing rate of food intake, and thus prey density. The mantid may

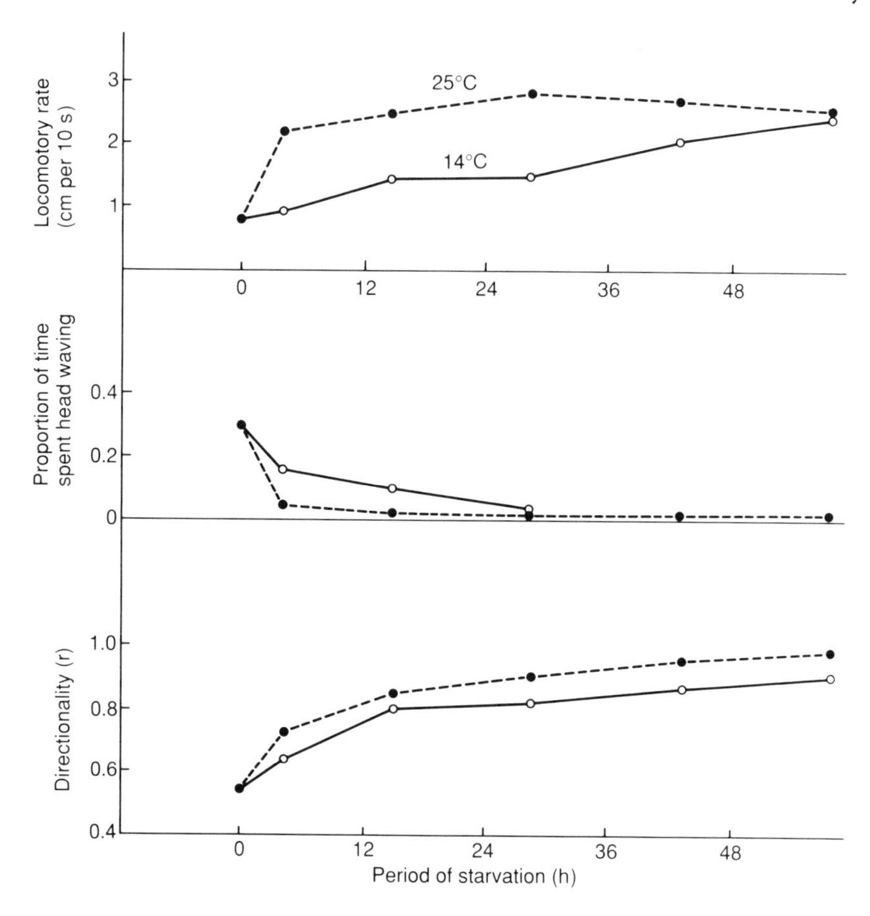

Figure 14.2 Changes in locomotory rate, proportion of time spent head waving, and directionality during starvation of butterfly larvae (*Pieris rapae*). (After Jones, 1977.)

not respond to more distant prey until it is hungry, because it may have more difficulty assessing prey quality when prey are distant; hence, when unstarved, strikes at distant prey are avoided. Its behaviour may also be affected by increased risk of predation while moving, and in fact its slow stalking manner may be a predator-avoidance adaptation. Indeed, its morphology, postures, and behaviour all may be influenced by predation risks.

14.1.4 Effects on choices of resources or patches

Time-dependent responsiveness can affect resource use in animal populations if unique time periods are required before particular

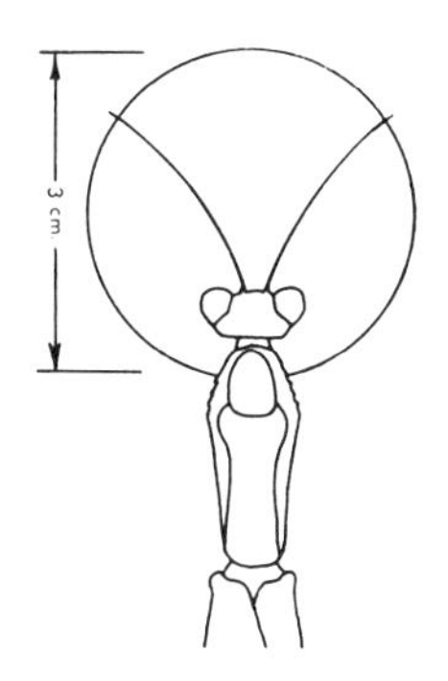

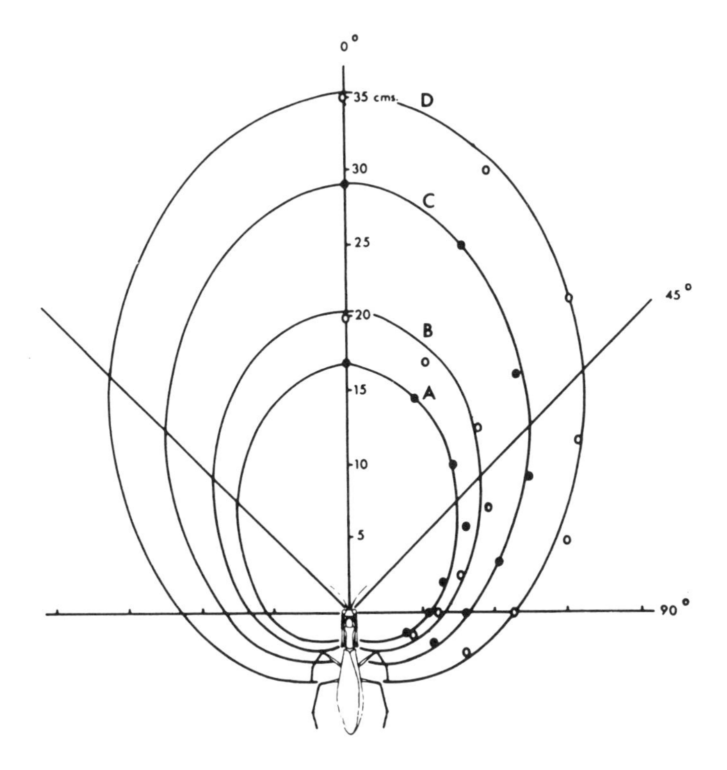

Figure 14.3 Shape of the reactive field of the praying mantis (*Hierodula crassa*) with hunger increasing from low (designated as A) to high (designated D) levels. Distances were measured along seven radii from 0° to 108°. Insert shows zone within which mantid can capture prey. (After Holling, 1966.)

resource types are accepted by individual animals. Changes such as these only affect searching behaviour indirectly, as an animal would be more likely to continue its searching when it encounters a resource that at the moment is unacceptable. Deprivation from a given resource could lead to

1. acceptance of a previously unacceptable resource,
2. decreased latency of acceptance, or
3. decreased number of 'errors' during selection.

Singer (1982) and Rausher *et al.* (1983), in field studies of ovipositing checkerspot butterflies (*Euphydryas editha*), found that for each potential species of host plant there is a corresponding acceptance threshold. Females will oviposit on an encountered plant only when the time elapsed since the last oviposition exceeds the threshold for that plant species. Singer proposed two quantitative measures of host specificity: rank-order of selectivity (i.e. the order in which host-types become acceptable over time since eggs were last laid), and the specificity of selectivity (i.e. the rate at which host types become acceptable over the time since eggs were last laid). For the simple case of two alternative host-plant species, Singer defines several phases through which a butterfly progresses during the time since its last oviposition. There is an initial refractory period after egg laying during which the butterfly will not oviposit on any plant. During the subsequent discrimination phase the insect accepts one plant species for oviposition, but not the other; if the animal has not yet oviposited, this phase is eventually followed by one in which both host species become acceptable.

Although Singer's observations on oviposition in checkerspot butterflies are restricted to examination of postalighting behaviour, his ideas are generally applicable to discrimination prior to host discovery as well, and are generalizable to other species. For example, after a shore crab (*Carcinus maenas*) eats a large mussel it is unlikely to accept the next two if they are small. But if the crab does not encounter another large mussel, it is likely to accept a smaller one on the third encounter (Elner and Hughes, 1978). This suggests a simple threshold mechanism based on mussel size (Elner and Hughes, 1978):

'each encounter with a large mussel raises (or maintains at a ceiling) the size threshold for accepting the next mussel, but the threshold decays rapidly to the level where small mussels are accepted'.

Hungry snails (*Achatina fulica*) orient nonselectively to food odours but with greater efficiency than do satiated individuals, whereas animals that have not been starved are much more selective in responding to odours (Croll and Chase, 1980). Hungry predators of many species accept prey indiscriminantly when hungry, whereas as they became satiated they accept only certain types. For example, as Fowler's toad (*Bufo fowleri*) becomes satiated the upper limit of prey size decreases (Heatwole and Heatwole, 1968). Similarly, starved salticid spiders show capture movements toward much larger prey items than they normally attack (Drees, 1952).

Deprivation can affect the number of errors made in selecting resource types. For example, hungry great tits (*Parus major*) made fewer errors in selecting food in that they took fewer unprofitable items or rejected fewer profitable items (Rechten *et al.*, 1983). They foraged more efficiently than did partially satiated birds. As the birds became increasingly satiated they appeared to 'strive' less and less toward the maximal energy intake rate.

Hunger also influences 'decisions' that animals make regarding variable versus constant food supplies. Caraco *et al.* (1980) (see also Caraco, 1981) offered yellow-eyed juncos (*Junco phaeonotus*), small granivorous birds, a choice between two feeding sites in an aviary. One site always offered a constant reward in number of seeds, whereas the other offered a variable reward, but with the same mean value over time. Thus, on some occasions the variable reward was greater than in the constant site and on some occasions it was less. Since over time the yield from the two types of patches would be the same, an animal maximizing energy gain should be indifferent to these options. However, although the juncos preferred the predictable site after one hour of starvation, after four hours they chose the unpredictable site. These findings may indicate that after one hour of starvation the constant site provides enough food to meet the animal's energy requirements, whereas the variable site has only a 50% chance of providing more than enough and a 50% chance of not providing enough. Choosing the predictable site therefore maximizes daily survival. After four hours of deprivation, however, the reverse argument applies: the constant site now has no chance of providing enough food, whereas there is at least a 50% chance that the unpredictable site will do so. In response to this interpretation we might ask why 50% of animals feeding at unpredictable sites do not die in nature, as might be expected from the above argument. Perhaps they do. It would seem that long-term experiments, monitoring behaviour of the animals, might answer this question.

14.1.5 Effect on search tactic

The effect of starvation on switching between ambush and active searching modes has been documented for several diverse predators.

Note, however, that it is never certain if an animal is responding to deprivation directly, or to some other indicator of resource density.

In naiads (immatures) of the damselflies *Ischnura verticalis* (Crowley, 1979) and *Anomalagrion hastatum* (Akre and Johnson, 1979) starvation led to a switch from ambushing to searching mode. Both studies used two cladocerans as test prey: *Simocephalus*, which swims in a smooth gliding action and often rests on the substrate, and *Daphnia*, which swims most of the time in a jerky motion. As starvation ensued, the naiads switched from ambush to active searching, and were more likely to capture the relatively sessile *Simocephalus*. Akre and Johnson (1979) suggest that hunger provides relevant information concerning the success of each searching mode, as well as triggering the switch from one mode to another. Crowley (1979) emphasizes that being motionless, as in the ambush mode, is adaptive for an animal that is preyed upon by visual predators such as fish, and that moving under exposed conditions is risky. Thus, only under starvation conditions would naiads be likely to switch from ambush to active searching.

Inoue and Matsura (1983) followed up on Holling's (1966) work on how hunger affects various components of the ambush activities of mantids. They found that *Paratenodera angustipennis* only used the ambush mode as long as prey were abundant. When the available prey decreased in numbers, the mantids moved to another place. When deprived of food for more than two days, fifth instar nymphs stopped ambushing and switched to active searching. In another experiment, Inoue and Matsura (1983) released either starved or fed adult mantids into an enclosed field plot. On average, adult females remained on plants containing food (tethered grasshoppers) for three days, but left plants not containing food. The three days of starvation before females switched to searching is a period that correlates well with time required for food to be dispelled from the digestive tract.

Finally, a deprived state and/or lack of abundant resources may stimulate ungulates to expand their range of food as abundance declines (Owen-Smith and Novellie, 1982; Belovsky, 1984). For example, moose (*Alces alces*) will use the most nutritious top twigs of birch when the plants are readily available, but will consume twigs of larger bite diameter and lower nutritional quality when the plants are scarce (Vivas and Säether, 1987).

14.2 STATUS-DEPENDENT CHANGES IN SEARCHING BEHAVIOUR

The requirements of parental care alter searching patterns or search duration. For example, the amount of time allocated to foraging in wild nonlactating herbivorous ungulates is only 85% of that for lactating females on the same range (Bunnell and Gillingham, 1985). Female pipistrelle bats (*Pipistrellus pipistrellus*) decreased foraging distances travelled during periods when lactating young (Racey and Swift, 1985).

Animals commonly switch from one complex kind of behaviour to another according to major shifts in their requirements. A good example is switching from search directed toward finding hosts to search directed toward finding oviposition sites. Females of the mosquito (*Aedes aegypti*) have two physiological mechanisms controlling the decision whether or not to approach a potential host (Klowden and Lea, 1979). Host-seeking behaviour is terminated by the distension-induced inhibition, which is mediated by abdominal stretch receptors when the abdomen is distended with blood. As yolk deposition proceeds, a second mechanism, ovary-induced inhibition, takes over at 30 hours when digestion, absorption, and excretion of the blood meal have reduced abdominal distension. In contrast to distention-induced inhibition involving neural feedback, ovarian inhibition terminates host seeking humorally; the release of the humoral host-seeking inhibitor, which is probably ecdysone, is dependent upon a factor released by the ovaries during the first 10–12 hours of yolk deposition. Ovarian-induced inhibition is apparently terminated when ovarioles no longer contain mature eggs. The proximate mechanism of the ovarian-induced inhibitor may be to inactivate peripheral sensory receptors rather than to act at the level of the central nervous system, since hemolymph transfused from gravid females reduces the sensitivity of antennal lactic acid receptors of unfed recipients (Davis and Takahashi, 1980).

A few studies demonstrate how parental care may influence searching behaviour of adult animals. For example, Carlson and Moreno (1981) compared the collecting times and loads of male and female breeding wheatears (*Oenanthe oenanthe*) early and late in the nesting cycle. For the most part the adults took significantly more prey and spent significantly less time per trip during the second period than during the first. Thus, the birds seemed to load faster when the nestlings were older than at the beginning of the study period. They also loaded more prey and spent significantly more time in distant than in the nearby patches. It would appear that the need to supply young with greater amounts of food drove the adults to increased foraging rates. Lack of male help is brought about in song sparrows (*Melospiza melodia*) when a male has two females to cope with and cannot serve either adequately, or when a female's mate dies and she has full responsibility of foraging for nestlings (Zach and Smith, 1981); under these conditions females took twice as many foraging

Table 14.3 Foraging trips of female song sparrows (*Melospiza melodia*) (from Zach and Smith, 1981)

	Feeding trips/young/h (mean ± SD)			
	n	*Early broods*	*n*	*Late broods*
Male help	(16)	262 ± 0.53	(15)	1.57 ± 0.55
No male help	(7)	4.24 ± 2.0	(3)	3.34 ± 0.20

trips as when a male was available to help (Table 14.3). One conclusion we can draw from these studies is that neither male nor female passerine birds maximize their foraging rate unless circumstances require them to do so. Females can be much more efficient, as shown by their activity when males are absent, and both males and females can be more efficient when food requirements of their young are demanding. One can only wonder about Cowie's (1977) great tits, whose performances were thought to be optimal.

Sex-related differences in time allocated to foraging have been observed in wild (herbivorous) ungulates. During the breeding season adult males frequently allocate less time to foraging than do females. Adult Asian elephants (*Elephas maximus*) normally spend about 76% of the day foraging, but when accompanied by a female in oestrus that value declines to 46% (Eisenberg *et al.*, 1971); in Soay sheep rams (*Ovis aries*), the percentage of time spent foraging declines from 68% before the rut to 28% during rutting (Grubb and Jewell, 1974). After the breeding season males of many such species are depleted nutritionally and they spend a great deal of their time foraging to recoup their body weight losses.

Dominant/subordinant status, presumably reflecting the internal state, can also affect how an animal searches. Hodapp and Frey (1982) studied the effects of status in dominant and subordinate pairs of juvenile firemouth cichlids (*Cichlasoma meeki*). In their experiments, three kinds of patches of Tubifex worms were placed in two different types of conditions: (1) food-poor with 2, 10, or 25 worms in low, medium, and high density, and (2) food-rich with 4, 20, or 50 worms in low, medium and high density. The results show that the dominant fish fed mainly in the best patch initially. When food was abundant, the subordinate fed predominantly in the second-best patch, minimizing costs associated with agonistic activity, but obtaining 33% fewer worms than the dominant. Both dominant and subordinate fish increased their sampling after the first few minutes, but the dominant did this more than the subordinate, regardless of whether the habitat was rich or poor. Overall, sampling was generally higher in the poor than in the rich habitat.

14.3 SUMMARY AND CONCLUSIONS

Deprivation may affect searching behaviour by: increasing reactive distance, increasing search duration, reducing specificity of resource preference, altering the motor pattern before or after an animal arrives in a resource patch, stimulating a predator to move its ambush site if one site turns out to be unprofitable, or increasing the probability of a response to a resource stimulus. Some components of searching behaviour are not affected by hunger: predators walk stealthfully regardless of their hunger level.

Depriving an animal of a given resource may lead it to

1. accept a previously unacceptable resource,
2. decrease latency of acceptance, or
3. decrease the number of 'errors' during selection.

Hunger increases the chances that an animal will gamble by choosing a variable patch that has at least some chance of providing more than enough food for survival, as compared to a predictable patch that has no chance of providing enough food.

Other influences of the internal state include behavioural changes associated with breeding and reproduction: during the breeding season adult male animals frequently allocate less time to foraging than do females, and parental care may alter searching patterns or search duration. Dominant/subordinant status, mediated through the internal state, can also affect how an animal searches, with dominant individuals feeding at the best patches.

15
Genetic factors

An evolutionary response to natural selection depends on phenotypic variation within a population, a genetic component to that variability, and selection according to phenotype (Lewontin, 1970). Although few examples dealing with searching behaviour include all three types of information, nearly all of the studies discussed in earlier chapters provide convincing evidence of phenotypic variation in behaviour. Is a significant portion of that variation heritable and therefore subject to selection?

Considering behavioural variability in a population, two kinds of variability may pertain. *Within-individual variability* represents the plasticity that an individual can apply to a given task or set of conditions during its lifetime. Such plasticity in behaviour could be embedded within a single genome as stochastic variability, so that phenotypic expression is determined by changing conditions, maternal or ontogenetic influences, or early adult experience (i.e. the variability discussed in Chapters 13, 14, and 16). *Among-individual variability* is the variation, genetic or otherwise, that leads to differences in searching behaviour of different individuals. The differences among individuals are keys to understanding the potential of a species for locating resources in new or differing spatial and temporal arrangements or to adapt searching tactics or alter preferences as resources change over the long or short term. Although among-individual variation may be genetically based, there is a limit to the amount of variability that can be stored this way. Variability among individuals could also be caused by various combinations of genetic factors and environmental (abiotic, maternal, ontogenetic, experiential) factors.

Heritability, the proportion of phenotypic variation that is genetically determined, tells us the potential response and rate of change of that response for a given trait to natural selection. Selection may act to reduce additive genetic variance in characters closely related to fitness (as searching behaviour is assumed to be), because alleles that are selected will become fixed in the population. Thus, heritabilities for characters closely associated with fitness are typically low

(Falconer, 1981). Several mechanisms, such as mutation, linkage disequilibrium, variable environments, and immigration, have been suggested to account for the persistence of additive genetic variance in characters such as foraging that are important to fitness.

In the following sections, we examine the kinds of phenotypic variability that characterizes search behaviour traits. This variability is shown to be in part genetic and in part environmental.

15.1 VARIABILITY IN SEARCHING TRAITS

The extreme condition contributing to variability is a mutation. No information is available on mutants in nature regarding search behaviour, and so mutations induced through technology are discussed instead. Note that individuals having such a mutation would not usually be found in nature.

Studies of mutant *Drosophila melanogaster* have not as yet provided direct evidence for genetic control of searching behaviour, but they do show that some of the traits influencing searching behaviour can be genetically controlled (review: Hall, 1985). The mutants fall into two categories: those that affect the timing of search and those that affect locomotory patterns and sensory or scanning abilities. Konopka and Benzer (1971) isolated three circadian-rhythm mutants which affect the temporal organization of searching behaviour: per^{0}, in which flies have with no circadian activity rhythm, per^{1}, in which flies have a long period of 28 hours, and per^{s}, in which flies have a short period of 19 hours. Rather than being absent, the rhythms of pero flies are actually shorter, less distinct and noisier than the wild type (Dowse *et al.*, 1987). Genetic mapping suggests that all three mutations are at the same functional gene, and mosaic analysis has traced the functioning of the gene to the head region (Konopka and Benzer, 1971). Recently, the abnormal phenotypes *per*^{0} and *per*$^{-}$ have been 'rescued' by P-element-mediated transformation of the germ line, using cloned fragments of DNA containing this region (e.g. Yu *et al.*, 1987). The *per* gene may either direct the coupling of some oscillator to the overt circadian activity rhythm, or it may direct the coupling of endogenous ultradian oscillators to produce the circadian rhythm.

Several motor and sensory mutations reveal genes involved in information gathering or generating locomotory patterns. For example, courtship mutations that influence wing vibration, following behaviour, and copulatory attempts (Hall, 1979), may also affect searching locomotory patterns. Sensory mutations that affect thresholds for smell or taste (Tanimura *et al.*, 1982) would be expected to affect resource assessment. In fact, a single mutation in an herbivorous insect could conceivably change it from monophagous to polyphagous and vice versa, since surgical removal of certain gustatory receptors can transform a lepidopterous larva from a specialist into a generalist (e.g. Dethier, 1953). Memory mutants (Dudai *et al.*, 1976; Tempel *et al.*, 1983) indicate that perhaps sliding memory windows used in foraging behaviour have a genetic basis. There are, in addition, many pleiotropic effects, where genes for morphological or physiological traits also affect behaviour. For example, eye-colour mutations in the honeybee also affect their dance communication and orientation abilities (Kuz′mina, 1977).

15.1.1 Resource preference/selection

More work has actually been accomplished on genetic aspects of resource preference and selection than on searching itself. For example, Arnold (1981) found geographic variance in the slug- feeding tendency among populations of garter snakes (*Thamnophis elegans*).

Singer (1983) provides one of the few examples in which the effects of genetic variation on host preference can be localized to particular components of host selectivity by individuals. Populations of the checkerspot butterfly (*Euphydrayas editha*) differ markedly in host use, although most populations are highly monophagous. Even though the same or similar set of host species is available, the primary host species may differ among populations (White and Singer, 1974). Singer has shown that much of this difference in host use is due to differences in the postalighting response to host plants. In addition, differences in specificity exist among populations exhibiting the same rank-ordering of host species. In other words, in some populations the second most preferred species becomes acceptable soon after the first, whereas in other populations the intervening discrimination phase is much longer. In a single population there are individuals with distinct preferences (monophagous) and individuals with nearly no preference at all (polyphagous). Since the butterflies used in Singer's experiments were reared in a common environment, it seems likely that the differences in ranking and the differences in specificity among populations are due to underlying genetic differences.

In some examples, where genetically-based host preferences would be predicted, the predictions do not hold. For example, Tabashnik (1980, 1983) studied Colorado populations of *Colias philodice* eriphyle from agricultural fields ('pest populations') and from nonagricultural fields ('nonpest populations'), and found that pest larvae that were fed on alfalfa had higher survivorships and shorter development times than did nonpest larvae fed on alfalfa (Figure 15.1). There also were

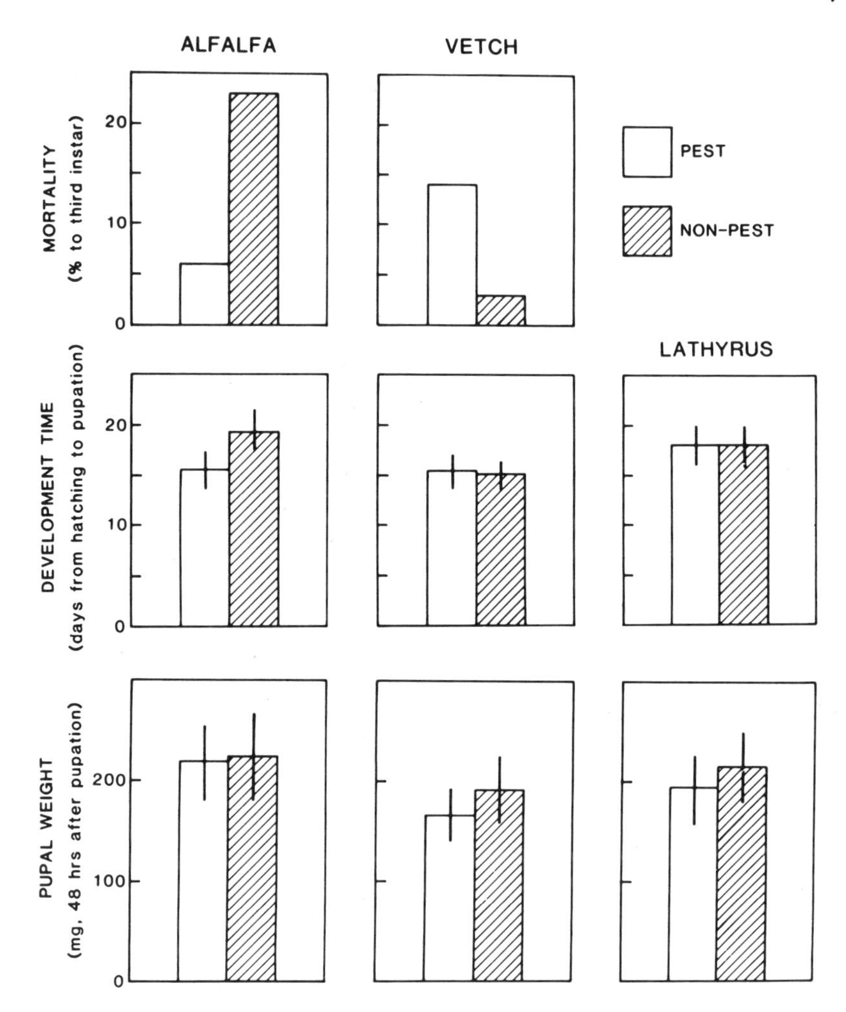

Figure 15.1 Mortality, development time, and pupal weight of pest and nonpest butterfly larvae (*Colias philodice eriphyle*) fed on alfalfa or native hosts. Vertical bars are standard deviations. From Tabashnik, 1983. Copyright B. Tabashnik, reprinted by permission.

significant interactions (host plant × population) in both survivorship and development time, indicating that each population grew best on its 'own' host. Tabashnik's evidence suggests that each population has become adapted to its own host and the observed phenotypic differences between pest and nonpest larvae are genetically based. However, when females were given an opportunity to oviposit upon the host plant on which their larvae have the best survival chances, they failed to do so (Tabashnik, 1983). The lack of a positive phenotypic correlation between larval growth on alfalfa and oviposition preference for alfalfa versus vetch suggests that larval growth and oviposition preference are controlled by different sets of genes.

Tabashnik offers the following explanations for why *C. p. eriphyle* females may not oviposit on the plants for which their larvae are best suited:

1. Host suitability may be affected by ecological factors such as host phenology, parasitism, or predation, so that the most nutritional host may be unsuitable in the field due to ecological factors that might not be evident in laboratory feeding tests.
2. Host-finding and acceptance behaviour may be influenced by foraging constraints not directly related to host suitability. For example, suboptimal hosts may be accepted because habitats containing suitable hosts are avoided. If time available for searching is a limiting factor, suboptimal hosts that are abundant may be used more frequently than more suitable hosts that are scarce. Host choice behaviour also may be affected by energy requirements, such that oviposition choice in butterflies might be influenced by the proximity of larval host plants to nectar sources.
3. Ovipositing females may not be capable of discriminating between plants that differ in suitability for larval success, but are similar in most other respects. This explanation seems to apply to Tabashnik's *Colias* example.
4. Poor correspondence between host preference and host suitability may occur when an herbivore does not have sufficient time to evolve appropriate responses to changes in host quality or availability. For instance, Chew (1977) suggests that crucifer-feeding *Pieris* butterflies lay eggs on a crucifer that is toxic to larvae because this plant was introduced recently and there has not been enough time for discrimination against it to evolve.

Contrasting Singer's *Euphydrayas* study and Tabashnik's *Colias* study, it seems clear that the diversification of populations depends critically on the existence of genetic variation for host preference among females. Physiological adaptations allowing animals to exploit habitats (as in *Colias* on alfalfa) are important, but have no consequence unless females oviposit in areas that would correlate with physiological adaptations of their larvae. Genetic variation in host preference also may allow species to evolve ecological differences that make coexistence possible, and it may facilitate the tracking of temporal and spatial variation in abundance of different resources by a single species. Tabashnik (1986) suggests that the magnitude of genetic variation for host preference within and among populations should have some effect on possible modes of speciation based on use of alternative hosts.

Another approach to the genetics of foraging behaviour has been to select for lines with particular resource preferences. For example, Wallin (1988) performed artificial selection for food choice in *D. melanogaster*. Significant responses were found in a line selected for high energy food and another selected for high protein food. Diet choices paralleled increases in fitness measures, such as larval weight, pupation time, and adult size.

15.1.2 Dispersal and migration tendency

Genetic variability has been revealed in various components of dispersal and migration tendency. An obviously important trait is locomotory ability. For example, several morphs of aphids differ in their ability to fly (Lees, 1966; Hardie, 1980), and gregarious and solitary morphs differ behaviourally and morphologically among migratory locusts (Kennedy, 1975).

Clones of cereal aphids (*Sitobion avenae*) differ in tendency to move between wheat plants (Lowe, 1981, 1984). Aphids from a clone labeled CASS were more mobile than aphids from other clones, and were best able to locate and colonize other plants after dispersal. They also differed from some of the other clones in their greater tendency to colonize the more susceptible host-plant genotypes (Lowe, 1981; Watt and Dixon, 1981). Their tendency to leave established colonies and to colonize adjacent plants could be adaptive in a monoculture cereal crop, leading to early dispersal and possibly greater ultimate exploitation of the habitat. Clones that tend to form more enduring groupings, but colonize neighbouring plants less often and produce more winged offspring, might be better adapted to a grassland environment with a patchy distribution of suitable hosts.

Population size thresholds responsible for bursts of migration differ among strains of *D. melanogaster* (Sakai *et al.*, 1957). Also in *D. melanogaster*, Narise (1962) was able to select for higher migratory activity, and McInnis and Schaffer (1984) selected for dispersal rate in the field. Genetically-determined differences in migratory tendency also exist among Colorado, Arizona and New Mexico populations of the grasshopper, *Melanoplus sanguinipes* (McAnelly, 1985). Dingle (1968a) applied intense selection to milkweed bugs (*Oncopeltus fasciatus*) for long flight duration of tethered animals, and increased the percentage of bugs flying more than 30 minutes from 25 to 60% in one generation. This trait presumably correlates with migration tendency or migratory persistence. Caldwell and Hegmann (1969) calculated heritability estimates for duration of tethered flight in another milkweed bug (*Lygaeus kalmii*) and found a significant genetic component to the trait.

Migration in birds also has a genetic component (see Chapter 13.4.2). Individual cormorants (*Phalacocorax carbo*) in the eastern north Atlantic differ in direction of migration and in tendency to fly over large bodies of water and/or land masses (Coulson and Brazendale, 1968). Similarly, Berthold and Querner (1981) crossed African and German blackcaps (*Sylvia atricapilla*) and found that the timing of migration restlessness of the hybrids was intermediate between their parents. Subsequently, Berthold (1988) selected for either migratory or nonmigratory tendency, and found that within three or four generations a population of partial migrant captive blackcaps could be channelled into one line of highly migratory individuals and another line of totally nonmigratory individuals. The rapid change in response to selection suggests that migratory behaviour is highly heritable. Calculation of the realized heritability (response to selection divided by the selection differential) gave a rather high value of 0.6.

15.1.3 Searching patterns

Several foraging traits in *D. melanogaster* larvae have been shown to be heritable, including mechanisms of feeding and locomotion, and pupation site preference (reviews: Sokolowski, 1985, 1986). Various types of adult orientation (review: Heisenberg, 1980), components of their local search (Nagle and Bell, 1987), olfactory perception (Kikuchi, 1973), and olfactory preferences (Fuyama, 1978), are also heritable. Capacity for conditioning behaviour has been shown to be genetically determined in blowflies (McGuire and Hirsch, 1977). The two most relevant examples for which substantial information is available are *rover/sitter* traits in *D. melanogaster* and *mover/stayer* traits in species of charr (*Salvelinus*). Two foraging types, *movers* and *stayers*, have been demonstrated within and among species of lake and brook charr. Lake charr (*Salvelinus namaycush*) actively forage on relatively dispersed food (*movers*), whereas brook charr (*S. fontinalis*) wait for food to drift by and thus spend more time stationary (*stayers*) (Ferguson *et al.*, 1983). Ferguson and Noakes (1983) examined the characteristics of F_1 and backcross hybrids for the two species of charr. The hybrids showed a closer affinity to brook charr for mobility measures in that they spent more time stationary than did lake charr, suggesting directional dominance. Grant and Noakes (1987) identified *movers* and *stayers* among individuals in populations of young-of-the-year *S. fontinalis*: *movers* swim slowly while foraging and they are rare in currents, whereas *stayers* remain stationary and swim only to intercept food items flowing by in the current.

Phenotypic variation in the searching tactic of *D. melanogaster* was first discerned in larvae by differences in the area traversed while feeding on a yeast slurry (review: Sokolowski, 1986). The two searching morphs, *rover* and *sitter*, were distinguished using an assay for larval trail-length: *rover* larvae have longer foraging trails and traverse a larger area than do *sitter* larvae. *Rover* larvae also dig deeper into and pupate higher above the feeding substrate than do sitters. Crosses between lines with extreme phenotypes suggest a relatively simple genetic system, with *rover* phenotype completely dominant over the *sitter*, and no significant sex-linked or maternal effects. A chromosomal analysis demonstrated that differences between *rover* and *sitter* traits can be attributed to genes on the second pair of chromosomes.

Phenotypic variation in search behaviour of adult *D. melanogaster* was observed in a study of wild-type flies initiated from field trapping, and in colonies of Sokolowski's (SOK) larval *rover/sitter* lines (Nagle and Bell, 1987). The bioassay indirectly measured search duration of adult flies by quantifying displacement from an ingested sucrose drop after 30 s. Nagle and Bell (1987) selected for adult *rover* and *sitter* traits, and were able to shift displacement scores of *sitters* and *rovers* significantly from the control line (Figure 15.2a). By generation 25 the median displacement of *rovers* was 37 cm, and *sitter* displacement was less than 2 cm (Figure 15.2c). Progeny of crosses based on the adult trait indicate a complete lack of dominance (F1 hybrids are intermediate between *rover* and *sitter* lines) (Bell and Nagle, 1987). This difference is corroborated by the continuous distributions observed in displacement scores of populations of adults trapped in the field, as compared to the discontinuous distribution in larvae. Thus, different sets of genes seem to distinguish searching traits in larvae and adults.

Neither locomotory rate nor turn bias differed significantly among unfed adult *rover*, *sitter*, and control line flies that were allowed to move across an experimental arena. Thus, the *rover/sitter* genetic influence has no substantial effect on the orientation of unfed flies. The traits can only be expressed after a fly ingests sucrose. There is a transition, abrupt in *rovers* and gradual in *sitters*, in locomotory rate from the low values immediately following feeding on a sucrose drop toward the higher prefeeding levels. Turning rates of *sitters* are significantly higher than *rovers* after feeding, and remain higher for a longer period of time during local search in *sitters* (Figure 15.3). The genetic effect would therefore seem to be exerted on (1) the extent to which locomotory rate is decreased and turning rate is increased after

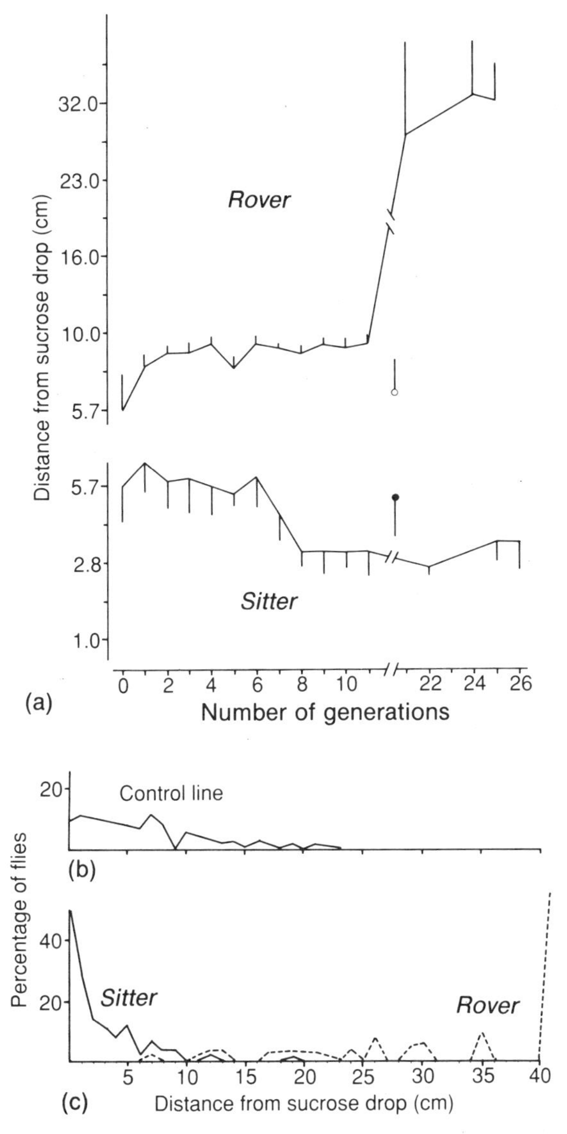

Figure 15.2 (a) Selection for change in displacement scores in *rover* and *sitter* lines of *Drosophila melanogaster* over 26 generations. (b) Control line flies after 25 generations. (c) Frequency distribution of displacement values for *rover* and *sitter* flies tested after 25 generations. (After Nagle and Bell, 1987.)

ingestion, and (2) the rate of return to the locomotory and turning rates of the prefeeding condition. Turn bias, which is higher during the first 5 s after feeding, accounts for the tight looping orientation of *sitters*, keeping the fly near the sucrose drop, as compared to *rovers* which walk relatively straight away from the drop. In addition to locomotory and turning traits, *rovers* tend not to stop for lengthy periods, whereas *sitters* spend approximately 50% of their time not moving; the control line is intermediate between the two extreme phenotypes (Figure 15.3b). The differences between *rover* and *sitter* search displacement seem to be dependent upon the additive effects of slow movement and periodic stopping in *sitters*, and on continuous and rapid movement in *rovers*.

Table 15.1 Heritability estimates for components of search behaviour in houseflies* (from Collins *et al.*, in press).

Search component	*Heritability* (h^2)	*(± SE)*
Locomotary rate (mm/s)	−0.416 ± 0.118	$P < 0.05$
Turning rate (°/s)	0.097 ± 0.039	$P < 0.01$
Meander (°/mm)	0.022 ± 0.037	NS†
Pivoting (°/mm)	0.129 ± 0.038	$P < 0.01$
Path length (mm)	0.091 ± 0.038	$P < 0.01$
Stop duration (s)	−0.063 ± 0.108	NS
Movement index	0.096 ± 0.115	NS

*Males and females combined.
†NS: not significant.

Heritabilities for various behavioural components of local search were estimated by parent/offspring regression in houseflies (*Musca domestica*) (Table 15.1) (Collins *et al.*, in press). For each of 136 families examined, the mean phenotypic value for the parents was compared to the mean value for their offspring for each locomotory parameter. An average of 5.75 offspring were assayed per family. Of the search components that were measured in houseflies, those related to locomotion (path length, which reflects locomotory rate, turning rate, pivoting and distance from origin) are heritable, whereas heritabilities for other components such as frequency or duration of stopping and direction taken are not significant. Thus, different aspects of local search appear to be inherited independently. This kind of system, where various search parameters are inherited independently may allow more flexibility in the response to varying selection pressures, which will ensure that a population is able to respond appropriately to shifting selection pressures.

In a related study on blowflies (*Phormia regina*), McGuire and Tully (1986) showed that high central excitatory state (CES), which is characterized by an increase in proboscis extensions to water after a fly has been stimulated with sucrose, is associated with longer search

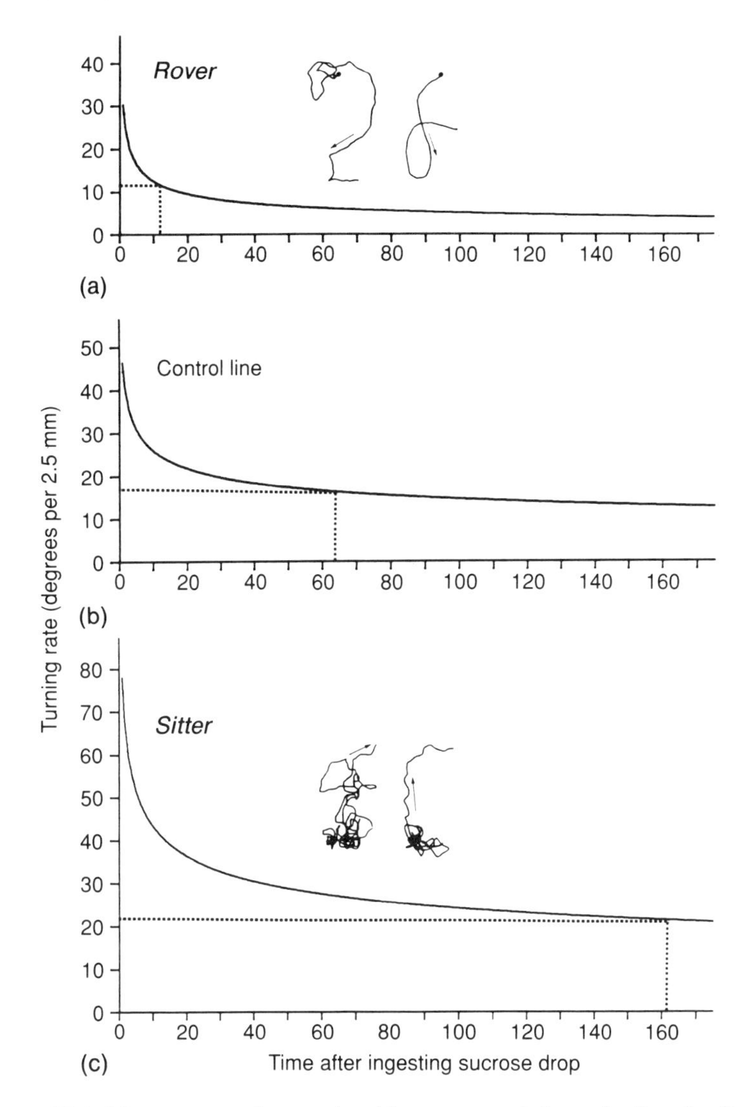

Figure 15.3 Mean rates of meander (degrees per 2.5 mm) after feeding on sucrose measured in (a) *rover*, (b) control, and (c) *sitter* lines of *Drosophila melanogaster*. Horizontal dashed lines represent the pre-feeding turning rate value. Insets are tracings of typical *rover* and *sitter* pathways. (After Nagle and Bell 1987.)

duration than in the low-CES condition. However, this phenotypic correlation disappeared in the F_2 CES hybrid, indicating that the observed variations in CES and search duration were not controlled by the same set of genes. Further, biometrical analyses showed that several linked, autosomal genes with digenic epistatic interactions and a complex pattern of maternal inheritance were responsible for the differences in search duration between the high- and low-CES lines. These results indicate that the *rover/sitter* trait may be commonly found in various species of flies, and perhaps in other animals as well, but that it may be controlled by vastly different genetic systems.

Sokolowski (1985) investigated the ecological correlates of the larval *rover/sitter* traits. Larvae of stocks originating from individuals collected from 'on-fruit' pupal sites (M1) or from 'in the soil' pupal sites (M4) showed different preferences for pupal microhabitat. This pattern was consistent with that expected from the type of microhabitat from which they were collected at an orchard site: larvae of the M1 stock had a greater tendency to pupate on fruit, and had shorter path lengths and lower pupal heights than larvae of the M4 stock. M1 larvae tended to pupate more on fruit than did M4, even when the soil water content was varied. Varying the soil water content had two important effects relevant to habitat selection by the M1 and M4 stocks:

1. when the soil water content was increased, the proportion of larvae in both stocks that pupated on the fruit decreased; and
2. the M1 larvae still tended to pupate more on the fruit than did M4 in all soil water content conditions.

A third effect of varying the soil water content was that the relative suitability of the two microhabitats for pupation changed. At low levels of soil water pupal survivorship is better on the fruit, whereas at high levels survivorship is better in the soil. Thus, there is a reversal in which the microhabitat is a better site for pupation as the habitat changes from dry to wet.

While the adaptive significance of larval *rover/sitter* traits seem to be clearcut, the ecological correlates of *rover/sitter* traits in adults are less clear. We know that adult *D. melanogaster* fly between resource habitats, and that their foraging strategy is influenced by allelochemicals (e.g. Flugge, 1934) and visual stimuli (review: Götz, 1980). Once a resource such as decaying fruit has been localized, however, search continues by walking. Tortorici and Bell (1988) tested adult *rover*, *sitter*, and control lines of *D. melanogaster* in single resource patches varying in sucrose drop density and in a multiple-patch array (Figure 15.4) in order to determine how resource dispersion and genetic differences in searching tactics influence searching success. The results indicate that the adult *rover* phenotype is adaptive for finding resources that are not spaced close together, or for walking away

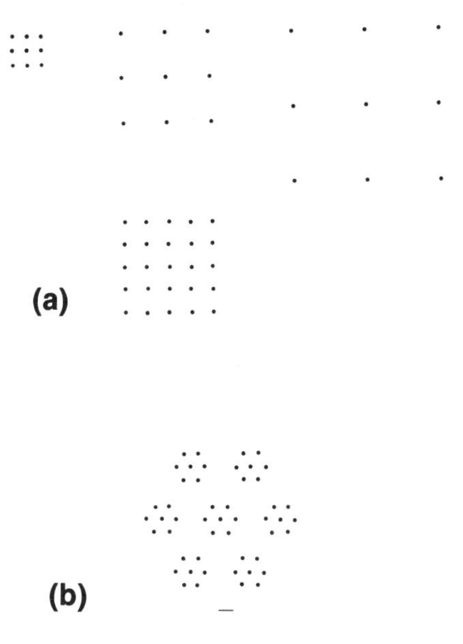

Figure 15.4 Diagrams of spatial patterns of sucrose drops used in *Drosophila melanogaster* searching experiments. (a) Single patches, (b) multiple-patch array. Distance between drops, 1 cm. (After Tortorici and Bell, 1987.)

from a resource rapidly and directly, instead of remaining near it. The adult *sitter* phenotype is more adaptive for locating closely-spaced resources, even at the expense of remaining at a feeding site for lengthy periods of time.

15.1.4 Habitat preference/selection

Mark/release/recapture experiments showed that *D. persimilis* and *D. pseudoobscura* tend to return to their area of origin or to ecologically similar areas when transplanted (Taylor and Powell, 1978). However, when flies preferring two different habitats were crossed, the F_1 progeny showed no significant habitat choice, indicating a lack of a strong genetic component for habitat choice. Alternative explanations include the possibility that habitat conditions may have changed during the nearly three weeks between the time that adults were collected and the time when the F_1 flies were released. Another possibility is that recombination and Mendelian assortment may have obscured the genetic differences which did exist in the parental generation; this implies that the genetic variation predisposing flies to behave differently is largely non-additive. Perhaps some of the habitat preferences are age-dependent, and perhaps the laboratory-reared F_1s lacked sufficient experience or the expertise to sort themselves as did the parents.

In other experiments, habitat preference seems to have a genetic component. For example, Jaenike (1985) observed that female *D. melanogaster* raised in the laboratory on tomatoes or mushrooms tended to be captured on their natal food. In addition, the different strains showed differential association with the two resources independent of experience. This strain difference and the observation that F_2 progeny from these strains had nearly intermediate association suggest a genetic basis for the behaviour involved. Jaenike (1986a,b) has also shown that oviposition behaviour and habitat fidelity appear to be determined by different genes.

A good demonstration of how environmental heterogeneity can maintain a polymorphism dependent on habitat selection is an experiment conducted by Jones and Probert (1980) using white-eye (*w*) mutants of *D. simulans*. In a uniform habitat of either normal or dim red light, the *w* mutant is lost; in 30 weeks its frequency changed from 0.5 to 0.01 in normal light and 0.5 to 0.06 in dim red light. However, in a cage divided so that half of the population of flies had normal light and half had dim red light, the frequency of *w* after 30 weeks was 0.32. Moreover, the frequency of *w* was much higher in the dim red light sector. Because in the control experiments *w* was always lost, these results indicate that heterogeneity in the environment is necessary for the maintenance of the polymorphism. The apparent basis for the polymorphism is habitat preference: wild-type flies are positively phototactic, whereas the white-eyed flies prefer (or perhaps are neutral to) the dim red light.

Do genotypes actually choose the habitat in which they have the highest fitness? Jones (1982) developed a technique using a paint that fades upon exposure to the sun, to determine the proportion of time snails, *Cepaea nemoralis*, spend in sunlight. An equal number of marked yellow unbanded snails and yellow banded snails in each of 10 natural populations were released into field cages, and 60 days later recaptured and measured for extent of fading. In all of the populations the banded individuals appeared to spend more time in the sun, most probably because the banded snails have lower tolerance to heat. Since lower positions on plants near the soil are hotter, the banded snails spent more time higher on the plants and were more exposed to the sun. Because in cages these snails tend to spend much of the day buried, the differences between banded and unbanded snails may be due principally to their time of activity (J.S. Jones, unpublished, in Hedrick, 1986). A possible interpretation is that the snails move until

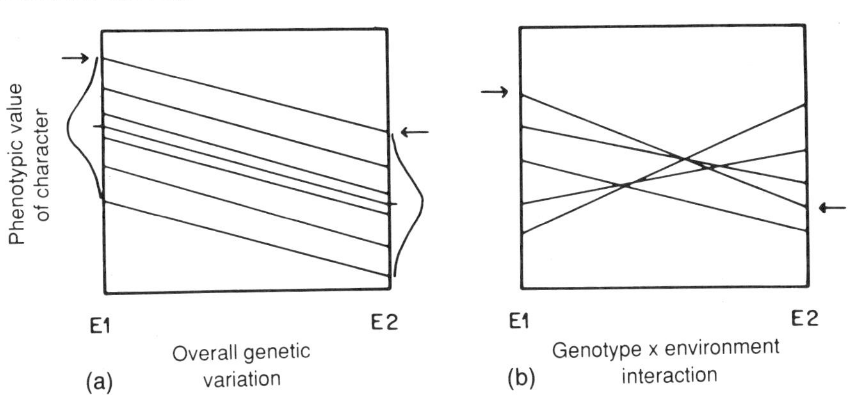

Figure 15.5 Graphical representation of the difference between overall genetic variation and genotype–environment interaction. The mean character state for a given family in environment 1 (E1) is plotted on the left axis, the mean in environment 2 (E2) is plotted on the right axis, and the two family means are connected by a line. The slopes of the lines denote the phenotypic response of each family to a change in the environment. Within environment variation among families is expected to be normally distributed. (a) Parallel lines indicate that all families respond similarly to a change in the environment. No changes in genotypic rank occur. (b) The variation among families within environments is the same as in (a), but crossing of the lines indicates that families vary in their responses to the environment. Arrows marking certain families show that changes in rank can occur. From Via, 1986. Copyright Plenum Press, reprinted by permission.

they find their optimum temperature, and then remain in this location. This is corroborated by observations that the behaviour of snails can be changed by painting the shells (J.S. Jones unpublished, in Hedrick, 1986).

15.2 INTERACTIONS BETWEEN INTERNAL AND EXTERNAL ENVIRONMENT AND GENES

The environmental noise of natural conditions could so strongly affect individuals that genetically-based behavioural variation among them may actually be negligible (Papaj and Rausher, 1983; Jaenike, 1985). For example, the genotype of an individual *D. tripunctata* can have a strong effect on its food preference in nature, but previous experience can modify that preference (Jaenike, 1985), and although smart- and dull-maze strains of rats have been observed, their performances are markedly affected by the characteristics of the test environment (Cooper and Zubek, 1958).

15.2.1 Environmental conditions

When the mean for a given behavioural character for one family in environment 1 is plotted on the left axis and the mean in environment 2 is plotted on the right axis (Figure 15.5), then the slope of the line connecting each pair of means illustrates the type of responses of families to a change in the environment (Via and Lande, 1985; Via, 1986). In Figure 15.5a, the parallel lines indicate that all families experience change in the environment in a similar way. The fact that all the phenotypes of all families can be modified (and thus differ substantially in average character value over environments) attests to the presence of a strong environmental component to the overall phenotypes. In contrast, the families in Figure 15.5b vary in their responses to the environment, suggesting a genotype–environment interaction. The change in rank of genotypes in the different environments indicates a differential response among families to the change in environment.

(a) Previous experience

In a study of the vegetable leafminer (*Liriomyza sativae*), an analysis of variance for the effect of larval experience (reared on either tomato or cowpea) on adult feeding preferences of individuals from full sib families (Table 15.2) illustrates no significant effect of either larval rearing plant (Rplant) or family on the percentage of adults feeding on cowpea (Via, 1986). In other words, when averaged over all genotypes, larval experience had no effect on adult feeding preference, nor was there significant variation among genotypes when averaged over both larval rearing plants. However, the presence of a significant interaction term (Rplant × family) suggests that responses to the larval host plant do occur but that they are genotype-specific.

Table 15.2 Analysis of variance (ANOVA) on feeding preference (transformed as arcsin of square root of percentage of individuals feeding on pea) (from Via, 1986). Rplant = larval rearing plant; Family = full sib family.

Source	*Df*	*SS*	*F*	$P > F$	R^2
Model	40	9.73	1.67	0.018	0.37
Rplant	1	0.00	0.01	>0.90	
Family	23	6.07	1.05	>0.50	
Rplant × family	16	4.03	1.73	0.50	
Error	115	16.78			
Total	155	26.51			

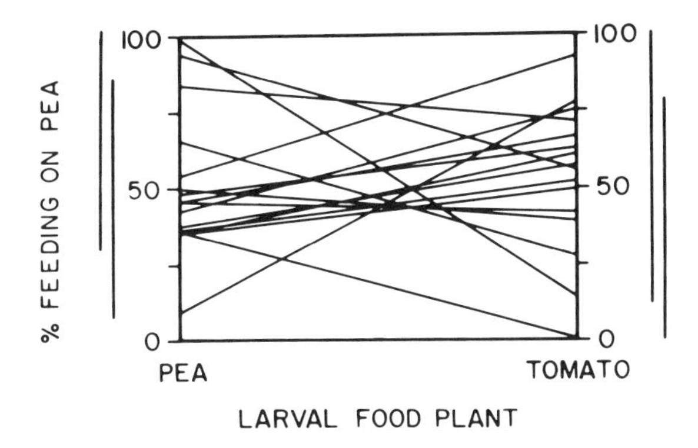

Figure 15.6 Variation in the proportion of feeding of individual leafminers (*Liriomyza sativae*) which occurred on cowpea for siblings reared on either cowpea or tomato. Means for each full sib family are plotted on the vertical axes and connected by lines. Genotype-environment interactions are indicated when lines are not parallel. Bars beside each axis are the homogeneous subsets. From Via, 1986. Copyright Plenum Press, reprinted by permission.

In other words, families vary in their susceptibility to larval induction of adult preference.

Genotype–environment interaction in which families change rank in the different environments means that no genotype in the experimental population has the highest value of a given character in all environments (Figure 15.6). For example, the genotypes that fed most on cowpea when reared on cowpea would not be the ones showing the highest preference for cowpea when reared on tomato. This change of rank of genotypes in different environments is what Haldane (1946) called 'crossing' genotype–environment interaction, suggesting a potential for genetic differentiation among populations under directional selection in different environments (Via and Lande, 1985).

(b) *Abiotic conditions*

Grant and Noakes (1987), in their study of young-of-the-year brook charr (*Salvelinus fontinalis*), examined the foraging behaviour of *movers* and *stayers*. They tested the hypothesis that sit-and-wait foragers (*stayers*) specialize on fast-moving prey (drift), whereas active foragers (*movers*) feed more on relatively sedentary prey (benthos). The results showed that *movers* have a higher foraging rate than do *stayers* in zero current, and that up to a point the foraging rate of *stayers* on drifting invertebrates increases with current velocity. *Movers* have a higher benthic foraging rate than *stayers* in running water, and individual fish that had been observed acting as both *movers* and *stayers*, tended to behave as movers most of the time when foraging on benthos. Substrate characteristics also affect the foraging rate of the fish: both *movers* and *stayers* are increasingly more successful in foraging on benthos from silt to sand to gravel, but the effect is much greater in *movers*.

15.2.2 Resource availability

Short-term changes in feeding history of the spider, *Agelenopsis aperta*, including periods of deprivation or satiation, have no effect on the distance at which spiders tolerate neighbouring web owners (Riechert, 1986). However, there is a significant difference in nearest neighbours between desert grassland (webs far apart) and riparian habitats (webs close together). With conditions of low prey availability (in desert), selection for long nearest-neighbour distances provide a more dependable system than would a facultative mechanism. In a communal spider, *Metepeira spinipes*, the mean nearest-neighbour distance increases as prey availability decreases in various habitats in central Mexico (Uetz *et al.*, 1987). If colonies are moved to another (poorer) site, nearest-neighbour distances increase. Although this response system seems to be facultative, there is also a genetic component. When spiderlings from three different habitats were reared under identical conditions in the laboratory, nearest-neighbour distances were still different.

15.2.3 Age

Graf and Sokolowski (1989) found that larvae of *D. melanogaster rover* and *sitter* moved further per unit time as they aged. *Rovers* showed a more dramatic increase in path length from 24 to 96 hours, as compared to smaller changes in the *sitters*. The *rover/sitter* trait is first expressed clearly at 48 hours (second instar); its expression is greatest in third instar larvae (72 and 96 hours post-hatch). At 24 hours post-hatch, food quality significantly affects path length, but strain and the interaction between strain and substrate quality are not significant. By the second and third instar, strain, substrate quality, and their interaction are all significant, indicating that the strains differ in both their path lengths and their response to different quality substrates.

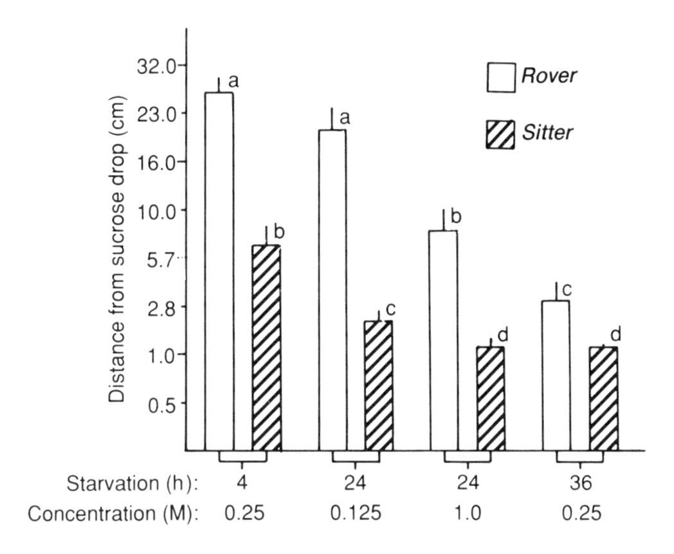

Figure 15.7 Displacement scores of *Drosophila melanogaster* adults that were starved for different periods of time and fed various concentrations of sucrose. Vertical lines are standard errors. Columns labelled with the same letter are not significantly different. (After Bell and Tortorici, 1987.)

15.2.4 Deprivation

Phenotypic plasticity in search duration of adult *rover* and *sitter D. melanogaster* is in part dependent on sensory perception of sucrose concentration and in part on physiological feedback responses to hunger (Bell and Tortorici, 1987). Figure 15.7 shows that search duration increases with period of starvation and with increase in sucrose concentration. The extent to which these variables affect search duration is dependent in part on the genotype of the fly. Under the same conditions of starvation and resource quality, *sitters* always search longer than *rovers* (e.g. pairs of columns in Figure 15.7). However, the search efforts of *rovers* and *sitters* could be equalized by specific combinations of starvation period and sucrose concentration: for example, the responses of 24-hour starved *rovers* fed 1 M sucrose were not significantly different from 4-hour starved *sitters* fed 0.25 M sucrose. Since both *sitters* and *rovers* engaged in at least some local search in all combinations of resource quality and deprivation, the genetic effect appears to be on the duration of search.

Graf and Sokolowski (1989) tested starved and unstarved *D. melanogaster rover* and *sitter* larvae for trail length. As shown in Figure 15.8, starvation significantly decreased path length in both the *rover*

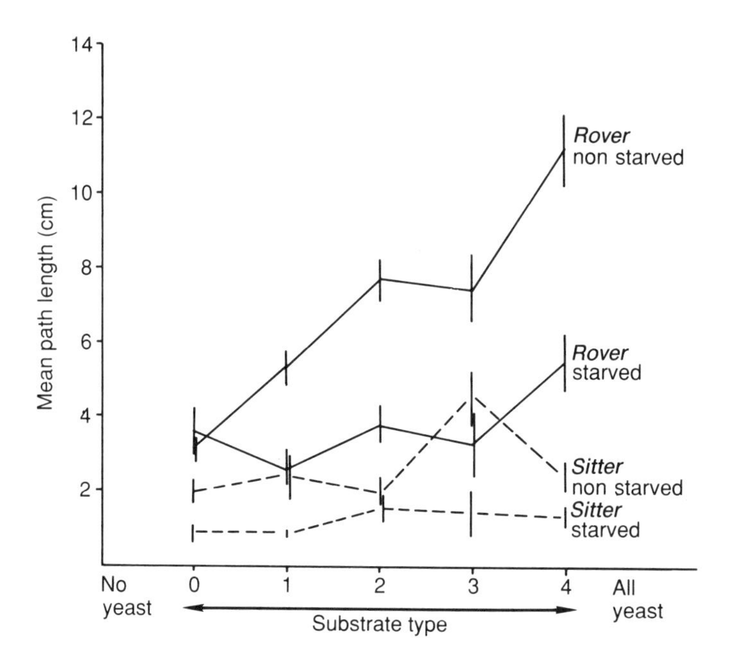

Figure 15.8 Effects of starvation and substrate type on search path displacement in *Drosophila melanogaster* larvae. Vertical lines are 95% confidence intervals. (After Graf and Sokolowski, 1989.)

and *sitter* strains, and secondly, starvation decreased the variability in response to substrates in both strains. The response curve for starved *rover* larvae is flatter than that for unstarved *rover* over the transition from low to high quality medium (Figure 15.8), suggesting a greater affect of medium quality on nonstarved flies. As with the adult phenotype, *rover* and *sitter* larval path lengths were still significantly different, regardless of whether the larvae were starved or unstarved.

The phenotypic variance in search duration of *D. melanogaster* seems to be dependent upon physiological state, as indicated by the influences of internal and external environmental factors. It is also influenced by genetic factors. Considering that the nutritional experiences of every wild fly would be unique, it may be that the genetic variance underlying behavioural characters involving food preference and search behaviour is not subject to severe selection. Selection may act primarily on variance in the behavioural phenotype, which is the product of substantial nongenetic factors as well as genetic factors. Thus, while the adaptive significance of foraging behaviours may be profound, the influence of variable internal and external factors may insulate the various genotypes from direct selection. This allows a population to remain more genetically variable than would be expected, regardless of the severity of selection on the phenotypes. This variation, cryptic in nature but demonstrable in the laboratory, should extend to any other traits subject to physiological control.

15.3 SUMMARY AND CONCLUSIONS

Within-individual variability in search behaviour is the plasticity that an individual can apply to a given task or set of conditions during its lifetime. Among-individual variability is the variation that leads to differences in searching behaviour of individuals.

Studies of mutant *Drosophila melanogaster* show that some traits influencing searching behaviour, such as locomotory patterns, scanning abilities, timing of search, and sensory thresholds, are heritable. In the laboratory, searching traits have been shown to be heritable through artificial selection and by examination of the progeny of hybrid crosses. Among heritable traits related to search behaviour are host-plant selection in insects, components of dispersal and migration tendency, population size thresholds responsible for bursts of migration, digging, pupation site preference, local search, olfactory perception, olfactory preferences, and capacity for conditioning behaviour.

That alternative searching tactics can be heritable is demonstrated by the two foraging types, *mover* and *stayer* charr, in which *stayers* wait for food to drift by and *movers* actively search for prey. The two searching morphs in flies, *rover* and *sitter*, occur in both larval and adult stages. *Rovers* have longer foraging trails and traverse a larger area than do *sitters*, which have shorter trails and area-restricted locomotory patterns.

External and internal environmental factors affect the phenotype, but always within the restrictions of an individual's genotype. Changes in rank of genotypes in the different environments indicate a differential response among genotypes to the changes in environment.

Selection may act primarily on variance in the behavioural phenotype which is the product of substantial nongenetic factors as well as genetic factors. Thus, while the adaptive significance of foraging behaviours may be profound, the influence of variable internal and external factors may insulate the various genotypes from direct selection. This might allow a population to remain more genetically variable than would be expected regardless of the severity of selection on the phenotypes.

16
Ontogenetic and maternal influences

Ontogenetic experiences and maternal influences can affect both search behaviour and search success, and individual adult animals of a given population may differ in their preferences or abilities because of these experiences. I differentiate here, rather arbitrarily, between experiences of animals that are categorized as juveniles or 'young' adults and those that are categorized as 'older' adults. (The importance of learning in older adults is discussed in other chapters.) It should also be noted that it is much more difficult to separate out ontogenetic from maternal effects than the simple categorization used in this chapter.

16.1 MATERNAL EFFECTS

The term maternal effect is defined narrowly in this chapter, referring specifically to female influences prior to birth or hatching. These influences include the differences in investment that may occur among the eggs or embryos of a single female, and differences between females in conversion of environmental resources into offspring. Other kinds of maternal effects, such as parental care and teaching the young to search for resources are discussed in other sections of this chapter.

A phenotypic characteristic often cited as due to maternal effects is body size, which in invertebrates is nearly always directly related to nutritional experiences while immature. Such experiences may be mediated either through vitellogenesis (e.g. Barbosa and Capinera, 1978) or larval provisioning (Alcock *et al.*, 1977). Other possible proximate causes of size variation in insect populations are temperature gradients, moisture or humidity, larval diet, and temporal fluctuation of resources and population density. Body size can have an important impact on searching behaviour.

Male dung flies (*Scatophaga stercoraria*) that receive poor food supplies as larvae, develop into small adults with limited fighting ability, and they search for females on the grass near a cow pad, whereas larger males fight for females on the cow pad itself (Parker,

1974). Borgia (1980, 1981, 1982) has shown that small males obtain more matings at high pat density and low fly density than at low pat and high fly density. Thus, an interaction between maternally-mediated male size and two external determinants – fly density and cow pat density – are also important factors in determining a male's behaviour. Large males are still more successful than are small males at all densities, largely because small males suffer proportionately more harm in interactions than do larger males, and success of small males in retaining captured females is very low, as shown by extreme differences in takeover rates relative to male size.

In a large anthophorid bee (*Centris pallida*) larger males patrol emergence sites, flying over emerging females, whereas smaller males tend to be hoverers, searching less profitable areas to locate females that escaped patrollers (Alcock *et al.*, 1977). The size dimorphism among males is probably the result of differential maternal provisioning of cells. Alcock *et al.* (1977) offer the following theory to explain how such differential investment could maintain both strategies in a population. First, under normal conditions females producing patrollers would be favoured because it is unlikely that a female could produce enough small males to offset the cumulative reproductive success of a brood of patrollers. However, if there were no counterbalancing selection, the population would soon become entirely patrollers and the only hoverers would be the unsuccessful large male patrollers making the best of a bad situation. It is possible, therefore, that high parasite densities periodically occur, and under such conditions females investing in small males would be favoured. This is because small males, tending not to search at nest sites, would be less likely to bring the parasites back into close association with the brood. Thus the periodic parasitic infections that favour investment in small males could maintain the two kinds of behaviours in a stochastic equilibrium. It would always be possible, however, for parasites to home in on females returning to their nests, since they must return regardless of the size of the males they produce. Rubenstein (1980) offered an alternative explanation, that the dimorphism could also be maintained as long as the 'expected value of each type of male decreased as its frequency increased in the population'; the dimorphism could then be maintained by 'females adjusting the relative frequency of each morph in her brood so as to maximize the expected value of the brood'.

Female gypsy moths (*Lymantria dispar*) raised on leaves of maple trees produce smaller and fewer eggs than those raised on oak (Barbosa *et al.*, 1981), suggestive of maternally regulated effects of host type on fecundity. Large early-instar larvae are more likely to disperse repeatedly in laboratory assays than are small larvae, indicating that the maternal effect influences later behaviour in a significant way (Barbosa *et al.*, 1981). Maternally-influenced differences in host-finding behaviour have been reported in larvae of two other lepidopterous species, the western tent caterpillar (*Malacosoma pluviale*) and the forest tent caterpillar (*Malacasoma disstria*). The caterpillars differed in their locomotory and orientational tendencies (e.g. Wellington, 1965; Greenblatt and Witter, 1976). Testing the effects of two populations of eastern cottonwood (*Populus deltoides*) on characteristics of aphids (*Pemphigus populitranversus*) raised on the trees, Bingham and Sokal (1986) found host-tree effects on the morphology of galls, stem mothers and the alate progeny of of stem mothers. Interestingly, one of two aphid morphs was more sensitive to the differences in the trees than was the other morph.

16.2 JUVENILE BEHAVIOUR

The comparative approach of this book dictates that juveniles must be defined simply as prereproductive, because there are such vast differences between baby mice, caterpillars, and tadpoles. The experiences of juveniles, whether self-initiated or parentally-guided, are extremely important for development of searching and resource assessment skills. Even among invertebrates, early experience can affect not only the behaviour of immatures but also the behaviour of the adults. In some cases we also know that prenatal behaviour has important consequences for successful future behavioural patterns.

16.2.1 Play

Many species of animals seem to require opportunities for direct interaction with components of their environment through exploration, manipulation and play (Fagen, 1982) in order to ensure later success in foraging and other search-related behaviours. In rats, for example, environments that are physically and socially stimulating (sometimes called 'enriched') enhance behavioural flexibility, the development of ability to solve novel problems, switching between alternative patterns of behaviour, and responses to changes in an animal's physical surroundings (e.g. Einon and Morgan, 1977).

Play behaviour includes diverse forms of social interaction, manipulation of inanimate objects, and solitary movement patterns. Bekoff and Byers (1979) define play as all motor activity performed which appears to be purposeless and in which motor patterns from other contexts may be used, often in modified forms and/or altered temporal sequencing. Bekoff (1974) suggests that play activity smooths behavioural sequences, and it may also allow acquisition of novel behaviours which might be used later in life as an alternative strategy for problem solving. At the very least, selection would seem to favour the early practising of motor patterns simply to enhance neuromuscular development.

If behaviours such as play shape an individual's future success in foraging, hunting or escape, as has been suggested (review: Symons, 1978), then depriving an animal of opportunities to engage in play should have a negative effect on its future abilities in these important kinds of behaviours. Some of the factors that might decrease play time or play possibilities include environmental variation, nutrition, illness, and hibernation. Baldwin and Baldwin (1974) found that group size in squirrel monkeys varied with habitat and food availability, and that in at least one sparse area, the young were not observed to play. In the laboratory, both squirrel monkeys (Baldwin and Baldwin, 1976) and rhesus monkeys from Cayo Santiago (Loy, 1970) showed significantly reduced play behaviour when food supply was lowered. Lambs growing up in lush environments play more than do lambs growing up in deserts, and the desert sheep end up with a more limited behavioural repertoire (Berger, 1979).

The value of play behaviour depends in part upon the kind of environment to which a species is subjected. Play and behavioural flexibility are less important for coping with a constant environment than a variable environment that fluctuates unpredictably over a given set of possible states (Fagen, 1982). In the variable environment, development should be plastic in order to adapt the phenotype to local circumstances, and behaviour should be rigidly adapted to these circumstances. In either case, however, if novel benefits can be expected to result from environmental change, play should evolve to have general effects on behavioural flexibility, and especially on activities related to habitat use, foraging, and social interaction.

16.2.2 Practice

When juveniles are about to be cast upon their own devices for survival, there is a premium on the frequency, quality and diversity of practice bouts in finding and handling resources. While many of the basic neural channels may be used during play behaviour, now the behaviours are linked to patterns of survival.

Experience is one factor capable of ensuring that cats (*Felis domesticus*) are able to catch, handle, and dispatch a prey item efficiently (Caro, 1980). Experiences with a bird or mouse prey between 4 and 12 weeks of age improved predation on some prey

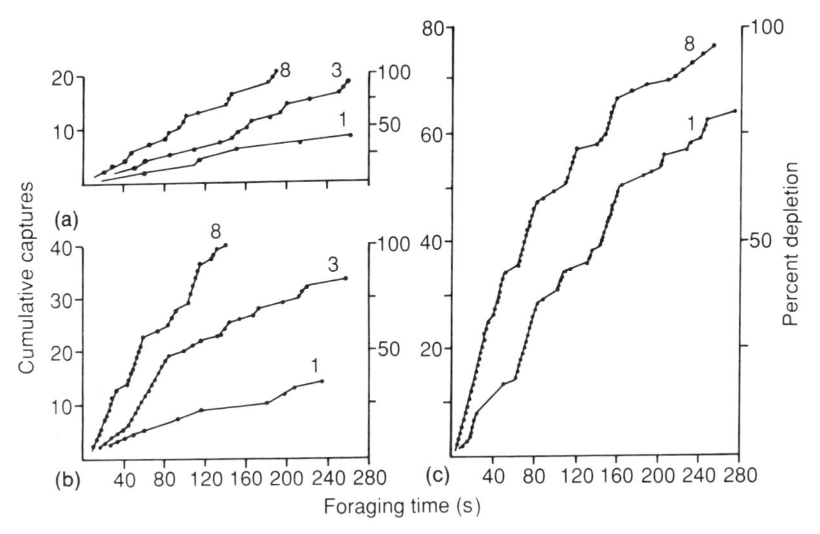

Figure 16.1 Examples of the accumulation of *Daphnia* captures within trials for 3 bluegill sunfish (*Lepomis macrochirus*) during trials 1, 3, and 8. (a) Low density, (b) medium density, (c) high density. Per cent depletion calculated as number of *Daphnia* captured in the trial up to that time divided by the total number of prey available in the arena at the beginning of the trial multiplied by 100. (After Ehlinger, 1986.)

types when tested at 6 months. The same was true for mouse-experienced cats tested on birds, but less so for bird-experienced cats tested on mice.

Juvenile pipistrelle bats (*Pipistrellus pipistrellus*) first begin to fly at the age of 4 to 5 weeks of age, and weaning is completed when the adults leave the roost about 2 weeks later (Racey and Swift, 1985). During the first 2 to 4 days after marking, juvenile bats flew clumsily within 100 m of the roost. They all returned to the roost after flying for less than 20 minutes, and many made up to ten attempts to land at the entrance hole before successfully entering the roost. During these 4 days of observations in, 1980, ten dead or injured juvenile bats were found in the vicinity of the roost. From 5 to 10 days after marking, young bats began to fly more strongly, and were sighted up to 1 km from the roost. At 11 to 15 days after marking, the flight of juveniles was indistinguishable from that of adults as judged by speed of turning and landing ability. Juveniles flew strongly and moved up to 2.5 km from the roost, and from 16 days onwards they covered the same area as adults and were sighted up to 5 km from the roost. The number of feeding buzzes emitted per minute by juvenile pipistrelles increased progressively over the 20 day period after marking. For up to 4 days after marking, juveniles made virtually no attempt to pursue insects.

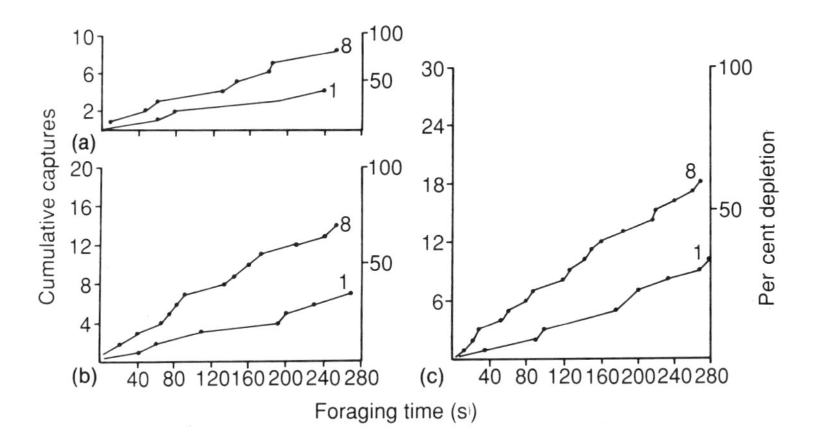

Figure 16.2 Examples of the accumulation of damselfly nymph captures within trials for 3 bluegill during trials 1, 3, and 8. (a) Low density, (b) medium density, (c) high density. (After Ehlinger, 1986.)

At 5 to 10 days they began to attempt to hunt, and there was then a steady incease in the rate of attempted feeding until 16 days, an average rate of 8.5 buzzes per minute was achieved. This is comparable with the maximum rate recorded for adults.

Ehlinger (1986) found that the time required per prey capture decreased with experience (shown here as successive trials) in bluegill sunfish (*Lepomis macrochirus*). Figures 16.1 and 16.2 show cumulative prey captures across time changed both within and between trials for individual fish. An analysis of variance of the mean time required per capture indicated significant effects of prey density and trial for two different prey types. Search time also decreased across trials for each prey density group, suggesting that naive bluegills increased searching efficiency as they gained experience with the prey. Of all the changes in searching behaviours across trials, the most visually striking was the increase in the tendency to change direction in the search paths taken by fish. Figures 16.3 and 16.4 show reconstructions of early and late trials for fish feeding on medium *Daphnia* and damselfly nymph densities. They illustrate the general patterns observed for all fish across trials:

1. a reduction in the number of times they crossed their search path, and
2. a reduction in the number of times they returned to the same location in the arena.

The qualitative patterns in the search paths observed suggest strongly

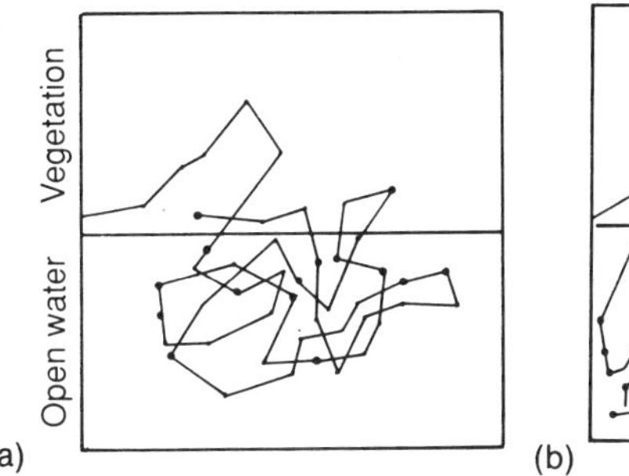

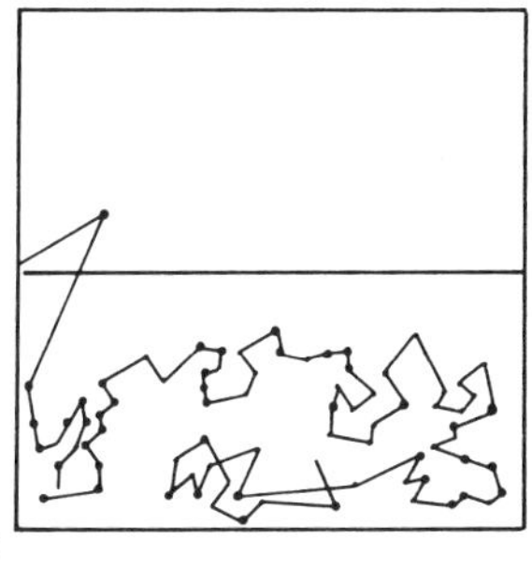

Figure 16.3 Maps of search paths taken by a bluegill feeding on *Daphnia* (0.5 per l) in an arena divided into 'open water' and 'vegetation'. Small dots indicate search hovers; large dots indicate captures. (a) Trial 2, (b) trial 8. (After Ehlinger, 1986.)

that changes in searching path across trials contributed to the overall increase in searching ability. Because prey did not remix in the arena during the experiments, the increase in change in direction of search paths reduced the effect of depletion on search time by reducing the number of times a forager revisited areas wherein prey were locally depleted.

16.2.3 Learning from parents

Some female cats promote the learning of skills connected with hunting in their young by creating situations where these skills can be observed or practised (Schaller, 1972). Cheetahs (*Actinonyx jubatus*), for example, may bring gazelle fawn to their cubs and release it. Tigresses (*Felis tigris*) may pull down a buffalo and then allow the cubs to kill it. Animals like lions (*Panthera leo*) may be dependent on older individuals for food for several years, attaining maximum searching proficiency only after considerable predation experience (Schaller, 1967, 1972). Young lions have opportunities to learn stalking techniques and killing methods by observing adults, and they accompany adults on hunts when only a few months old and have been observed watching adults kill prey.

The length of time required to become proficient often depends on the relative difficulty of handling prey. For example, European oystercatchers (*Haematopus ostralegus*) learn how to open bivalves by observing their parents and the behaviour is highly resistant to change. A given population uses a special pattern of opening bivalves and crabs, either by stabbing or hammering but not with both

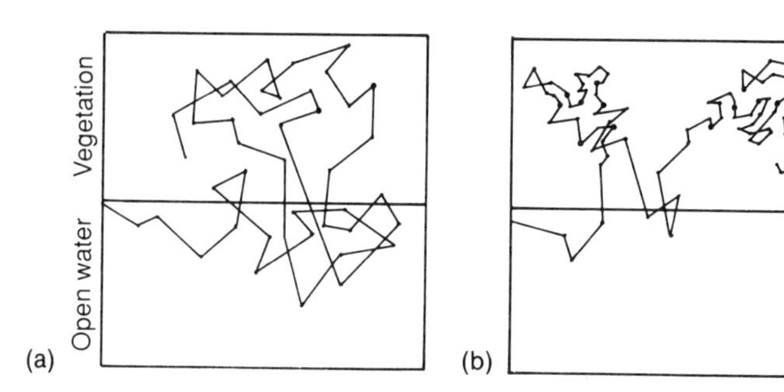

Figure 16.4 Maps of search paths taken by a bluegill feeding on damselfly nymphs (0.25 per l) in an arena divided into 'open water' and 'vegetation'. Small dots indicate search hovers; large dots indicate captures. (a) Trial 2, (b) trial 8. (After Ehlinger, 1986.)

techniques, and they often do not master the feat for many months (Norton-Griffiths, 1969).

16.2.4 Ontogenetic changes in tactics

Ontogenetic changes in the tactics of fish larvae have been observed in several species with quite different habits: bluegill sunfish (*Lepomis macrochirus*) – mainly facultative planktivores; pumpkinseed sunfish (*L. gibbosus*) – benthivores; rock bass (*Ambloplites rupestris*) – insectivores and piscivores; largemouth bass (*Micropterus salmoides*) – mainly piscivores (Brown and Colgan, 1984).

The time that each feeding act (fixating, lunging, orienting, biting) remained in the repertoire varied among the species and appears to be related to the physiological development of the fry. Fixating, for example, is an act whereby the 'prey is viewed stereoscopically before the larva darts forward'. This pause is necessary for fry as their visual system may not be fully developed. In the study of Brown and Colgan (1984), fixating was no longer part of the repertoire of the fry after their initial length had doubled. Once fixating is no longer present in the feeding repertoire, the visual fixation of prey is presumably incorporated into the act of orienting. Lunging and snapping also eventually dropped out of the feeding behaviour of all four species. The two acts that remained, orienting and biting, were the major later feeding acts. As the fry grew the act of orienting became less obvious, and the speed and angle of approach began to differ among the species. Rock bass, for example, were quite 'deliberate' in their orienting, whereas the other three species performed this act much faster. Perhaps in the substrate foraging of rock bass the prey are more difficult to find and capture as compared to finding prey in the water column.

The decline of the acts over time can be linked to a number of factors. The first may be related to feeding success, where in larval herring and northern anchovy, for example, initially only less than 10% of the feeding attempts were successful, but by three weeks of age the success rate had increased to 90% (review: Hunter, 1981). A second contributing factor is that prey selected by larvae increase in size as the larvae grow. Larger prey have more nutritive value per item than smaller prey, and thus fry feeding on small zooplankton would have to capture more than similarly sized fry feeding on large plankton.

16.3 EARLY ADULT EXPERIENCES

The kinds of experiences that young adults encounter that influence their later searching success are too numerous to include here in much detail. The amount of teaching provided by parents to offspring alone is enormous, especially since young of many species must be taught how to find all of the resources important in their lives. In the following examples we examine how experience leads to assessment of resources and to more efficient orientation, two important components of searching behaviour.

16.3.1 Adult experience mediates host recognition in insects

Variation in host selection by adult insects may result from

1. 'induction' during larval exposure to particular host plants (see above),
2. conditioning by early adult experience, or
3. congenital genetic variation among individuals in host plant preference.

Determining how preferences are acquired during ontogeny is critical because such predispositions can have significant implications for the extent of host restriction. For example, experience plays a significant role in the arrestment of female parasitoids (*Pseudocoila bochei*) on host patches (Van Lenteren and Bakker, 1975). Inexperienced females which showed no arrestment response to a patch containing the host odour were separated and exposed to their host *H. viriscens* larvae in association with the kairomone within 1 hour after the initial test. After parasitism of several larvae, the formerly inexperienced females were re-exposed to a kairomone-contaminated patch. All of the 33 inexperienced parasitoids which failed initially to recognize the patch were arrested on re-exposure.

16.3.2 Improvement of navigational skills in pigeons and honeybees

The use of the sun compass by homing pigeons (*Columbia livia*) is not under rigid genetic control. Rather, a pigeon must learn the relationship between the sun's position, its own internal clock, and the geographic direction. Birds homing in the southern hemisphere, for example, learn a different relationship and appear to compensate for changes in the sun's position with time by correcting in a counterclockwise rather than a clockwise direction. Wiltschko and Wiltschko (1984) reviewed the importance of ontogenetic experience in the formation of this relationship. They reared pigeons from the age of 24 days under permanently clock-shifted conditions. Young birds that had never seen the sun were placed in light-tight rooms with a photocycle of lights on 6 hours after sunrise and lights off 6 hours after sunset. Thus, their internal clocks were 6 hours out of phase with the real sun time. An attempt was then made to switch the clock-shifted birds to the natural day, that is to 'normalize' their clocks, by confining them for five days in a room where the lighting cycle was identical to that of the natural day. When these now 'normal' birds were first released to home, they behaved as if they had been temporarily clock-shifted, flying 90° in error. After a number of flights under normal sun, however, the birds gradually began to fly normally again. Moreover, following a year's normalization under real sun time, temporary 6-hour clock-shifting had comparable effects on both the formerly permanently clock-shifted birds and normal birds. The authors interpret their results as indicating that the pigeons synchronize their clocks on the basis of their early experience with relevant environmental cues. Experience, in a sense, teaches them how to use the sun compass by allowing for an accurate coupling of the internal clock with the position of the sun. Thus, pigeons apparently have no inherent ability to interpret the sun's azimuth at a given time as indicating a given direction; instead the sun compass must be established by experience (Wiltschko *et al.*, 1976).

Young honeybees (*Apis mellifera*), on their first flight from the hive, show no navigational ability even from distances as short as 20 m, unless by chance an individual arrives sufficiently close to the hive to recognize it by olfaction as the home site (Frisch, 1967). At the beginning of the first exploratory bout, bees examine the hive. They then fly in front of it for a few seconds before heading off in a straight line away from the hive. They return a few minutes later and hover in front of the hive before settling. During this flight it is assumed that by a combination of sun compass and distant landmarks the bee establishes the position of the hive relative to the surrounding landmarks. Even after the outward phase of the first exploratory bout the bee can return to the hive from a distance of several hundred metres and after several bouts she can successfully return to the hive from distances up to 8 km.

16.4 SUMMARY AND CONCLUSIONS

Individual adults may differ in their search behaviour and search success because of ontogenetic experiences and maternal influences.

Maternal influences, through which females affect the future behaviour of their offspring, include different amounts of investment in eggs or embryos and differences in conversion of environmental resources into offspring. The experiences of juveniles, whether self-initiated or parentally-guided, are important for the development of searching and assessment skills. Many species of animals seem to require opportunities for direct interaction with components of their environment through exploration, manipulation, and play, to ensure later success in foraging and other search-related behaviours.

Young adults practise searching behaviours, thereby improving their abilities in finding and handling resources. Acquisition of search-related skills is sometimes promoted by mothers who create situations where young can observe or practise these skills. The length of time required to become proficient often depends on the relative difficulty of handling prey.

PART FIVE
Methodology

a little knowledge is a dangerous thing, and how well that axiom applies to reading tracks

T. Hillerman, The Dark Wind, 1982

Just as a spider web provides a tangible record of web-building behaviour, the pathway of a walking animal is a record of the behaviours involved in search. Both the spider's web and an orientation path can be photographed or traced as a static record, or they can be monitored continuously during construction to obtain a record of the temporal sequences of behaviour. As discussed in Chapter 17, search pathways are amenable to analysis, especially those of walking animals, and with a bit more sophistication of swimming and flying animals. Chapter 18 describes how imaginative investigators have been able to delineate, through data collection and then computer simulation, how a given species searches. In previous chapters we examined locomotory search patterns and the effects of resource distribution on searching success, and so it is intriguing to match computer simulation attempts with the empirical results that have been arrived at through experimentation.

17
Analysing search tracks

Analysis of search orientation paths yields several general principles relating to the functional design of a search pattern.

First, the path may reflect the output of a motor pattern generator, and so a path may contain deterministic components or specific geometric configurations such as zigzags or spirals.

Second, the geometric figure toward which the path seems to average (e.g. a spiral or a zigzag) is imperfect in form because of variability in successive directions or manner in which an animal walks. This 'noise' could derive from imperfections in the neuromuscular operations of the animal during locomotion, it might represent a strategy by which an optimal degree of variableness improves searching ability, or it could mainly come from environmental heterogeneity in the medium or substratum.

Third, paths often reflect the attention that an animal gives to directional cues in the environment, such that a path may be quite straight and at some angle to an external cue.

Fourth, the path configuration is often directly related to the accompanying sensory scanning mechanisms. An animal might, for instance, improve its scanning sector by walking in a sinusoidal path, or it might scan to the left and to the right while stopping and look straight ahead while moving. In both cases the animal has an opportunity to scan a volume of space from straight ahead to at least 90° left and right, but the first path will be sinusoidal and continuous, and the second will be straight and punctuated by stops.

Animals may change their search paths, however, often as a direct result of locating and utilizing resources or perceiving resource-related cues (see Chapters 2 and 3). The objective here is to delineate the search track, compare methods for doing so, and to quantify several track measures.

7.1 DATA COLLECTION

The path of a searching animal can be described by recording the animal's position in space at successive points in time and then reconstructing the path. From this kind of data base we can ask to what degree the path is determined by stochastic processes, if it changes over time, and if it changes with respect to resource finding or various external cues. In addition, the scanning sector can be determined through successive observations of the distances in the animal's sphere at which it can detect resources.

In the laboratory, paths can be recorded by several means, including videotaping (e.g. Kuenen and Baker, 1982) and locomotion compensators (e.g. Weber *et al.*, 1981). A relatively simple method, described by Bond (1980), focuses the image of a searching insect onto the surface of a digitizing tablet; as the insect moves its path is traced with a cursor pen and stored in real time as a digitized sequence of x and y coordinates by a computer. A sonic digitizer, which does not require a pad, can be attached to a video monitor, and the cursor pen moved over the screen of the monitor to input video-taped sequences. Other video techniques digitize each video frame, either automatically using a 'frame-grabber' computer board, or manually with a VCR capable of frame-by-frame play.

Paths of male German cockroaches searching for females (Schal *et al.*, 1983) and desert isopods searching for their burrow (Hoffman, 1983a) were recorded with a locomotion compensator (Figure 17.1), which is a two-dimensional treadmill for maintaining an untethered animal on top of a 50 cm diameter sphere. Orthogonally-positioned optical encoders transmit the rotational movements of the sphere to a computer, thus indirectly recording the path of an animal (see Kramer, 1976; Weber *et al.*, 1981; Tobin and Bell, 1986).

The method most often used in the field is to record the position of an animal on a grid. Smith (1974a) recorded search paths of thrushes by relating their successive positions to a grid of marker pegs laid out in a meadow. This information, spoken into a tape-recorder, was later plotted onto scale maps which were then photographed and projected onto a digitizer. Zach and Falls (1976c) videotaped the movements of ovenbirds as they searched among patches of mealworms or flies in an enclosure, and later digitized the pathways. Roitberg *et al.* (1982) and Aluja *et al.* (1989) reported a solution to the difficult problem of monitoring a flying animal in three dimensions. As shown in Figure 17.2, each leaf of an apple tree was marked with a numbered label; as an apple maggot fly moved through a tree, observers called out the numbers of leaves flown past; later the sequences were organized into pathways for analysis. The other way to accomplish this task is to use two recording devices to obtain x, y, and z coordinates, as in records of butterflies moving among food plants in seminatural conditions (Zalucki and Kitching, 1982) and moths flying in a wind tunnel (Baker, 1985).

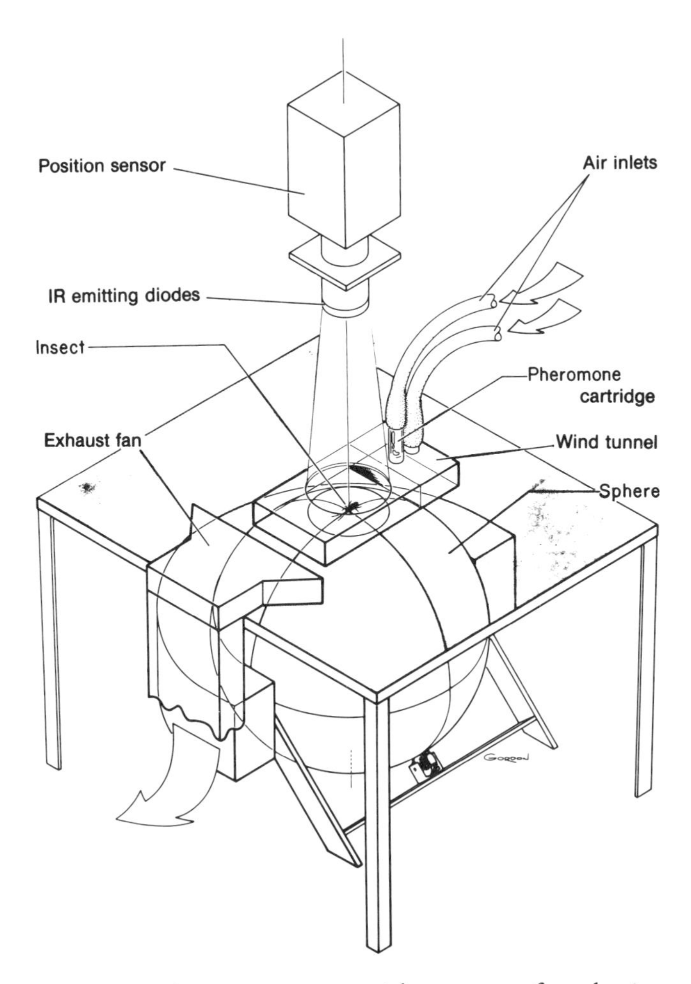

Figure 17.1 Locomotion compensator with apparatus for a horizontal wind current. Sphere is 0.5 m in diameter. Arrows indicate entry of air through tubes connected to valves (not shown). The column at the top of the apparatus contains a lens surrounded by infrared LEDs; the illuminated area is indicated as a beam shining down onto the sphere.

Animals can also be marked with tags or colour codes so that the release site of a captured individual can be determined, or so that the track of an individual can be compiled (see Opp and Prokopy, 1986; Brown and Downhower, 1988). On a more global scale, the movements of animals have been followed by telemetry and even by radar. Using radar, Riley and Reynolds (1983) showed that during migration, locusts in the Sahalian zone of Mali orient primarily

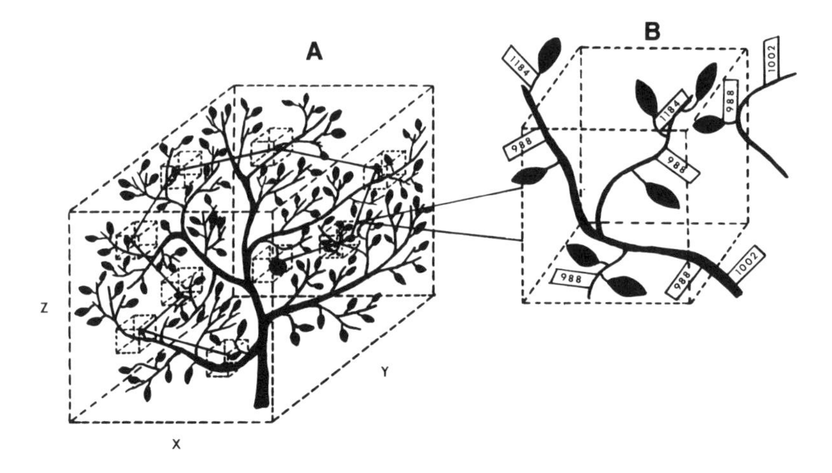

Figure 17.2 A small apple tree with each twig labelled for identifying movements of apple maggots flying through the tree. An observer calls out the numbers on labels as the insect flies by. The inset (diagram B) shows a close-up of branch labels. From Aluja *et al.*, 1989. Copyright Entomological Society of America, reprinted by permission.

downwind. Lingren *et al.* (1978), with the aid of night vision goggles, observed moths orienting to pheromone sources at night.

Each searching problem is unique and requires the appropriate method for accurate and efficient data collection. Put another way, each technique is appropriate for only certain types of searching problems. It would be inefficient to use Smith's method for recording an animal moving in a small arena where the entire scene could be videotaped; alternatively, it would be impossible to videotape elephants moving through a 50×50 km^2 area over a one-year period.

17.2 ANALYSIS OF TRACKS

The main idea in analysing search paths is to answer specific questions or to test specific hypotheses, as well as to delineate a search tactic. The following section, taken largely from Tourtellot *et al.* (1989), delves into analytical technique and motor pattern measures in some detail.

Cartesian coordinates can be used to describe each point in a recorded path. A value which provides time information may accompany each set of spatial coordinates, or each point may have been recorded at a constant rate. Analysis is accomplished by computer programs that read a file of consecutive pairs of x and y coordinates of the animal's position, as well as the coordinates of

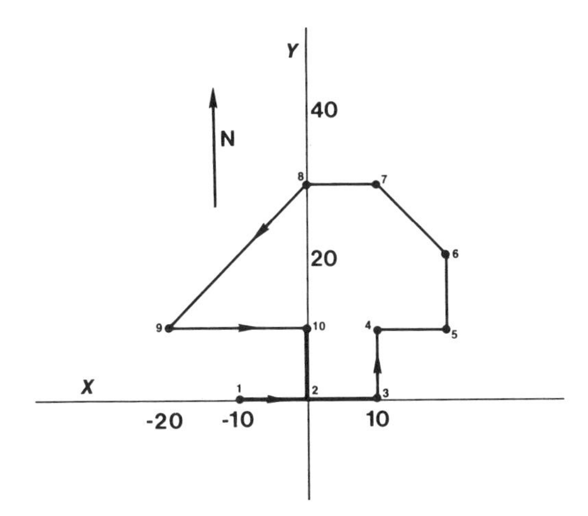

Figure 17.3 Cartesian coordinates of a hypothetical search pathway.

resources in the habitat or the vectors of cues such as wind currents carrying prey odour. The file usually provides other pieces of information, such as scale units, in the 'header'. Specific quantitative measures, as required by the investigator, can be calculated by a program that sequentially determines the distance and angle between two sets of x and y coordinates.

Figure 17.3 shows a pathway comprised of several coordinates recorded over a period of approximately 10 s. Table 17.1a lists the coordinates beginning with –10,0, and then shows the basic information extracted. Distance per point shows a span of 10 mm between points 1 and 2 as the animal moved along the x-axis. The compass angle of 90° shows the animal moving due east from point 1 to 3. There is no turn between points 1 and 3 (0°), whereas the animal turns left (–90°) at point 3 and then walks north (0°) toward point 4. Distance from target reflects distances in mm from the starting point at –10,0. From these basic values it is possible to calculate various measures as shown in Table 17.1b.

7.2.1 Useful measures for quantifying search behaviour

When files from several search paths, such as those of the housefly (*Musca domestica*) depicted in Figure 17.4a and b, are read individually by a computer program, the angle between each successive point and the distance between each point can be calculated. In this case a fly was fed a drop of sucrose and then its local search, which is released

Table 17.1 Analysis of the pathway resulting from the x, y coordinates in Figure 17.4

Trace: filename = DL19A.DAT

Line#	x	y	*Distance per point*	*Compass angle*	*Turn angle*	*Distance from target*
001	–10.0	0.0	[initial x, y]			
002	0.0	0.0	10.0	90	–	10.0
003	10.0	0.0	10.0	90	0.0	20.0
004	10.0	10.0	10.0	0	–90.0	22.4
005	20.0	10.0	10.0	90	90.0	31.6
006	20.0	20.0	10.0	0	–90.0	36.1
007	10.0	30.0	14.1	315	–45.0	36.1
008	0.0	30.0	10.0	270	–45.0	31.6
009	–20.0	10.0	28.3	225	–45.0	14.1
010	0.0	10.0	20.0	90	–135.0	14.1
011	0.0	0.0	10.0	180	90.0	10.1

Segment analysis. Data collected at 1.0 points/s

Parameter	*Mean*	*SD*	*Variance*	*n*
Trip distance (mm)	132.4	–	–	1
Stop duration (s)	0.0	0.0	0.0	9
Maximum distance to target (mm)	36.1	–	–	1
Turn bias (deg/s)	–27.0			
(deg/mm)	–2.2			
(deg/move)	–30.0	77.9	6075.4	9
Turning rate (deg/s)	63.0			
(deg/mm)	5.2			
(deg/move)	70.0	39.7	1575.2	9
Locomotory rate (mm/s)	13.6	6.5	41.8	9
Per cent time spent moving	100.0			1
Linearity index	0.08			1

Parameter	Mean	Circ. SD	*r*	*n*
Compass heading (deg)	120.4	75.6	0.1288	9

Trip distance: total distance moved (mm).
Stop duration: mean duration (s) of stationary periods delineated by two periods of continuous movement.
Maximum distance to target: the point in the path furthest from the starting point (mm).
Turn bias: amount of turning in the direction (left or right) that an animal tends on the average to turn (signed degrees).
Turning rate: amount of turn regardless of turn direction (unsigned degrees).
Locomotory rate: speed of movement (mm/s).
Per cent time spent moving: time spent moving as a fraction of the total time.
Linearity index: beeline distance divided by total path distance.
Compass heading: direction from 0° to 360° (degrees).

by feeding, was digitized at a rate of 5 points per s. The maximum resolution would therefore be 0.2 s.

(a) *Locomotory and turning rate*

For the study of the fly it was necessary to test for changes in locomotory and turning rate measures in order to determine if local search intensity wanes after resource utilization. Mean locomotory rate and turning rate were calculated for 10 flies for each consecutive 5 s period. As shown in Figure 17.5a and b, turning rate increases and locomotory rate decreases after a fly ingests a drop of sucrose (arrows in figure), and then both measures readjust over a period of 60 s until the prefeeding levels are reached (Murdie and Hassell, 1973; White *et al.*, 1984).

(b) *Compass heading*

Whereas the scale for calculating turning rate and turn bias is ±180°, the scale for real positions in space is the compass points 0 to 360°. In calculating the mean compass direction and angular deviation of an animal, one must use circular statistics (Batchelet, 1981), for the following reason. If an animal moves toward 5°, then 355°, 5°, and 355°, it is obviously heading nearly directly north, but if these four numbers are added and divided by four, the mean is 180°, or directly south. By taking the sines and cosines of these compass directions, adding these values, and then converting the mean sine and the mean cosine to the angle, one obtains a mean of 0°. For calculating mean and standard deviation for turning rate and turn bias with a ±180° scale, however, linear statistics should be employed.

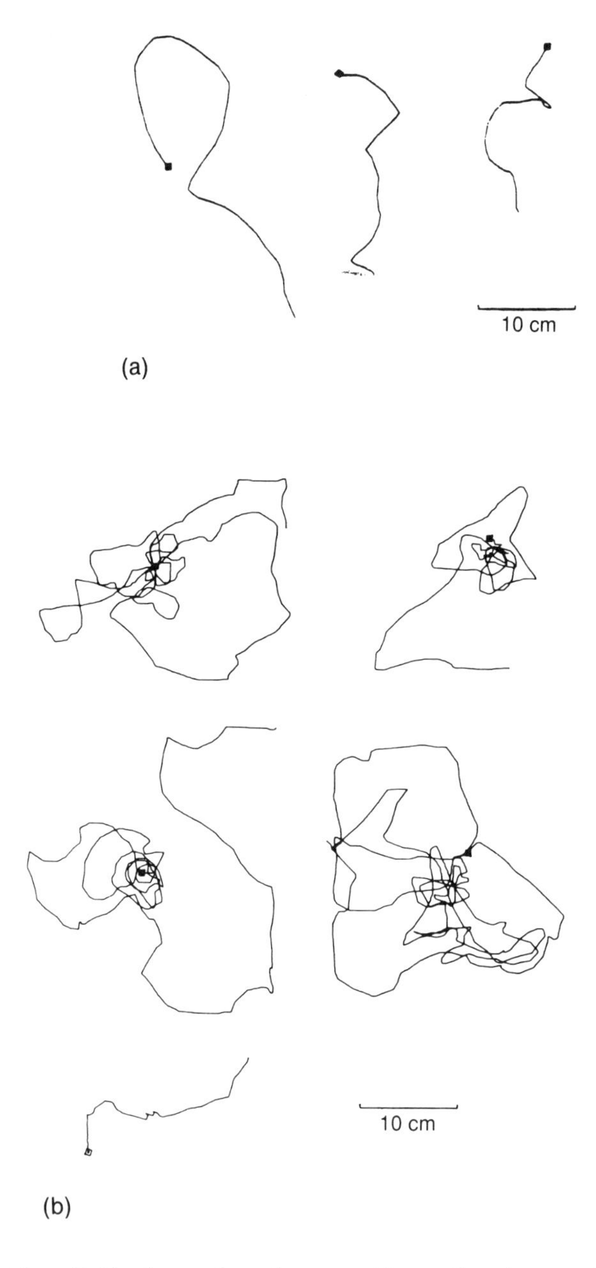

Figure 17.4 Individual search pathways of houseflies (*Musca domestica*). (a) After being offered water, or (b) after being fed a 2 μl drop of 0.25 M sucrose. Small square in each diagram shows the position of the sucrose drop. (After White *et al.*, 1984).

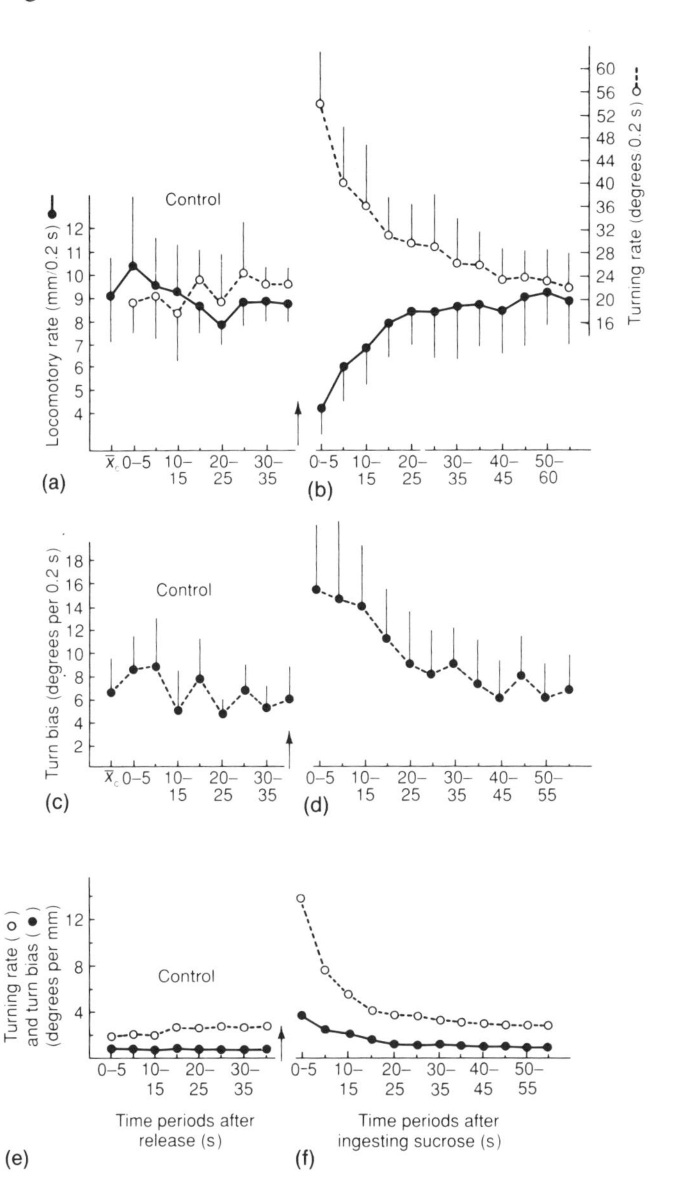

Figure 17.5 Quantitative parameters of search orientation of houseflies (*M. domestica*) at a resolution of 5 s (means of 5 s periods). (a) and (b) Turning rate, as degrees per 0.2 s (open circles), and locomotory rate, as mm per 0.2 s (closed circles); (c) and (d) turn bias, as degrees per 0.2 s; (e) and (f) turning rate, as degrees per mm. xc = mean control value. Vertical lines are standard deviations. Arrows indicate when flies were fed. (After White *et al.*, 1984.)

(c) Turn bias

The turn bias measure quantifies the looping characteristic of local search of flies. By designating left turn angles as (–) and right turn angles as (+), summing the signed values for each 5 s interval, and then taking the average, a measure of turn bias can be obtained. Thus, six consecutive turns of –10, –15, +20, +10, –20, and –20° would average out for a left-hand turn bias of –5.83°. For the same interval the average turning rate would be 15.83°. To combine and average the turn bias for several flies, the signs are removed (distributions are folded) and the mean turn bias is calculated (Figure 17.5c and d). From Figure 17.5d, the turning of the flies is shown to have a bias in one direction or another, although a fly does occasionally turn in the direction opposite to its bias direction. The data show that this measure also decays over time from looping to relatively straight locomotion.

(d) Turning rate per unit distance

Turning rate per unit distance was calculated by dividing turning rate (°/s) by locomotory rate (mm/s) (Figure 17.5e and f). In most studies, turning rate per unit distance generates a smoother curve than does turning rate per unit time, since the latter measure can reflect changes in locomotory rate as well as real turning rate. This conversion can be used for turn bias as well.

(e) Path straightness or directionality

The straightness of a path can be measured using a linearity index, calculated by the beeline distance divided by the total path length. Although indices of path straightness do not explain why a path is straight or not straight, they provide a quick measurement. Another similar measure of path straightness, directionality, is usually calculated as the mean vector, based on the relative directions of an animal at each time or distance move from the first to the last move. The vector (**r**) equals 1.0 if all moves are in the same direction, and 0.0 if all possible directions are equally represented.

(f) Displacement

Sometimes it is necessary to determine the displacement of the animal from a resource over time. If the x and y coordinates of the resource are known, displacement can be calculated for each x and y coordinate of the animal as it moves.

(g) *Thoroughness index*

One can calculate various indices of thoroughness of search, such as the T-index suggested by Bond (1980), based on the equation for a random walk. In this case:

$$T = 1 - \exp\left(\frac{-4\,l\,w}{\pi r^2}\right)$$

where
l = path length
w = path width
r = radius of minimum circumscriptive circle that will enclose path

As might be expected, the thoroughness index for local search is initially high and then decreases over time (see Figure 7.2a).

(h) *Autocorrelation*

Events or actions may be correlated with each other either positively or negatively and either in a first-order or nth-order manner. Thus, autocorrelation is the degree to which an animal's position x at time t is dependent on its position at some earlier time t-i. It may be expressed as:

$$K_t = \alpha X_{t-1} + \Sigma_x$$

In this expression, α is the autocorrelation coefficient and represents random error. When plotted as a function of time between successive observations, can permit an examination of the manner in which orientation is time-dependent during each search bout. A search path is a sequence of vectors, each of which is determined by the directional heading and speed of movement. Kareiva and Shigesada (1983) used speed of movement and change in direction (turn angle) to construct a correlated random walk model of insect movement. They provide equations for calculating the expected squared displacement from the origin of the search path under the assumption that the velocity and turn angle for each vector in the path are independent random variables. The expected values can then be compared to observed search paths. If nonrandom search is indicated, examination of autocorrelation functions of turns and velocities may provide insight into the cause of the nonrandomness. For example, Rausher (1979) found that alternation of left- and right-hand turns was one source of nonrandomness in movements of *Battus philenor* larvae.

(i) *Stop frequency and duration*

The periodicity and duration of stops may be important in relating search movements to function. For example, scanning pauses may be related to the presence of prey or may be determined by an 'internal timer' mechanism.

17.2.2 Analysis problems

Although these measures can be effective for characterizing the output of a search pattern generator or for deciphering the responses of an animal to changes in environmental stimuli, several problems such as the following must always be dealt with.

(a) *Time- and distance-based data*

An important consideration is dealing with time as it applies to spatial (x,y) data. The basic problem is that a track, as represented after it is completed by an animal, will not represent accurately the temporal sequence of moves made by the animal, unless the animal always moves at a constant rate. Thus, if one takes a pathway and divides it up into equal distance units, and calculates turning rate or some other measure per unit, these results may not reflect the true temporal sequence of the movements of the animal. The way that data are collected influences the extent to which this is a problem. For example, the data may contain only one reference to time (e.g. the beginning and ending points), as in tracks made in the sand by snails and recorded by Kitching and Zalucki (1982) as a continuous line which could be traced onto another surface and preserved. Although the length of time required for the snail to complete the entire path is known, it is not known when any particular (x,y) coordinate was reached by the snail. The same situation holds, but to a lesser degree, when data are recorded at a varying rate, as when sightings are made of an elephant whenever possible, but at intervals at least partially controlled by the elephant and not by the observer. However, if the data are videotaped or recorded from a locomotion compensator, then it is possible to determine when stops and changes in locomotory rate occur.

(b) Size of move

Another potential problem is determining the appropriate move size to be used in collecting and in analysing orientation data. For data collection, the solution is simple: always collect data at the highest rate practical. For data analysis, the solution requires some deliberation. McFarland (1971) has discussed the importance of move size when it must be arbitrarily determined by the investigator, suggesting that the high resolution gained by small unit moves serves only to emphasize the variability of the data, while important information can be lost by employing too large a move size. The data base can also be a limiting factor, since a sufficiently long pathway is required if large move sizes are to be used. When the animal has a pause/move/pause locomotory mode, such as in hopping birds, one can use real move length as the move size in analysing data.

The example shown in Figure 17.6a is a 10 s segment (100 data points) of search behaviour of a male cockroach (*Blattella germanica*), stimulated by antennal contact with female sex pheromone (Schal *et al.*, 1983). After the male loses contact with the female, it engages in local search. The diagrams show the results of stepping off the path in units from 0.1 to 10.0 s. The frequency distributions of turning rate and turn bias per time unit for each of the plots in Figure 17.6 were graphed in Figure 17.7 to illustrate the effects of changing the move size on turning rate and its deviation. The frequency distribution of turn bias values at a move of 0.1 s shows a narrow concentration around 0° with a slight turn bias to the left of –4.6°/move 16.8 (SD); only 15% of the turns exceeded 10°. At this resolution it appears as though the animal moved almost exactly straight most of the time, even though a cursory examination of the path (Figure 17.7a) reveals a strong left turn bias. The mean turning rate, 12.7°/move ±11.9, exceeds turn bias, since it represents the extent to which the cockroach turned in both the right and left directions as it maintained a left bias of 4.6°. The largest turn at this resolution was 69°/move, whereas 50% of the values are between 0 and ±10°/move.

For many sequences of 0.1 s moves, the cockroach oscillates between left and right turns with a period of approximately 1 move. High-resolution videotape analysis of *B. germanica* moving on a locomotion compensator or in an arena confirmed that the 'wobble' detected at a move size of 0.1 s is indeed due to lateral translational movements when the cockroach runs relatively quickly. A substantial component of the turning recorded at a move of 0.1 s can therefore be ascribed to the characteristic 'gait' or 'wobble' of the cockroach, caused by the way in which extensions and retractions of the six legs

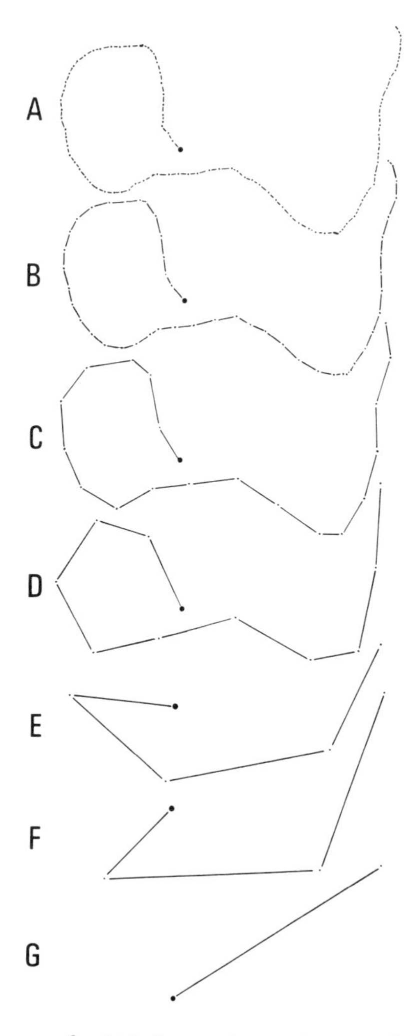

Figure 17.6 Diagrams of a 100-data point pathway of a German cockroach plotted at different resolutions. Time intervals used to 'step off' the path (from A to G): 0.1, 0.2, 0.5, 1.0, 2.0, 3.0, 10.0 s.

bring about lateral translational movements of the body. The amount of variation due to body wobble, revealed by analysis at a move representing less than one body length, would be expected to differ depending on the characteristic gait of the animal used.

As the move size is increased, turn bias per move increases proportionally toward the left direction (Figure 17.7). At the larger move sizes the angular deviation also increases, indicating an

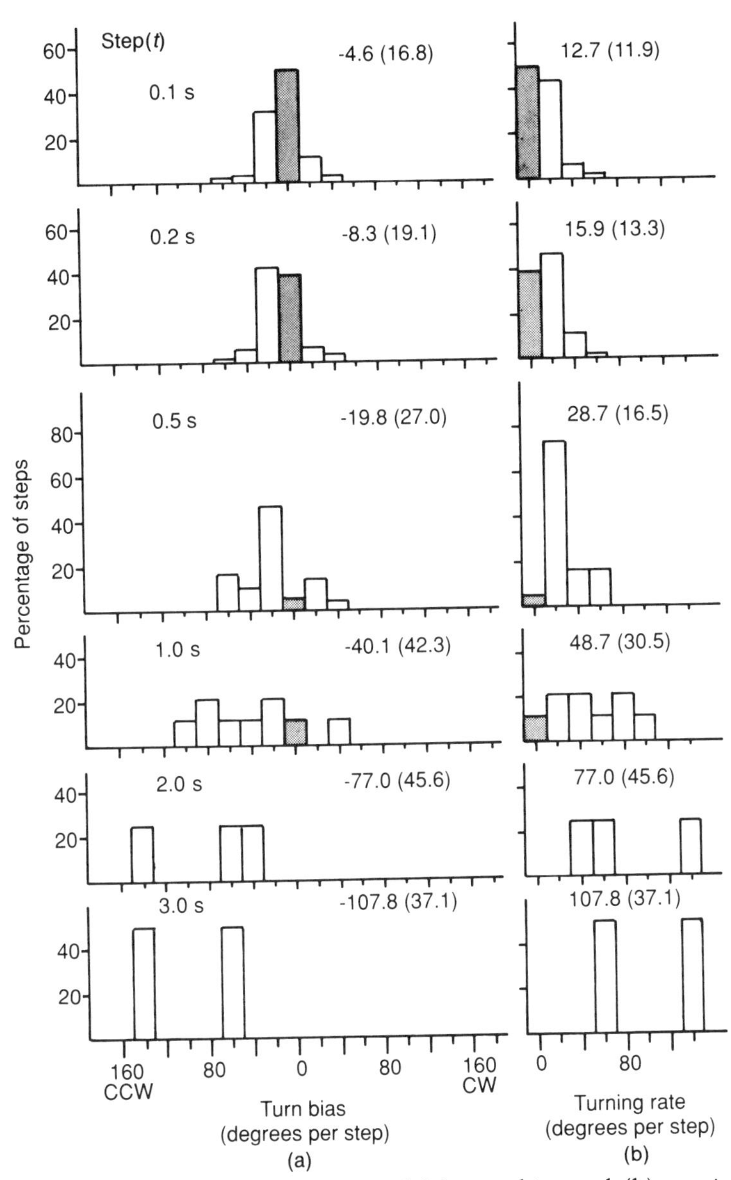

Figure 17.7 Frequency distributions of (a) turn bias and (b) turning rate values from the paths shown in Figure 17.6. Step refers to the time interval used in the analysis. Values are means and standard deviations.

increased tendency of the cockroach to turn in both the right and left directions rather than in one direction. At a move size of 1.0 s, the pathway begins to deviate from its real structure (Figure 17.6d), and from Figure 17.7 it would seem as though the cockroach orients by turning in units of 40°! Higher resolution data described above shows, however, that the cockroach does not run straight with intervening turns of 40°, but instead it turns in a much smoother manner. This is an important distinction, since some arthropods do exhibit a run/stop/run locomotory mode with sharp turns executed during each stop period.

At even lower resolutions, from move sizes of 2.0 to 5.0 s, turns are exclusively in the left direction, cancelling out the right turning phase in the middle portion of the path. At the 3.0 s move size (Figure 17.6f), the two turns are not only very large, but are very similar, and hence the relatively low variance. The animal seems to have turned 107.8° per 3.0 s move, a value that has little biological meaning, because a cockroach running at 7.2 cm/s could not possibly execute such large turns without stopping first.

The compromise move size for analysing these data seems to be between 0.2 and 0.5 s, at which the orientation pattern of the animal is emphasized realistically. The 'body wobble' is minimal, and the sinusoidal or zigzag pattern characteristic of a running cockroach is still evident.

(c) *Minimum move and the problem of stops*

A further consideration in an analysis program is determining a minimum move, and to use this minimum move to eliminate the calculation of measures for extremely short segments. Minimum move is therefore a special case of the move size, and must be considered in addition to the move size used in an analysis. The problem relates to errors that arise in digitizing body movements of animals that are not engaged in locomotion. Swaying movements or turning without translation (pivoting on the spot) are examples of moves that should not be included in calculation of locomotory or turning rate. To eliminate these artifacts, the analysis program must be set to ignore all moves less than some value which includes such movements.

A related problem is deciding when a stop is a stop. Some animals tend to turn primarily after a stop, during which information from the environment is gathered. When calculating turning rate per s, for example, problems would be encountered if a 5 s stop and the next 0.2 s of movement including a 160° turn were summed to give a turning rate of 30.8°/s. This value does not reflect accurately the animal's movements. A better method might be to separate post-stop turns, called pivots, from turns generated while moving. The two populations can be logically analysed separately because it is likely that they are caused by different neural generators.

(d) Defining a turn

The third problem is defining a turn. The analysis of high resolution pathways shows the presence of translatory movements due to the gait of the animal. These movements are not associated with a change of heading, and raise the question of how to define a turn (see for example, Weston and Miller, unpublished. One is never certain as to whether turns of smaller dimensions should be filtered out by arbitrarily defining a minimum turn or if all turn values extractable from the data should be included in the computation of a mean turn value. The case of an animal which is moving steadily through a large curve, making very small turns with each move, illustrates the problem of setting a minimum turn value. With each move, the turn is composed of both noise and intentional course change.

17.2.3 Comparing data sets

Among the various reasons for comparing turning rate values are comparisons of orientation of an individual or population under different stimulus conditions, and comparisons between species. A technique is described below that standardizes orientation data by defining a move in units of body length, and in units of the time required for the animal to move one body length. This technique reduces variation resulting from lateral translational movements and prevents the use of move sizes that lead to artifactual or unrealistic turning values per move.

(a) Comparisons between experimental variables

Data were collected from male *Trogoderma variabile* beetles on a locomotion compensator at 10 data points per s during a 10 s control period (a-b) and a 110 s period of sex pheromone exposure (b-c) (refer to Figure 2.1a) (Tobin and Bell, 1986).

The data set was analysed at a move of 0.625 mm (0.25 body lengths), 2.5 mm (approximately 1.0 body length), and 10.0 mm (4 body lengths). The frequency distributions of turn bias values (Figure 17.8) show that even after exposure to sex pheromone the cockroach seems to be running almost exactly straight at a move of 0.625 mm. Figure 2.1, however, clearly substantiates a change from relatively straight movements (A-B) to intensive right-hand looping (B-C). At

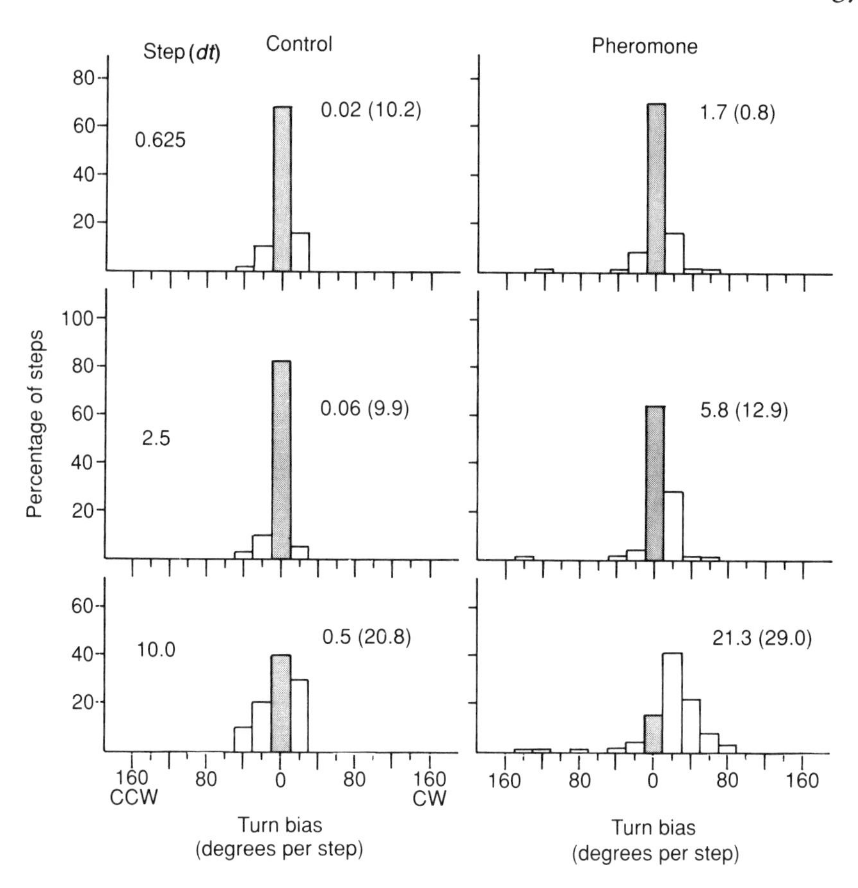

Figure 17.8 Frequency distributions of turn bias values from experiments with grain beetles before and after exposure to sex pheromone. Step refers to the distance interval used in the analysis. Values are means and standard deviations.

a move of 10.0 mm the data suggest that the beetle has executed turns greater than 80° per move, which it is not capable of making, and that the turn magnitudes per move are highly variable. Thus the result using 2.5 mm, a distance equal to its body length, seems to be a compromise that realistically compares the path before and after pheromone presentation.

From the analysis using a distance move, the mean time required for the beetle to move one body length was calculated, and this move could be used for a time-based analysis.

Table 17.2 Comparison of turn bias in search orientation of the grain beetle (*Trogoderma variabile*) and the German cockroach (*Blattella germanica*) at different analysis step sizes

Step size (mm)	*Turn bias (°/step ± SD)*			
	T. variabile	*n*	*B. germanica*	*n*
0.25	7.9 ± 11.9	40	4.6 ± 13.7	84
0.50	14.2 ± 13.1	20	4.7 ± 12.9	82
0.75	20.7 ± 18.4	13	6.2 ± 13.8	60
1.00	27.6 ± 21.2	10	8.3 ± 16.5	43
1.25	34.8 ± 27.7	8	8.9 ± 17.3	40
1.50	41.3 ± 33.9	6	11.3 ± 19.7	31
1.75	49.8 ± 38.8	5	13.1 ± 20.8	28
2.00	56.6 ± 44.4	4	13.5 ± 20.5	27

(*b*) *Comparisons between species*

The search orientation patterns of sex pheromone-stimulated male *B. germanica* and *T. variabile* are both characterized by looping search. The results of analysing the paths at various move sizes are shown in Table 17.2. The objective is to compare the orientation of the two species. Some adjustments must be made in the analysis because the cockroach runs more quickly (7.2 cm/s) than the beetle (1.2 cm/s), and the cockroach is larger (1.25 cm) than the beetle (0.25 cm). If the same move size were used for both insect species the turn bias of the beetle is much larger than that of the cockroach (Table 17.2). By using a move size that is equal to the body length of each species (0.25 cm for the beetle, 1.25 cm for the cockroach), the data are comparable. Moreover, at appropriate move sizes the turn bias values of the beetle and cockroach are not significantly different.

17.3 SUMMARY AND CONCLUSIONS

Movement is the means by which an animal visits various sites and exposes its sensory organs systematically for efficient scanning of successive portions of the environment. Delineating the locomotory actions that generate a search orientation pattern is often important for understanding the way that an animal uses its stored and perceived orientation information, and also to provide the basis for generating computer models of the locomotory components of search orientation. The search path may reflect the output of a motor pattern generator that is endogenously stored. Paths are also characterized by variability, which might reflect a strategy by which an optimal degree of variableness improves searching ability, or it could mainly derive from environmental heterogeneity in the medium or substratum. Paths often reveal the attention that an animal gives to directional cues in the environment, such that a path may be quite straight and at some angle to the external cue. The path configuration may be directly related to the accompanying sensory scanning mechanisms.

To obtain a record of the behaviours involved in search, an orientation path can be photographed or traced as a static record, or it can be monitored continuously to include the temporal sequences. The path of a searching animal can then be described by delineating the animal's position in space at successive points in time and then reconstructing the path. Each technique for recording pathways has certain advantages and disadvantages, and it is important to select the method best suited to the animal, the environment, and the objective of the research. Analysis of search paths is accomplished by computer programs that read a file of consecutive pairs of x and y coordinates of the animal's position, as well as the coordinates of resources in the habitat or the vectors of cues such as wind currents carrying prey odour.

18

Computer simulations of search behaviour locomotory patterns

The locomotory components of searching behaviour can be simulated using computer programs, with the idea of achieving a better understanding of the relative effects of factors involved in searching success. Among the variables that can be included are

1. reactive distance (at which an animal detects the resource),
2. move length over which the simulated animal does not change direction,
3. turn angle concentration or directionality, and
4. changes in locomotory patterns over time.

A simulated animal can be run in a 'habitat' with specific types of resource dispersion, density, and stand size.

The simplest means of representing a path, which takes into account the distribution of compass directions, is to allow a simulated animal to move only exactly forward, backward, left or right (e.g. Rohlf and Davenport, 1969). A better way of representing the turn distribution is to take the circular normal (von Mises) distribution with the mean angle equal to 0° and a parameter of concentration, k, ranging between 0 ($r = 0$) and infinity ($r = 1$). Because the von Mises distribution is difficult to compute, however, most simulators have used a linear normal distribution, wrapped around a trigonometrical circle (Mardia, 1972; Batchelet, 1981). Successive turn angles during a simulated path are then drawn from the distribution, independently of each other, either randomly or with some kind of assigned weight. The mean vector length, r, of the turn angle distribution can then be expressed as a function of the standard deviation, σ:

$$\mathbf{r} = \exp(-\sigma^2/2)$$

Figure 18.1 shows three distributions of turn angles with a mean of 0°, and examples of pathways that might result from such distributions (Siniff and Jesson, 1969).

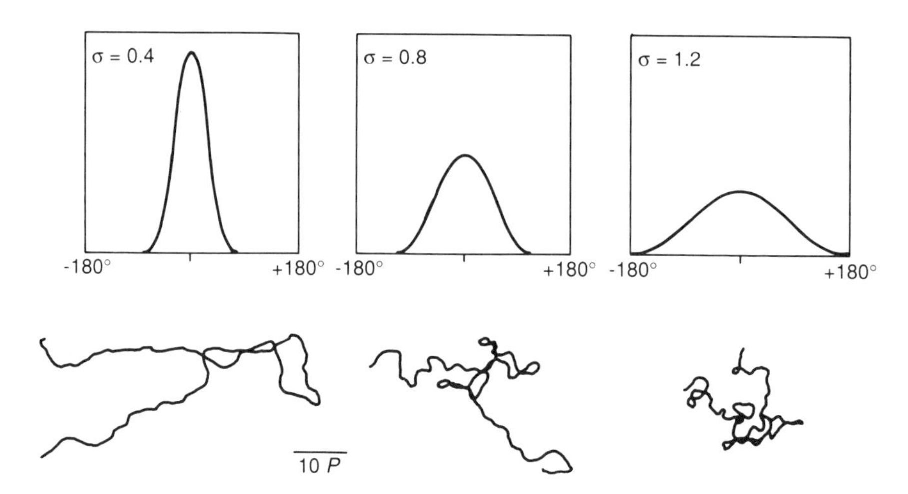

Figure 18.1 Examples of 200-move paths obtained with three different values of the standard deviation of the turn angle distribution. The distributions from which the turn angles were drawn are shown in the graphs. *P* = move length. (After Bovet and Benhamou, 1988.)

In some studies the move length is held constant, whereas in others it varies. Although long moves are sometimes used to signify a higher rate of locomotion then short moves, in reality these differences are simply a function of the scale imposed upon the simulation area. Reactive distance can be set so that the simulated animal detects a resource when it arrives within the set distance. Resources are distributed according to some index of density, and they may be clumped, uniform (regular), or random.

Cain (1985) devised a simulation model to represent butterflies searching for host plants upon which to oviposit. The host plants varied in density from 0.04 to 4 plants/m^2, and were dispersed either uniformly or clumped. Stand sizes were 10 × 10, 20 × 20 or 40 × 40 m plots. Move length varied from 5 cm to 5 m, and reactive distance was changed from 1 cm to 1 m. Turn angle concentration varied from 0.5 (intensive, spatially restricted path) to 4.0 (nearly a straight path). Turn angle concentration was estimated from data recorded from several species of adult (flying) insects. The objective was to determine how each of the plant distribution factors and the insect motor patterns influence searching success.

The results of Cain's (1985) simulation showed that for all stand sizes and plant dispersion patterns, insect searching success declined as plant density declined. Success also increased as stand size increased.

These two outcomes simply mean that any searching tactic will work better on higher than on lower density resources, and that insects are less likely to leave (and therefore decrease success rate) when stands are large than when they are small. At a high plant density, insects located them more easily when they were uniform than when clumped; at low plant densities, however, this difference was not evident. An increase in directionality did not alter the ability of an insect to locate plants in simulations with small reactive distances, short move length, or small stand, but did alter success in simulations with large reactive distance, long move length, or large stand. In simulations where host-plant density was varied from 0.04 to 0.25 to 1.0 plants/m^2, and directionality was varied from 0.5 to 4.0, differences in directionality affected the searching success of butterflies in both clumped and uniform distributions when density was low (0.04 plants/m^2). Success was greater when directionality was low. Short reactive distance and short move length markedly reduced success in finding plants at all plant densities and dispersions, as compared to long reactive distance and long move length.

Based on data collected from larvae of the small white butterfly (*Pieris rapae*) (see Chapter 18), Jones (1976) delineated three common searching modes: low directionality and low locomotory rate (LD), high directionality and high locomotory rate (HD), and intermediate (ID) with characteristics between the two extremes. The LD individuals characterize unstarved larvae, which tend to walk straight, and HD individuals characterize starved larvae, which tend to walk in more restricted pathways. The results of the simulations suggested that in high density plant stands both HD and LD larvae readily found plants, but HD individuals found them faster. At lower densities, in small stands, the most successful type was LD which tended to search intensively; ID individuals did nearly as well, but HD ones, which tended to move straight ahead, were less successful. As stand size was increased, with a concomitant decrease in plant density, HD larvae were most successful, and LD larvae were least successful. Over an intermediate range of stand sizes the ID larvae were more successful than either HD or LD larvae. In all cases LD larvae produced a more aggregated post-movement distribution than HD ones. At densities greater than 0.2 plants/m^2 (i.e. patch widths less than 5, with 5 plants per patch, and less than 16, with 50 plants per patch), ID larvae redistributed themselves like LD larvae. At lower plant densities they redistributed themselves like HD larvae. In all cases, successful HD types found plants more rapidly than either LD or ID ones.

Differences in the success of the various search types in simulations by Jones (1976) seem due to the slow movement, frequent turning, and high perceptual ability of LD individuals; these attributes combine to give a slow and thorough search of a small area, and reduce the chances of a larva leaving the patch. If plants are dense, this tactic works; if not, it fails. In less dense stands, the faster and straighter movement of HD individuals brings greater success; fast search allows a larva to travel far enough to find a new plant before it dies. In a larger stand there is little chance of leaving, and searching a small area thoroughly, as do LD individuals, is inefficient.

Most simulation models, such as those described above, assume that the turn angle distribution from which turns are drawn remains constant throughout a given search bout, and the appropriate mean of the turn angle distribution to use is always 0°. However, changes in turn angle distribution might be expected if an animal alters its search intensity after encountering prey; and, an animal might be expected to have a turn bias for at least short periods of time if it searches in a looping or spiralling manner.

Bornbusch and Conner (1986) showed that looping patterns, generated by a turn bias, were the most effective pattern for locating an odour source in a linear chemical gradient. In simulations without a turn bias, routes to the odour source were longer than with a turn bias. Recall also that turn bias was a significant component in local search of many vertebrates and invertebrates (see Chapter 7).

Yano (1978) recognized the importance of changes in behaviour immediately following oviposition, and this parameter was incorporated into a search simulation of a parasitoid wasp (*Trichogramma dendrolimi*) searching for eggs of its host, the almond moth (*Cadra cautella*). In real experiments with wasps and hosts, a brief period of intensive search followed oviposition, and then the wasp began to walk faster and in a straighter path. Consequently, Yano divided each search bout of the simulation into two phases: the first, immediately after leaving the host, had a high turning rate, and the second, after several moves, had a low turning rate. The change from the first phase to the second was abrupt, rather than gradual, and this apparently matched the transition observed in the wasp. The relative success of Yano's simulations approximated the data collected from experiments with real wasps and hosts in both clumped and uniform host distributions.

Roitberg (1985) simulated individual insects foraging within a resource-containing stand consisting of 2500 cells, 125 of which contained resources. The prey were either clumped or not clumped. Giving-up time was incorporated into the simulation, such that the forager had to encounter consecutive resources within a 10-move interval, otherwise it left the patch. Further, the forager could exhibit

Table 18.1 Comparison of foraging performance for three hypothetical foraging types (Roitberg, 1985)†

Clumping level	*Foraging type*	*Number of resources encountered*	*Time in patch*	*Resource encounter rate*	*Probability of success*
Low	1	2.8a	20.0a	0.12a	0.78*
Low	2	6.1b	29.4b	0.18a	0.93*
Low	3	8.8c	38.1c	0.21b	0.98*
High	1	4.3A	21.9A	0.12A	0.55
High	2	4.5A	20.8A	0.13A	0.52
High	3	8.2B	32.9B	0.22B	0.83

Values followed by different letters are significantly different ($P < 0.005$).
*Signifies significantly different within a foraging type between the two clumping levels ($P < 0.005$).
†Time in patch = number of moves; encounter = encounters per move; probability of locating a resource before leaving the patch.

one of three types of search behaviour in which the reactive distance and turning parameters were altered in the following ways:

Type-1 foragers moved randomly through the patch. Their reactive envelope had a radius of one cell, so that they only discovered prey if they entered a prey-containing cell or a cell adjacent to a prey-containing cell. After resource utilization, they increased their turning rate, thereby restricting their next two moves to areas immediately surrounding the cell in which the prey was found. This type of search pattern was expected to enhance efficiency in distributions with clumped prey.

Type-2 foragers were similar to type 1, except that they had a larger reactive envelope (radius = 2.25 cells). In addition, type-2 foragers did not increase their turning rate following resource utilization, thus reducing the probability of re-encounter with prey that remain in the patch following utilization.

Type-3 foragers moved through the simulated resource field along straight paths. Following each move, they 'assessed' the patch and moved to prey-containing cells that fell within a reactive envelope with a radius of 2.25 cells. After resource utilization they increased their turning rate in the same manner as type-1 foragers. Such behaviour was thought to enable foragers to 'allocate more search time in prey-rich areas'.

Roitberg (1985) drew several conclusions from the results of these simulations (Table 18.1).

1. Although not all values are statistically significant, the three foraging types tend to have different performance levels. Type-3 foragers, the individuals with both a large reactive envelope and local search after resource utilization, were most successful in both resource clumping levels.
2. All three foraging types experience far higher variation in resource encounter rates at the high than at the low resource clumping level, and there is considerable variation in performance within individuals (not shown in table) as they move through the patch.
3. During various segments of search, even successful individuals had low encounter rates, and in nature they might have altered their behaviour by modifying their prey preferences. Further, since patch residence time in these simulations is a function of prey encounter rate and the giving-up time, foragers at any point in time would be expected to emigrate as readily from a high prey density patch as from a low density patch, so long as the current encounter rate exceeds the giving-up time.
4. Finally, the results shown here and observations of others (e.g. Kareiva, 1983) suggest that foraging success is often the function of a first- or second-order Markovian process.

Thus, foraging movements and the probability of host encounter are dependent upon recent past and present events. For example, position in a resource stand is dependent upon previous moves. When the probability of host encounter is dependent upon present and past circumstances, foraging models make very different predictions than when such events are considered independent of them. Roitberg (1985) suggests that detailed analysis of search behaviour provides a means of assigning probability values to such processes in simulations.

Roggero and Bell (in press) based their simulations on data collected and analysed from search orientation of a variety of terrestrial arthopods. Data from walking flies (White *et al.*, 1984; Bell *et al.*, 1985) indicated that turning parameters, including turn bias, increased immediately after feeding, and then decayed geometrically over time. Further, these data indicated that flies and other arthropods do not have a mean forward tendency of 0° immediately after locating a resource, but in fact have a tendency to make large turns (high absolute turning) coupled with a high tendency of turning to the same direction (high turn bias). The combination of high absolute turn and high turn bias, and their decay through time, results in the characteristic tight looping pattern of local search after

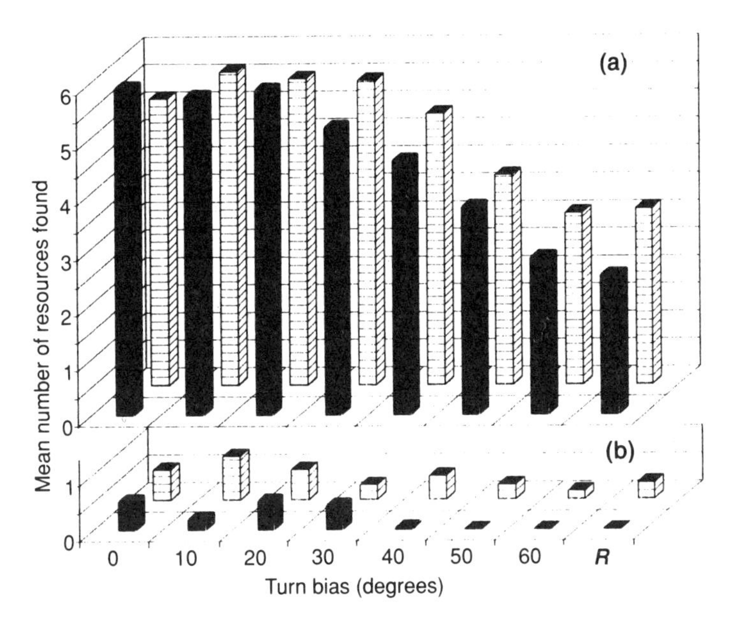

Figure 18.2 Mean number of resources found by simulated insects in clumped (dark bars) and regular (open bars) distributions as a function of turn bias. (a) High density stand (clumps of nine resources separated by 30 mm, clumps separated by 90 mm; resources in regular distribution separated by 49 mm), (b) low density stand (clumps of nine resources separated by 30 mm, clumps separated by 440 mm; resources in regular distribution separated by 157 mm). *R* = random turn angle distribution. (After Roggero and Bell, in press.)

feeding, and then a gradual return to straighter movements over time. In these simulations, turn angles and turn bias were calculated from an equation in which these values decreased geometrically as a function of move number. Thus, initially turning rate and turn bias were high, and then these values decreased as a search bout progressed. Three resource densities, each with clumped or evenly distributed resources, were used to compare the efficiency of a particular search tactic in locating resources in patches of different resource densities and distribution.

As with the simulation results described above, search success was highest in dense resource stands, regardless of the search tactic employed. This idea is demonstrated in Figure 18.2a and b, in which 5 to 10 times as many resources were located in high (a) than in low (b) densities. In these distributions, resources within clumps were 30 mm apart, and the clumps were separated by 90 mm in (a) and 440 mm in

Table 18.2 Mean number of resources located in clumped distributions as a function of turn bias and rate of change (slope) of turn bias (from Roggero and Bell, in press)*

Turn bias slope	*Turn bias* 0°	10°	20°	30°	40°	50°	60°	*Random*
0.4	1.32	1.32	0.88	0.40	0.00	0.00	0.00	0.00
0.3	1.36	1.24	0.96	1.32	0.92	0.32	0.00	0.00
0.2	1.32	1.60	1.44	1.64	1.00	0.96	1.36	0.00

*Resource parameters: 30 mm between nine resources in each clump; 240 mm between clumps. Note that the simulated animal is 8 mm in length and has a reactive distance of 4 mm. Absolute turn rate equals 70° at the beginning of a search bout. Each combination was run 25 times.

(b); in the regular distributions, resources were 49 mm apart in (a) and 157 mm in (b). Note that the length of the animal (also the move length) was 8 mm, and the reactive distance was 4 mm. Figure 18.2 also shows that particular search tactics enhance resource finding in certain resource distributions. The different tactics shown here are essentially variability in turn bias (from 0° to 60°), or random turns. For the resource distributions used in this study, success decreased as turn bias increased, and random turn distributions were least successful. Interestingly, stand size is extremely important in determining the extent to which turn bias affects success in locating resources. In this data set the stand size was infinite, but when the stand contained only 36 resources (four clumps of nine resources, or 36 resources distributed in a 6 × 6 array), turn bias was important for maintaining the simulated animal within the stand of clumps or within the stand of regularly distributed resources. For example, in the finite stand, animals with no turn bias located 1.8 resources, whereas they found 6.6 resources if they had a turn bias of 50°.

Roggero and Bell (in press) also observed an effect of the rate of decay from high to low turn bias during a search bout. In simulations with clumped resources, and where the turn bias decayed quickly (slope = 0.2), animals were more successful finding resources (Table 18.2). This was especially true for those with turn biases between 10° and 60°. These conditions allow the animal to explore resources in the vicinity of the last found resource more efficiently, but at the same time enable the animal to leave the area more quickly after unsuccessful search. Of course the success relative to decay rates would be a function of the distances between resources within a patch.

18.1 SUMMARY AND CONCLUSIONS

Computer simulations examine search success relative to the effects of reactive distance, move length, turn angle concentration, and changes in locomotory patterns over time. Simulations can be run in a 'habitat' with specific types of resource dispersion, density, and stand size. Although simulation algorithms may differ, their combined results indicate that the various factors (search components and resource distributions) must be tested in combinations, such that the interactions can be revealed. Thus, one search component, such as turn bias, may influence search sucess in finite stands, but to a lesser degree in infinite resource stands. It is important to incorporate into simulations dynamic changes in search parameters, since it is unlikely that any search component remains constant over time.

References

Akre, B.G. and Johnson, D.M. 1979. Switching and sigmoid functional response curves by damselfly naiads with alternative prey available. *J. Anim. Ecol.* **48**, 703–20.

Alatalo, R.V., Carlson, A. and Lundberg, A. 1988. The search cost in mate choice of the pied flycatcher. *Anim. Behav.* **36**, 289–91.

Alcock, J. 1973. Cues used in searching for food by red-winged blackbirds (*Agelaius phoeniceus*). *Behaviour*, **46**, 174–87.

Alcock, J., Eickwort, G.C. and Eickwort, K.R. 1977. The reproductive behavior of *Anthidium maculosum* (*Hymenoptera: Megachilidae*) and the evolutionary significance of multiple copulations by females. *Beh. Ecol. Sociobiol.* **2**, 385–96.

Alder, H.E. 1964. Sensory factors in migration. *Anim. Behav.*, **11**, 566–77.

Alphen, J.J.M., van and Boer, H. (1980). Avoidance of scramble competition between larvae of the spotted asparagus beetle, *Crioceris duodecimpunctata* by discrimination between unoccupied and occupied asparagus berries. *Neth. J. Zool.* **30**, 136–43.

Altmann, S.A. and Altmann, J. 1970. *Baboon Ecology. African Field Research.* The University of Chicago Press, Chicago.

Aluja, M., Prokopy, R.J., Elkinton, J.S. and Laurence, F. 1989. Novel approach for tracking and quantifying the movement patterns of insects in three dimension under seminatural conditions. *Environ. Entomol.*, **18**, 1–7.

Ambrose, H.W. 1972. Effect of habitat familiarity and toe-clipping on rate of owl predation on *Microtus pennsylvanicus*. *J. Mammal.*, **53**, 909–12.

Andersen, D.C., Armitage, K.B. and Hoffmann, R.S. (1976). Socioecology of marmots: female reproductive strategies. *Ecology* **57**, 552–60.

Arak, P.A. 1983. Sexual selection by male–male competition in natterjack toad choruses. *Nature*, **306**, 261–62.

Arnold, S.J. 1981. The microevolution of feeding behavior. In: *Foraging Behavior* (eds. A.C. Kamil and T.D. Sargent), Garland and STPM Press, New York, pp. 409–55.

Aronson, L.R. 1971. Further studies on orientation and jumping behavior in the gobiid fish, *Bathygobius soporator*. *A. New York Acad. Sci.*, **188** 378–92.

Aschoff, J. 1962. Spontane lokomotorische aktivitat. *Handb. Zool.*, **8**, 1–76.

Aschoff, J. Ed. 1981. Biological rhythms. In: *Handbook of Behavioral Neurobiology*, Vol 4. Plenum Press, New York.

Atsatt, P.R. and O'Dowd, D.J. 1976. Plant defense guilds. *Science*, **193**, 24–9.

Ayal, Y. (1987). The foraging strategy of diaeretiella rapae. I. The concept of the elementary unit of foraging. *J. Anim. Ecol.*, **56**, 1057–68.

Bailey, W.J. and Thomson, P. 1977. Acoustic orientation in the cricket *Teleogryllus oceanicus* (Le Guillou). *J. Exp. Zool.*, **67**, 61–75.

Baker, R.R. 1972. Territorial behaviour of the nymphalid butterflies, *Aglais urticae* (L.) and *Inachis io* (L.). *J. Anim. Ecol.* **41**, 453–69.

Baker, R.R. 1978. *The Evolutionary Ecology of Animal Migration*. Hodder and Stoughton, London.

Baker, T.C. 1985. Chemical control of behavior. In: *Comprehensive Insect Physiology, Biochemistry and Pharmacology*, Vol. 9 (eds G.A. Kerkut and L.I. Gilbert), 621–771. Pergamon Press, London.

Balda, R.P. 1980. Recovery of cached seeds by a captive *Nucifraga caryocatactes*. *Z. Tierpsychol.*, **52**, 331–46.

Balda, R.R. and Bateman, G.C. 1971. Flocking and annual cycle of the pinyon jay, *Gymnorhinus cyanocephalus*. *Condor* **73**, 287–302.

Baldwin, J.D. and Baldwin, J.I. 1974. Exploration and social play in squirrel monkeys (Saimiri). *Amer. Zool.*, **14**, 303–15.

Baldwin, J.D. and Baldwin, J.I. 1976. Effects of food ecology on social play: a laboratory simulation. *Z. Tierpshchol. 40*, 1–14.

Banks, C.J. 1957. The behaviour of individual coccinellid larvae on plants. *Brit. J. Anim. Behav.* **5**, 12–24.

Bänsch, R. 1964. On prey-seeking behaviour of aphidophagous insects. In *Proc. Prague Symp. Ecol. Aphidophagous Insects* (ed. I. Hodek), pp. 123–28. Academia, Prague.

Barbosa, P. and Capinera, J.L. 1978. Population quality, dispersal and numerical change in the gypsy moth, *Lymantria dispar* (L.) *Oecologia* **35**, 203–09.

Barbosa, P., Cranshaw, W. and Greenblatt, J.A. 1981. Influence of food quantity and quality on polymorphic dispersal behaviors in the gypsy moth, *Lymantria dispar*. *Can. J. Zool.*, **59**, 293–96.

Barker, J.F. and Herman, W.S. 1976. Effect of photoperiod and temperature on reproduction of the monarch butterfly, *Danaus plexippus*. *J. Insect Physiol.*, **22**, 1565–68.

Barnard, C.J. 1978. Aspects of winter flocking and food fighting in the house sparrow (*Passer domesticus domesticus* L.). D. Phil. thesis, University of Oxford.

Barnard, C.J. 1980. Equilibrium flock size and factors affecting arrival and departure in feeding house sparrows. *Anim. Behav.* **28**, 503–11.

Barnard, C.J. 1984. The evolution in food-scrounging strategies within and between species. In: *Producers and Scroungers, Strategies of Exploitation and Parasitism* (ed. by C.J. Barnard), pp. 95–126. Croom Helm, London.

Barnard C.J. and Sibly, R.M. 1981. Producers and scroungers: a general model and its application to captive flocks of house sparrows. *Anim. Behav.* **29**, 543–50.

Barnes, R.F.W. 1982. Mate searching behaviour of elephant bulls in a semi-arid environment. *Anim. Behav.*, **30**,1217–23.

Barnett, S.A., Dickson, R.G., Marples, T.G. and Radha, E. 1978. Sequences of feeding sampling and exploration by wild and laboratory rats. *Behav. Proc.*, **3**, 29–43.

Barrows, E.M. 1975. Individually Distinctive Odors in an invertebrate. *Behav. Biol.* **15**, 57–64.

Barry, W.J. 1976. Environmental effects on food hoarding in deermice (*Peromyscus*). *J. Mammalogy* **57**, 731–46.

Bateson, P.P.G. 1983. Genes, environment and the development of behaviour. In *Genes, Development and Learning* (eds. T.R. Halliday and P.J.B. Slater), p. 52–81. Blackwell, Oxford.

Batchelet, E. 1981. *Circular Statistics in Biology*. Academic Press, London.

Baum, W.M. 1982. Choice, changeover, and travel. *J. Exp. Anal. Behavior* **38**, 35–49.

Baum, W.M. 1983. Studying foraging in the psychological laboratory. In: *Animal cognition and Behavior* (eds. R.L. Mellgren), p. 253–84. North-Holland, The Netherlands.

Beevers, M., Lewis, W.J., Gross, H.R. and Nordlund, D.A. 1981. Kairomones and their use for management of entomophagous insects. X. Laboratory studies on manipulations of host-finding behavior of *Trichogramma pretiosum* Riley with a Kairomone extracted from *Heliothis zea* (Boddie) moth scales. *J. Chem. Ecol.*, **7**, 635–48.

Bekoff, M. 1974. Social play and play-soliciting by infant canids. *Amer. Zool.* **14**, 323–40.

Bekoff, M. and Byers, J. 1979. A critical reanalysis of the ontogeny and phylogeny of mammalian social and locomotor play: an ethological hornet's nest. In *Behavioural Development in Animals and Man* (eds. K. Immelmann, G. Barlow, M. Main and L. Petrinovich). p. 296–337. Bielefield Interdisciplinary Conference, Cambridge University Press.

Beling, I. 1935. Uber das Zeitgedachtnis bei Tieren. *Biol. Rev.*, **10**, 18–41.

Bell, W.J. 1984. Chemo-orientation in walking insects. In *Chemical Ecology of Insects* (eds W.J. Bell and R.T. Card), pp. 93–106. Chapman and Hall, London.

Bell, W.J. 1985. Sources of information controlling motor patterns in arthropod local search orientation. *J. Insect Physiol.* **31**, 837–47.

Bell, W.J. and Nagle, K.J. 1987. Pattern of inheritance of adult rover/sitter traits. *Dros. Inform. Serv.*, **66**, 22–23.

Bell, W.J. and Tobin, T.R. 1982. Chemo-orientation. *Biol. Rev. Cambr. Phil. Soc.* **57**, 219–260.

Bell, W.J. and Tortorici, C. 1987. Genetic and non-genetic control of search duration in adults of two morphs of *Drosophila melanogaster*. *J. Insect Physiol.* **33**, 51–4.

Bell, W.J., Tobin, T.R., Vogel, G. and Surber, J. 1983. Course control and visual orientation in cockroaches. *Physiol. Entomol.* **8**, 121–132.

Bell, W.J., Tortorici, C., Roggero, R.J., Kipp, L.R. and Tobin, T.R. 1985. Sucrose-stimulated searching behaviour in *Drosophila melanogaster* in a uniform habitat: modulation by period of deprivation. *Anim. Behav.*, **33**, 436–448.

Belovsky, G.E. 1984. Herbivore optimal foraging: a comparative test of three models. *Amer. Nat.* **124**, 97–115.

Bernays, E.A. and Chapman, R.F. 1974. The regulation of food intake by acridids. In *Experimental Analysis of Insect Behaviour* (eds. by L. Barton Browne). pp. 48–59, Springer-Verlag, New York.

Bernstein, R. 1975. Foraging strategies of ants in response to variable food density. *Ecology* **56**, 213–219.

Berthold, P. and Querner, U. 1981. Genetic basis of migratory behavior in European Warblers. *Science* **212**, 77–79.

Beukema, J.J. 1968. Predation by the three-spined stickleback (*Gasterosteus acculeatus* L.): the influence of hunger and experience. *Behaviour.*, **31**, 1–126.

Bider, J.R., Thibault, P. and Sarrazin, R. 1968. Schémes dynamiques spatio-temporels de l'activité de *Procyon lotor* en relation avec le comportement. *Mammalia* **32**, 137–63.

Birch, M.C. 1984. Aggregation in bark beetles. *Chemical Ecology of Insects.* (eds W.J. Bell and R.T. Cardé), pp. 331–53. Chapman and Hall, London.

Blaxter, J.H.S. 1968. Visual threshold and spectral sensitivity of herring larvae. *J. exp. Biol.*, **48**, 39–53.

Bobisud, L.B. and Neuhaus, R.J. 1975. Pollinator constancy and survival of rare species. *Oecologia*, **21**, 263–72.

Bolles, R.C. 1962. The readiness to eat and drink: the effect of deprivation conditions. *J. Comp. Physiol. Psychol.*, **55**, 230–4.

Bond, A.B. 1980. Optimal foraging in a uniform habitat: the search mechanism of the green lacewing. *Anim. Behav.*, **28**, 10–19.

Borgia, G. 1982. Experimental changes in resource structure and male density: size-related differences in mating success among male *Scatophaga stercoraria. Evolution* **36**, 307–15.

Borgia, G. 1980. Size- and density-related changes in male behaviour in the fly *Scatophaga stercoraria. Behaviour* **75**, 185–206.

Borgia, G. 1981. Strategies of mate selection in the fly *Scatophaga stercoraria. Anim. Behav.* **29**, 71–80.

Bornbusch, A.H. amd Conner, W.E. 1986. Effects of self-steered turn size and turn bias upon simulated chemoklinotactic behavior. *J. Theor. Biol.* **122**, 7–18.

Bovet, P. and Benhamou, S. 1988. Spatial analysis of animals' movements

using a correlated random walk model. *J. Theor. Biol.* **131**, 419–33.
Bradbury, J.W. 1977. Social organization and communication. In *Biology of Bats*, vol. 3 (ed. Wimsatt, W.), pp. 1–72. New York: Academic Press.
Brady, J. 1981. Behavioral rhythms in invertebrates. In *Handbook of Behavioral Neurobiology* (ed. J. Aschoff), pp. 125–40. Plenum Press, New York.
Brant, D.H. and Kavanau, J.L. 1965. Exploration and movement patterns in the canyon mouse *Peromyscus cranitus* in an extensive laboratory enclosure. *Ecol.* **46**, 452–61.
Brown, J.A. and Colgan, P.W. 1984. The ontogeny of feeding behavior in four species of centrarchid fish. Behav. Process. **9**, 395–411.
Brown, L. and Downhower, J.F. 1988. Analyses in Behavioral Ecology, A Manual for Lab and Field. Sinauer Press, Sunderland, MA.
Bunnell, F.L. and Gillingham, M.P. 1985. Foraging behavior: dynamics of dining out. In *Bioenergetics of Wild Herbivores* (ed. by R.J Hudson and R.G. White), pp. 53–79. CRC Press, Boka Raton, Florida.
Cade, W.H. 1979. Effects of male-deprivation on female phonotaxis in field crickets (Orthoptera: Gryllidae; *Gryllus*). *Can. Entomol.* **111**, 741–44.
Cain, M.L. 1985. Random search by herbivorous insects: a simulation model. *Ecology* **66**, 876–88.
Cain, M.L., Eccleston and Kareiva, P.M. 1985. The influence of food plant dispersion on caterpillar searching success. *Ecol. Entomol.* **10**, 1–7.
Caldwell, R.L. and Hegmann, J.P. 1969. Heritability of flight duration in the Milkweed Bug *Lygaeus kalmii. Nature* **223**, 91–2.
Caraco, T. 1981. Energy budgets, risk and foraging preferences in dark eyed juncos *Junco hyemalis. Behav. Ecol. Sociobiol.* **8**, 213–17.
Caraco, T. and Wolf, L.L. 1975. Ecological determinants of group sizes of foraging lions. *Amer. Natur.* **109**, 343–52.
Caraco, T. and Martindale S. and Pulliam, H.R. (1980). Avian flocking in the presence of a predator. *Nature* **285**, 400–1.
Cardé, R.T. 1984. Chemo-orientation in flying insects. In *Chemical Ecology of Insects* (ed. by W.J. Bell and R.T. Cardé), pp. 111–21. Chapman and Hall, London.
Carl, E.A. 1971. Population control in arctic ground squirrels. *Ecol.* **52**, 395–413.
Carlson, A. and Moreno, J. 1981. Central place foraging in the wheatear *Oenanthe oenanthe*: an experimental test. *J. Anim. Ecol.* **50**, 917–24.
Caro, T.M. 1980. The effects of experience on the predatory patterns of cats. *Behav. Neural. Biol.* **29**, 1–28.
Caro, T.M. and Bateson, P. 1986. Organization and ontogeny of alternative tactics. *Anim. Behav.* **34**, 1483–99.
Carter, M.C. and Dixon, A.F.G. 1982. Habitat quality and the foraging behaviour of coccinellid larvae. *J. Anim. Ecol.* **51**, 865–78.
Chance, M.R.A. and Mead, A.P. 1955. Competition between feeding and

investigation in the rat. *Behaviour* **8**, 174–82.
Chandler, A.E.F. 1969. Locomotory behaviour of first instar larvae of aphidophagous Syrphidae (Diptera) after contact with aphids. *Anim. Behav.* **17**, 673–78.
Chapman, R.M. and Levy, N. 1957. Hunger drive and reinforcing effect of novel stimuli. *J. Comp. Physiol. Psychol.* **50**, 233–38.
Charnov, E.L. 1976. Optimal foraging: The marginal value theorem. *Theoret. Pop. Biol.* **9**, 129–36.
Charnov, E.L., Orians, G.H. and Hyatt, K. 1976. Ecological implications of resource depression. *Amer. Nat.* **110**, 247–59.
Chase, R., Pryer, K., Baker, R. and Madison, D. 1978. Responses to conspecific chemical stimuli in the terrestrial snail Achatina fulica (Pulmonata: Sigmurethra) *Behav. Biol.* **22**, 302–15.
Cheverton, J., Kacelnik, A. and Krebs, J.R. 1985. Optimal foraging: constraints and currencies. In *Experimental Behavioral Ecology and Sociobiology* (Ed. by B. Hölldobler and M. Lindauer), pp. 109–26. Sinauer Associates, Sunderland MA.
Chew, F.S. 1977. Coevolution of pierid butterflies and their cruciferous foodplants. II. The distribution of eggs on potential foodplants. *Evolution* **31**, 568–79.
Cloarec, A. 1969. Etude de'scriptive et experimentale du comportement des capture de *Ranatra linearis* au cours de son ontogene'se. *Behaviour* **35**, 83–113.
Cock, M.J.W. 1977. Searching behaviour of polyphagous predators. Ph.D. thesis, University of London.
Cody, M.L. 1968. On the methods of resource division in grassland bird communities. *Amer. Nat.* **102**, 107–47.
Cody, M.L. 1971. Finch flocks in the Mohave desert. *Theor. Pop. Biol.* 2, 142–48.
Cody, M.L. 1974. Optimization in ecology. Science 183, 1156–64.
Cole, G.F. 1972. Grizzly bear-elk relationships in Yellowstone National Park. *J. Wildl. Management.* **36**, 556–61.
Cook, R.G., Brown, M.F. and Riley, D.A. 1985. Flexible mimory processing by rats, use of prospective and retrospective information in the radial maze. *J. Exp. Psychol., Anim. Behav. Process.* **3**, 453–69.
Cooper, R.M. and Zubek, J.P. 1958. Effect of enriched and restricted early environments on the learning ability of bright and dull rats. *Can. J. Psychol.* **12**, 159–64.
Coulson, J.C. and Brazendale, M.G. 1968. Movements of cormorants ringed in the British Isles and evidence of colony-specific dispersal. *Br. Birds* **61**, 1–21.
Cowan, P.E. 1977. Systematic patrolling and orderly behaviour of rates during recovery from deprivation. *Anim. Behav.* **25**, 171–84.
Cowie, R.J. 1977. Optimal foraging in great tits (*Parus major*). *Nature, Lond.* **268**, 137–9.

Cowie, R.J., and Kerbs, R.J. 1979. Optimal foraging in patchy environments. In *Population Dynamics*, (Ed. by R.M. Anderson, B.D. Turner and L.R. Taylor). Blackwell Scientific Publications, Oxford, England.
Cowie, R.J., Krebs, J.R. and Sherry, D.F. 1981. Food storing by marsh tits. *Anim. Behav.* **29**, 1252–59.
Croll, R.P. and Chase, R. 1980. Plasticity of olfactory orientation to foods in the snail *Achatina fulica. J. Comp. Physio.* **136**, 267–77.
Crowcroft, P. 1954. The daily cycle of activity in British shrews. *Proc. Zool. Soc. Lond.* **123**, 715–29.
Crowley, P.H. 1979. Behavior of Zygopteran nymphs in a simulated weed bed. *Odonatologica*, **8**, 91–101.
Croze, H. 1970. Searching image in carrion crows. *Z. Tierpsychol.* Suppl. **5**, pp. 1–85.
Curio, E. 1976. The Ethology of Predation. Springer-Verlag, Berlin.
Daan, S. and Slopsema, S. 1978. Short-term rhythms in foraging behaviour of the common vole, *Microtus arvalis. J. Comp. Physiol.* **127**, 215–27.
Daan, S. 1981. Adaptive daily strategies in behavior. In *Handbook of Behavioral Neurobiology*, Vol 4, *Biological Rhythms* (Ed. by J. Aschoff), pp 275–98. Plenum Press, New York.
Dalquest, W.W. and Walton, D.W. 1970. Diurnal retreats of bats. In *About Bats. A Chiropteran Biology Symposium* (ed. Slaughter, B.H. and Walton, D.W.), pp. 162–87. Dallas: Southern Methodist University Press.
Darwin, C. 1876. The Effect of Cross- and Self-Fertilization in the Animal Kingdom. Murray, London.
Daumer, K. 1958. Blumenfarben, wie sie die bienen sehen. *Z. Vergl. Physiol.* **41**, 49–110.
David, W.A.L. and Gardiner, B.O.C. 1962. Oviposition and the hatching of eggs of *Pieris brassicae* in a laboratory culture. *Bull. Entomol. Res.* **53**, 91–109.
Davies, N.B. 1977. Prey selection and the search strategy of the spotted flycatcher (*Muscicapa striata*), a field study on optimal foraging. *Anim. Behav.* **25**, 1016–33.
Davies, N.B. 1978. Territorial defence in the speckled wood butterfly (Pararge aegeria), the resident always wins. *Anim. Behav.* **26**, 138–47.
Davies, N.B. 1982. Behaviour and competition for scarce resources. In *Current Problems in Sociobiology* (Ed. by King's College Sociobiology Group), pp. 363–80. Cambridge University Press, Cambridge.
Davies, N.B. and Halliday, T.R. 1979. Competitive mate searching in common toads, *Bufo bufo. Anim. Behav.* **27**, 1253–67.
Davies, N.B. and Houston, A.I. 1981. Owners and satellites: the economics of territory defence in pied wagtail, *Motacilla alba. J. Anim. Ecol.* **50**, 157–80.
Davis, D.E. 1967. The annual rhythm of fat deposition in woodchucks *(Marmota monax). Physiol. Zool.* **40**, 391–402.

Davis, E.E. and Takahashi, F.T. 1980. Humoral alteration of chemoreceptor sensitivity in the mosquito. In *Olfaction and Taste VII* (ed. H. van der Starre), pp. 139–42. IRL Press Ltd., London.

Davis, W.J. Mpitsos, G.J. Pinneo. J.M. and Ram, J.L. 1977. Modification of the behavioral hierarchy in Pleurobranchaea I. Satiation and feeding motivation. *J. Comp. Psychology.* **117**, 99–125.

Dawkins, M.E. 1971. Perceptual changes in chicks: another look at the 'search image' concept. *Anim. Behav.* **19**, 566–74.

de Boer, G. and Hanson, F.E. 1987. Differentiation of roles of chemosensory organs in food discrimination among host and non-host plants by larvae of the tobacco hornworm, *Manduca sexta. Physiol. Entomol.* **12**, 387–98.

Dethier, V.G. 1953. Host plant reception in phytophagous insects. *Proc. 9th Int. Congr. Entomol.* **2**, 81–8.

Dethier, V.G. 1957. Communication by insects: physiology of dancing. *Science* **125**, 331–6.

Dethier, V.G. 1959. Chemical factors determining the choice of food plants by papilio larvae. *Amer. Nat.* **75**, 61–73.

Dethier, V.G. 1976. *The Hungry Fly*. Harvard University Press, Cambridge.

Dethier, V.G. 1982. The contribution of insects to the study of motivation. In *Changing Concepts of the Nervous System* (ed. by A.R. Morrison and P.L. Strick), pp. 445–55. Academic Press, New York.

Dingle, H. 1968a. The influence of environment and heredity of flight activity in the milkweed bug *Oncopeltus. J. Exp. Biol.* **48**, 175–84.

Dingle, H. 1968b. Life history and population consequences of density, photo-period, and temperature in a migrant insect, the milkweed bug, *Oncopeltus. Amer. Nat.* **102**, 149–63.

Dingle, H. 1980. Ecology and evolution of migration. In *Animal Migration, Orientation, and Navigation.* (ed. S.A. Gauthreaux, Jr.), pp. 78–83. Academic Press, New York.

Dixon, A.F.G. 1959. An experimental study of the searching behaviour of the predatory coccinellid beetle *Adalia decempunctata* (L.). *J. Anim. Ecol.* **28**, 259–81.

Dixon, A.F.G. 1969. Population dynamics of the sycamore aphid, *Drepanosiphum platanoides* (Schr.) (Hemiptera: Aphididae): migratory and trivial flight activity. *J. Anim. Ecol.* **38**, 585–606.

Dolnik, V.R. and Blyumental, T.I. 1967. Autumnal premigratory and migratory periods in the chaffinch (*fringilla coelebs coelebs*) and some other temperate-zone passerine birds. *Condor* **69**, 435–68.

Dolnik, V.R. and Gaurilov, V.M. 1972. Photoperiodic control of annual cycles in the chaffinch, a temperate-zone migrant (in Russian, with English summary.) *Zool. Zh.* **51**, 1685–96.

Dominey, W.J. 1984. Alternative mating tactics and evolutionary stable strategies. *Amer. Zool.* **24**, 385–96.

Dorst, J. 1962. *The Migration of Birds*. Heinemann, London.
Douwes, P. 1968. Host selection and host finding in the egg-laying female *Cidaria albulata* L. (Lepidoptera: Geometridae). *Opusc. Entomol.* **33**, 233–79.
Downes, W.J. and Dahlem, G.A. 1987. Keys to the evolution of diptera: role of homoptera. *Envir. Entomol.* **16**, 847–54.
Dowse, H.B., Hall, J.C. and Ringo, J.M. 1987. Circadian and ultradian rhythms in period mutants of *Drosophila melanogaster. Behav. Genet.* **17**, 19–35.
Drees, O. 1952. Untersuchungen Über die angeborenen Verhaltensweisen bei Springspinen (Salticidae). *Z. Tierpsychol.* **9**, 169–207.
Dryl, S. 1973. Chemotaxis in ciliate protozoa. In *Behaviour of Micro-Organisms* (ed. by A. Pérez-Miravete). Plenum Press, N.Y.
Dudai, Y., Jan, Y.-N., Byers, D., Quinn, W.G. and Benzer, S. 1976. *Dunce*, a mutant of *Drosophila* deficient in learning. *Proc. Nat. Acad. Sci. USA* **73**, 1684–88.
Dunbrack, R.L. and Dill, L.M. 1984. Three dimensional prey reaction field of the juvenile coho salmon (*Oncorhyncus kisutch). Can. J. Fish. Aquat. Sci.* **41**, 1176–82.
Dunn, R.M. 1982. Choice, relative reinforcer duration, and the changeover ratio. *J. Exper. Anal. Behav.* **38**, 313–19.
Ehlinger, T.J. 1986. Learning, sampling and the role of individual variability in the foraging behavior of bluegill sunfish. Ph.D. Dissertation, Michigan State University.
Einon, D.F. and Morgan, M.J. 1977. A critical period for social isolation in rats. *Devel. Psychobiol.* **10**, 123–32.
Eisenberg, J.F., McKay, G.M. and Jainudeen, M.R. 1971. Reproductive behaviour of the Asiatic elephant (*Elephas maximus* L.) *Behaviour* **38**, 193–225.
Ellen, P. and Thinus-Blanc, C. (eds) 1987. Cognitive Processes and *Spatial Orientation in Animal and Man.* Volume I, *Experimental Animal Psychology and Ethology*. Martinus Nijhoff Publishers, Dordrecht.
Elliott, J.P. Cowan, Mc. and Holling, C.S. 1977. Prey capture by the African lion. *Can. J. Zool.* **55**, 1811–28.
Elner, R.W. and Hughes, R.N. 1978. Energy maximization in the diet of the shore crab, *Carcinus maenas. J. Anim. Ecol.* **47**, 103–16.
Emlen, S.T. and Ambrose, H.W., III. 1970. Feeding interactions of snowy egrets and red-breasted mergansers. *Auk* **87**, 164–5.
Enders, F. 1975. The influence of hunting manner on prey size, particularly in spiders with long attack distances (Araneidae, Linyphiidae, and Salticidae). *Amer. Nat.* **109**, 737–63.
Enright, J.T. 1975. The circadian tape recorder and its entrainment. In *Physiological Adaptation to the Environment* (ed. F.J. Vernberg) pp. 465–76. Intext, New York.

Ens, B.J. and Goss-Custard, J.D. 1984. Interference among Oystercatchers, *Haematopus ostralegus*, feeding on mussels, *Mytilus edulis*, on the Exe Estuary. *Anim. Ecol.* **53**, 217–31.

Etienne, A.S. 1969. Analyse der schlagauslösenden Bewegungsparameter einer punktförmigen Beuteattrappe bei der *Aeschnalarve. Z. Vergl. Physiol.* **64**, 71–110.

Evans, B.I. and O'Brien, W.J. 1986. An analysis of the feeding rate of white crappie. *Contemporary studies on fish feeding* (ed. Charles A. Simenstad and Gregor M. Cailliet) pp. 299–306. Dr. W. Junk Publishers, The Netherlands.

Evans, H.F. 1976. The searching behaviour of *Anthocoris confusus* (Reuter) in relation to prey density and plant surface topography. *Ecol. Entomol.* **1**, 163–9.

Fagen, R. 1982. Evolutionary issues in development of behavioral flexibility. In *Perspective in Ethology*, Vol. 5, Ontogeny (ed. P.P.G. Bateson and P.M. Klopfer). pp. 365–83. Plenum Press, New York.

Falconer, D.S. 1981. *Introduction to Quantitative Genetics*. Longman, New York.

Fantino, E. 1969. Choice and rate of reinforcement. *J. Exper. Anal. Behav.* **12**, 723–30.

Fantino, E. and Abarca, N. 1985. Choice, optimal foraging and the delay-reduction hypothesis. *Behav Brain Sci.* **8**, 315–30.

Feeny, P.P. 1975. Biochemical coevolution between plants and their insect herbivores. In *Coevolution of Animals and Plants* (ed. by L.E. Gilbert and P.R. Raven) pp. 3–19. University of Texas Press, Austin.

Ferguson, M.M. and Noakes, D.L.G. 1983. Movers and stayers: genetic analysis of mobility and positioning in hybrids of Lake Charr, *Salvelinus namaycush*, and Brook Charr, *S. fontinalis* (Pisces, Salmonidae). *Behav. Gen.* **13**, 213–22.

Ferguson, M.M., Noakes, D.L.G. and Romani, D. 1983. Restricted behavioural plasticity of juvenile lake charr, *Salvelinus namaycusch. Env. Biol. Fish.* **8**, 151–6.

Fink, J.B. and Patton, R.M. 1953. Decrement of a learned drinking response accompanying changes in several stimulus characteristics. *J. Comp. Physiol. Psychol.* **46**, 23–7.

Fitzgerald, T.D. and Peterson, S.C. 1983. Elective recruitment by the eastern tent caterpillar (*Malacosoma americanum*). *Anim. Behav.* **31**, 417–23.

Flügge, C. 1934. Geruchliche Raumorientierung von *Drosophila melanogaster. Z. vergl. Physiol.* **20**, 464–99.

Formanowicz, D.R., Jr. and Bradley, P.J. 1987. Fluctuations in prey density: effects on the foraging tactics of scolopendrid centipedes. *Anim. Behav.* **35**, 453–61.

Fowler, H. 1965. *Curiosity and Exploratory Behaviour*. Macmillan, New York.

Frankie, G.W., Opler, P.A. and Bawa, K.S. 1976. Foraging behavior of solitary bees: implications for outcrossing of a neotropical forest tree species. *J. Ecol.* **64**, 1049–57.

Franks, N.R. 1986. Teams in social insects: group retrieval of prey by army ants (*Eciton burchelli*, Hymenoptera: Formicidae). *Behav. Ecol. Sociobiol.* **18**, 425–9.

Fretwell, S.D. 1972. *Populations in a Seasonal Environment.* Princeton University Press, Princeton.

Fretwell, S.D. and Lucas, H.L. 1970. On territorial behaviour and other factors influencing habitat distribution in birds. *Acta Biotheoretica* **19**, 16–36.

Frisch, K. von 1967. *The Dance Language and Orientation of Bees.* Harvard University Press, Cambridge, MA.

Fromm, J.E. 1988. Search orientation in *Musca domestica.* Ph.D. Thesis, Univ. Kansas.

Fromm, J.E. and Bell, W.J. 1987. Search orientation of *Musca domestica* in patches of sucrose drops. *Physiol. Entomol.* **12**, 297–307.

Futuyma, D.J. 1983. Selective factors in the evolution of host choice by phytophagous insects, In *Herbivorous Insects Host-seeking Behavior and Mechanisms* (Ed. by S. Ahmad), pp. 227–44. Plenum Press, New York.

Fuyama, Y. 1978. Behavior genetics of olfactory responses in Drosophila. II. An odorant-specific variant in a natural population of *Drosophila melanogaster. Behav. Gen.* **8**, 399–414.

Gandolfi, G., Mainardi, D. and Rossi, A.C. 1968. La reazione di paura d lo svantaggio individuale dei pesci allarmisti (esperimenti con modelli). *Zoologia* **102**, 8–14.

Gass, C.L. 1978. Experimental studies of foraging in complex laboratory environments. *Amer. Zool.* **18**, 729–38.

Gass, C.L. 1985. Reaching for an integrated science of behavior. *Behav. Brain Sci.* **8**, 337–8.

Gass, C.L., and Montgomerie, R.D. 1981. Hummingbird foraging behavior, decision-making and energy regulation. In *Foraging Behavior. Ecological, Ethological, and Psychological Approaches* (ed. A.C. Kamil and T.D. Sargent), pp. 159–94. Garland STPM Press, New York.

Gendron, R.P. 1986. Searching for cryptic prey: evidence for optimal search rates and the formation of search images in quail. *Anim. Behav.* **34**, 898–912.

Gendron, R.P. and Staddon, J.E.R. 1983. Searching for cryptic prey: the effect of search rate. *Am. Nat.* **121**, 172–86.

Gentry, A.H. 1974. Coevolutionary patterns in central Americal Bignoniaceae. *Ann. Missouri Botan.* Gardens **61**, 728–59.

Gerhardt, H.C. and Rheinlaender, J. 1980. Accuracy of sound localization in a miniature dendrobatic frog. *Naturwissen.* **67**, 362–3.

Getz, L.L. 1970. Influence of vegetation on the local distribution of the meadow vole in southern wisconsin. *Occas. Papers, Biol. Sci. Ser. University of Connecticut*, **1**, 213–41.

Gibb, J.A. 1954. Feeding ecology of tits, with notes on the treecreeper and

goldcrest. *Ibis* **96**, 513–43.
Gibb, J.A. 1958. Predation by tits and squirrels on the eucosmid *Ernarmonia conicolana* (Heyl.) *J. Anim. Ecol.* **27**, 275–96.
Gibb, J.A. 1962. L. Tinbergen's hypothesis of the role of specific search images. *Ibis* **104**, 106–11.
Gilbert, L.E. 1975. Ecological consequences of a coevolved mutualism between butterflies and plants. In *Coevolution of Animals and Plants* (eds. L. E. Gilbert, and P. H. Raven,, pp.210–40. University of Texas Press, Austin.
Gilbert, L.E. 1980. Ecological consequences of a coevolved mutualism between butterflies and plants. In *Coevolution of Animals and Plants* (Ed. by L.E. Gilbert and P.H. Raven), pp. 210–40. University of Texas Press, Austin.
Gill, F.B. and Wolf, L.L. 1975. Economics of feeding territoriality in the golden-winged sunbird. *Ecology* **56**, 333–45.
Giraldeau, L.-A. and Lefebvre, L. 1987. Scrounging prevents cultural transmission of food-finding behaviour in pigeons. *Anim. Behav.* **35**, 387–94.
Glen, D.M. 1974. Searching behaviour and prey-density requirements of Blepharidopterus angulatus (Fall.) (Heteroptera:Miridae) as a predator of the lime aphid, *Eucallipterus tiliae* (L.), and leafhopper, Alnetoidae alneti (Dahlbom) *J. Anim. Ecol.* **44**, 115–34.
Glickman, S.E. and Sroges, R.W. 1966. Curiousity in zoo animals. *Behaviour* **24**, 151–88.
Glickman, S.E., Higgins, T.J. and Isaacson, R.L. 1970. Some effects of hippocampal lesions on the behavior of mongolian gerbils. *Physiol. Behav.* **5**, 931–8.
Godin, J.J. and Keenleyside, M.H.A. 1984. Foraging on patchily distributed prey by a cichlid fish (Teleostei, Cichlidae): a test of the ideal free distribution theory. *Anim. Behav.* **32**, 120–31.
Gossard, T.W. and Jones, R.E. 1977. The effects of age and weather on egg-laying in *Pieris rapae* L. *J. Appl. Ecol.* **14**, 65–71.
Goss-Custard, J.D. 1977. Optimal foraging and the size selection of worms by redshank, *Tringa totanus*, in the field. *Anim. Behav.* **25**, 10–29.
Götmark, F., Winkler, D.W. and Andersson, M. 1986. Flock-feeding of fish schools increases individual success in gulls. *Nature* **319**, 589–91.
Götz, K.G. 1980. Visual guidance in *Drosophila*. In *Development and Neurobiology of Drosophila* (ed. O. Siddiqi, P. Babu, L.M. Hall and J.C. Hall) pp. 391–407, Plenum Press, New York.
Gould, S.J. and Lewontin, R.C. 1979. The spandrels of San Marco and the Panglossian paradigm: A critique of the adaptationist programme. *Proc. R. Soc. London Ser. B.* **205**, 147–64.
Graf, S.A. and Sokolowski, M.B. 1989. Rover/sitter *Drosophila melangoster* larval foraging polymorphism as a function of larval development, food

patch quality and starvation. *J. Insect Behav.* **2**, 301–13.
Grant, J.W.A. and Noakes, D.L.G. 1987. Movers and stayers: Foraging tactics of young-of-the-year brook charr, *Salvelinus fontinalis*. *J. Anim. Ecol.* **56**, 1001–13.
Gray, R.D. 1987. Faith and foraging A: Critique of the "Paradigm Argument from Design" In *Foraging Behavior* (Kamil, A.C., Krebs, J.R. and Pulliam, H.R.), pp. 69–140. Plenum Perss, New York.
Green, G.W. 1964. The control of spontaneous locomotor activity in *Phormia regina* Meigen – I. Locomotor activity patterns in intact flies. *J. Insect Physiol.* **10**, 711–26.
Greenblatt, J.A., and Witter, J.A. 1976. Behavioral studies on *Malacosoma disstria* (Lepidoptera: Lasiocampidae). *Can. Entomol.* **108**, 1225–28.
Griffin, D.R. 1944. The sensory basis of bird navigation. *Quart. Rev. Biol.* **19**, 15–31.
Griffin, D.R. 1958. *Listening in the Dark*. Yale University Press, New Haven, Conn.
Griffiths, D. 1980. The feeding biology of ant-lion larvae: prey capture, handling and utilisation. *J. Anim. Ecol.* **49**, 99–125.
Gross, G.C. 1968. General activity. In *Analysis of Behavioral Change* (ed. L. Weiskrantz). pp. 91–106. Harper and Row, New York.
Grubb, P. and Jewell, P.A. 1974. Movement, daily activity, and home range of Soay sheep. In *Island Survivors: The Ecology of the Soay Sheep of St. Kilda* (ed. P.A. Jewell, C. Milner and J. Morton Boyd), pp. 160–94. The Athlone Press, University of London, London.
Guilford, T. and Dawkins,M.S. 1987. Search images not proven: a reappraisal of recent evidence. *Anim. Behav.* **35**, 1838–45.
Gwinner, E. 1977. Circannual rhythms in bird migration. *A. Rev. Syst. Ecol.* **8**, 381–405.
Hafez, M. 1961. Seasonal fluctuations of population density of the cabbage aphid, *Brevicoryne brassicae* (L.) in the Netherlands and the role of its parasite, *Aphidus (Diaretiella) rapae* (Curtis). *Tijdschrift over Plantenzeikten* **67**, 445–548.
Hagen, O.V. 1967. Nachweis einer kinasthetischen Orientierung bei *Uca rapax*. *Z. Morph. Okol. Tiere* **58**, 301–20.
Haldane, J.B.S. 1946. The interaction of nature and nurture. *Ann. Eugen. (Lond.)* **13**, 197.
Haley, K.B and Stone, L.D.(eds.) 1980. *Search Theory and Applications*. Plenum Press, New York.
Hall, J.C. 1979. Control of male reproductive behavior by the central nervous system of *Drosophila*: dissection of a courtship pathway by genetic mosaics. *Genetics* **92**, 437–57.
Hall, J.C. 1985. Genetic Analysis of Behavior in insects. In *Comprehensive Insect Physiology Biochemistry and Pharmacology* Vol. 9 (ed. G.A. Kerkut and L.I. Gilbert), p. 287–373. Pergamon Press, New York.

Hall, K.R.L. 1965. Behaviour and ecology of the wild patas monkey. *J. Zool.*, Lond. **148**, 15–87.

Hanski, I. 1980. Movement patterns in dung beetles and in the dung fly. *Anim. Behav.* **28**, 953–64.

Hanson, F.E. 1983. The behavioral and neurophysiological basis of food plant selection by lepidopterous larvae. In *Herbivorous Insects. Host-Seeking Behavior and Mechanisms* (ed. S. Ahmad). pp. 3–21. Academic Press, New York.

Hansson, L. 1971. Small rodent food, feeding and population dynamics. *Oikos* **22**, 183–98.

Hardie, J. 1980. Behavioural differences between alate and apterous larvae of the black bean aphid, *Aphis fabae*: dispersal from the host plant. *Entomol. Exp. Appl.* **28**, 338–40.

Harkness, R.D. and Maroudas, N.G. 1985. Central place foraging by an ant (*Cataglyphis bicolor* Fab.) a model of searching *Anim. Behav.* **33**, 916–28.

Harper, D.G.C. 1982. Competitive foraging in mallards: 'ideal free' ducks. *Anim. Behav.* **30**, 575–84.

Harris, M.O. and Miller, J.R. 1982. Synergism of visual and chemicals stimuli in the oviposition behavior of *Delia antiqua* (Meigen)(Diptera: Anthomyiidae). *Proc. 5th Int. Symp. Insect– Plant Relationships*, Pudoc Wageningen.

Harris, M.O. and Miller, J.R. 1984. Foliar form influences ovipositional behaviour of the onion fly. *Physiol. Entomol.* **9**, 145–55.

Hartnoll R.G. and Wright, J.R. 1977. Foraging movements and homing in the limpet *Patella Vulgata* L. *Anim. Behav.* **25**, 806–10.

Haskell, P.T., Paskin, M.W.J. and Moorhouse, J.E. 1962. Laboratory observations on factors affecting the movements of hoppers of the desert locust. *J. Insect Physiol.* **8**, 53–78.

Hassell, M.P. 1978. *Arthropod Predator-Prey Systems.* Princeton University Press, Princeton, New Jersey.

Hassell, M.P. and Southwood, T.R.E. 1978. Foraging strategies of insects. *A. Rev. Ecol. Syst.* **9**, 75–98.

Havukkala, I. and Kennedy, J.S. 1984. A programme of self-steered turns as a humidity response in *Tenebrio*, and the problem of categorizing spatial manoeuvres. *Physiol. Entomol.* **9**, 157–64.

Hawkes, C. and Coaker, T.H. 1979. Factors affecting the behavioural responses of the adult cabbage root fly, *Delia brassicae*, to host plant odour. *Entomol. Exp. Appl.* **25**, 45–58.

Heatwole, H. and Heatwole, A. 1968. Motivational aspects of feeding behavior in toads. *Copeia* **4**, 692–8.

Hedrick, P.W. 1986. Genetic polymorphism in heterogeneous environments: A decade later. *A. Rev. Ecol. and Syst.* **17**, 535–66.

Hegner, R.E. 1982. Central place foraging in the white-fronted bee-eater. *Anim. Behav.* **30**, 953–63.

Heinrich, B. 1976. The foraging specializations of individual bumblebees. *Ecol. Monographs* **46**, 105–28.
Heinrich, B. 1978. The economics of insect sociality. In *Behavioral Ecology. An Evolutionary Approach.* (ed. R.J. Krebs and N.B. Davies) p. 97–128, Sinauer Assoc., Sunderland MA.
Heinrich, B. 1979a. *Bumblebee Economics.* Harvard University Press, Cambridge, MA.
Heinrich, B. 1979b. Resource heterogeneity and patterns of foraging in bumblebees. *Oecologia* (Berl.) **40**, 234–45.
Heinrich, B. 1979c. "Majoring" and "minoring" by foraging bumblebees, *Bombus vagans*: an experimental analysis. *Ecol.* **60**, 245–55.
Heinrich, B. 1983. Do bumblebees forage optimally, and does it matter? *Amer. Zool.* **23**, 273–81.
Heinrich, B., Mudge, P.R. and Deringis, P.G. 1977. Laboratory analysis of flower constancy in foraging bumblebees, *Bombus ternarius* and *B. terricola*. *Behav. Ecol. Sociobiol.* **2**, 247–65.
Heisenberg, M. 1980. Mutants of the brain structure and function: what is the significance of the mushroom bodies for behavior? In *Development and Neurobiology of Drosophila* (ed. O. Siddiqi, P. Babu, L.M. Hall, and J.C. Hall), pp. 373–390. Plenum Press, New York.
Herbert, H.J. 1972. The Population Dynamics of the Waterbuck, *Kobus ellipsiprymnus* (Ogilby, 1833) in the Sabi-sand Wildtuin. Verlag Paul Parey, Hamburg.
Herrnstein, R.J. 1970. On the law of effect. *J. Exp. Anal. Behav.* **13**, 243–66.
Herrnstein, R.J. and Vaughan, W. 1980. Melioration and behavioral allocation. In *Limits to Action: The Allocation of Individual Behavior* (ed. J.E.R. Staddon) Academic Press, New York.
Hodapp, A. and Frey, D. 1982. Optimal foraging by firemouth cichlids, *Cichlasoma meeki*, in a social context. *Anim. Behav.* **30**, 983–9.
Hodges, C.M. 1981. Optimal foraging in bumblebees, hunting by expectation. *Anim. Behav.* **29**, 1166–71.
Hoffmann, G. 1983a. The random elements in the systematic search behavior of the desert isopod *Hemilepistus reaumuri. Behav. Ecol. and Sociobiol.* **13**, 81–92.
Hoffmann, G. 1983b. The search behavior of the desert isopod *Hemilepistus reaumuri* as compared with a systematic search. *Behav. Ecol. Sociobiol.* **13**, 93–106.
Holling, C.S. 1959. Some characteristics of simple types of predation and parasitism. *Can. Entomol.* **91**, 385–98.
Holling, C.S. 1963. An experimental component analysis of population processes. *Mem. Entomol. Soc. Canada* **32**, 22–32.
Holling, C.S. 1966. The functional response of invertebrate predators to prey density. *Mem. Entomol. Soc. Can.* **48**, 1–86.
Horn, H.S. 1968. The adaptive significance of colonial nesting in the

Brewer's blackbird (*Euphagus cyanociphalus*). *Ecol.* **49**, 682–94.
Huffaker, C.B. and Matsumato, B.M. 1982. Group versus individual functional responses of *Venturia* (=*Nemeritis*) *cansescens* (Grav.). *Res. Pop. Ecol.* **24**, 250–69.
Hulscher, J.B. 1976. Localization of cockles (*Cardium Edule* L.) by the oystercatcher (*Haematopus ostralegus* L.) in darkeness and daylight. *Ardea* **64**, 292–310.
Humphries, D.A. and Driver, P.M. 1970. Protean defence by prey animals. *Oecologia* (Berl.) **5**, 285–302.
Hunter, J.R. 1981. Feeding ecology and predation of marine fish larvae. In *Marine Fish Larvae* (ed. R. Lasker), pp. 33–71. University of Washington Press, Seattle.
Immelmann, K. 1975. Ecological significance of imprinting and early learning. *A. Rev. Ecol. Syst.* **6**, 15–37.
Innes, R.R. and Mabey, G.L. 1964. Studies on browse plants in Ghana. III. Browse/grass ingestion ratios. (a) Determination of the free-choice griffonia/grass ingestion ratio for west African shorthorn cattle on the accra plains using the 'simulated 'shrub' technique. *J. Exper. Agri.* **32**, 180–90.
Inoue, T. and Matsura, T. 1983. Foraging strategy of a mantid, *Paratenodera angustipennis* S.: Mechanisms of switching tactics between ambush and active search. *Oecologia (Berl.)* **56**, 264–71.
Jaeger, R.G., Nishikawa, K.C.B., and Barnard, D.E. 1983. Foraging tactics of a terrestrial salamander: costs of territorial defence. *Anim. Behav.* **31**, 191–8.
Jaenike, J. 1985. Genetic and environmental determinants of food preference in *Drosophila tripunctata. Evolution* **39**, 362–9.
Jaenike, J. 1986a. Intraspecific variation for resource use in *Drosophila. Biol. J. Linn. Soc.* **27**, 47–56.
Jaenike, J. 1986b. Genetic complexity of host-selection behavior in *Drosophila. Proc. Natl. Acad. Sci. USA* **83**, 2148–51.
Jander, R. 1957. Die optische Richtungsorientierung der roten Waldameise (*Formica rufa* L.). *Z. vergl. Physiol.* **40**, 162–238.
Jander, R. 1965. Die Hauptentwicklungsstufen der lichtorientierung bei den tierischen Organismen. *Naturwiss.* **18**, 318–24.
Jander, R. 1975. Ecological aspects of spatial orientation. *A. Rev. Syst. Ecol.* **6**, 171–88.
Janzen, D.H. 1971. *Cassia grandis* L. beans and their escape from predators: A study in tropical predator satiation. *Ecol.* **52**, 964–79.
Janzen, D.H. 1972. Escape in space of *Sterculia apetala* seeds from the bug *Dysdercus fasciatus* in a Costa Rican deciduous forest. *Ecol.* **53**, 350–61.
Jermy, T., Hanson, F.E. and Dethier, V.G. 1968. Induction of specific food preference in lepidopterous larvae. *Entomol. Exp. Appl.* **11**, 211–30.
Johnson, R.P 1973. Scent marking in mammals. *Anim. Behav.* **21**, 521–35.

Jones, R.E. 1976. Search behavior: a study of three caterpillar species. *Behaviour* **60**, 237–59.

Jones, J.S. 1982. Genetic differences in individual behavior associated with shell polymorphism in the snail Cepaea nemoralis. *Nature* **298**, 749–50.

Jones, J.S. and Probert, R.F. 1980. Habitat selection maintains a deleterious allele in a hetrogenous environment. *Nature* **287**, 632–3.

Jonkel, C.J. and Cowan, I. McT. 1971. The black bear in the spruce-fir forest. *Wildlife Monographs* **27**, 5–57.

Kacelink, A., Houston, A.I. and Krebs, J.R. 1981. Optimal foraging and territorial defence in the great tit *Parus major. Ecol. Sociobiol.* **8**, 35–40.

Kamil, A.C. and Balda, R.P. 1985. Cache recovery and spatial mamory in clark's nutcrackers (*Nucifraga columbiana) J. Exper. Psychol.* **11**, 95–111.

Kamil, A.C. and Roitblat, H.L. 1985. Foraging theory: Implications for animal learning and cognition. *Ann. Rev. Psychol.* **36**, 141–69.

Kamil, A.C., Krebs, J.R. and Pulliam, H.R. (eds.) 1987. *Foraging Behavior.* Plenum Press, New York.

Kareiva, P. 1983. Influence of vegetation texture on herbivore populations: resource concentration and herbivore movement. In *Variable Plants and Herbivores in Natural and Managed Systems* (eds R.F. Denno and M.S. McClure), pp. 259–89. Academic Press, New York.

Kareiva, P.M. and Shigesada, N. 1983. Analyzing insect movement as a correlated random walk. *Oecologia (Berl.)* **56**, 234–8.

Kasuya, E. 1982. Central place water collection in a Japanese paper wasp *Polistes chinensis antennalis. Anim. Behav.* **30**, 1010–14.

Kayser, C. and Heusner, A.A. 1967. Le rhythme nycthemeral de la depense d'energie. *J. Physiol.* **59**, 3–116.

Keeton, W.T. 1979. Avian orientation and navigation. *A. Rev. Physiol.* **41**, 353–66.

Kennedy, J.S. 1975. Insect dispersal. In *Insects, Science, and Society* (ed. D. Pimentel), pp. 103–19. Academic Press, New York.

Kenward, R.E. 1978. Hawks and doves: factors affecting success and selection in goshawk attacks on woodpigeons. *J. Anim. Ecol.* **47**, 449–60.

Kettle, D. and O'Brien, W.J. 1978. Vulnerability of arctic zooplankton species to predation by small lake trout *Salvelinus namaycush. J. Fish. Res. Board Can.* **35**, 1495–1500.

Kikuchi, T. 1973. Genetic alteration of olfactory functions in Drosophila melanogaster. *Japan J. Genetics* **48**, 105–18.

Kipp, L.R. 1984. Movement rules of honeybees on real and artificial flowers: locomotion turning theory. Ph.D. thesis, University of Kansas.

Kipp, L.R. 1987. The flight directionality of honeybees foraging on real and artificial inflorescences. *Can. J. Zool.* **65**,587–93.

Kitching, R.L. and Zalucki, M.P. 1982. Component analysis and modelling of the movement process: analysis of simple tracks. *Res. Pop. Ecol.* **24**, 224–38.

Kleber, E. 1935. Hat das Zeitgedachtnis der Beinen biologische Bedeutung? *Z. vergl. Physiol.* **22**, 221–62.

Kleerekoper, H. 1972. Orientation through chemoreceptor in fishes. In *Animal Orientation and Navigation*, (eds S.R. Galler, K. Schmidt-Koeing, C.J. Jacobs and R.E. Beeleville). pp. 459–68. NASA Publ., Washington.

Kleerekoper, H., Matis, J., Gensler, P. and Maynard, P. 1974. Exploratory behaviour of goldfish Carassius auratus. *Anim. Behav.* **22**, 124–32.

Klowden, M.J. and Lea, A.O. 1979. Humoral inhibition of host seeking in *Aedes aegypti* during oocyte maturation. *J. Insect Physiol.* **25**, 231–5.

Koltermann, R. 1971. 24-Std-Periodik in der Langzeiterinnerrung and Duft- und Farbsignalen bei der Honigbiene. *Z. vergl. Physiol.* **75**, 49–68.

Konishi, M. 1973. How the owl tracks its pray. *Amer. Scient.* **61**, 414–24.

Konopka, R.J. and Benzer, S. 1971. Clock mutants of *Drosophila melanogaster. Proc. Nat. Acad. Sci., USA* **68**, 2112–16.

Kovac, M.P. and Davis, W.J. 1977. Behavioral choice: neural mechanisms in *Pleurobranchaea. Science* **198**, 632–4.

Kovac, M.P. and Davis, W.J. 1980. Reciprocal inhibition between feeding and withdrawal behaviors in *Pleurobranchaea. J. Comp. Physiol.* **139**, 77–86.

Kramer, E. 1976. The orientation of walking honeybees in odour fields with small concentration gradients. *Physiol. Entomol.* **1**, 27–37.

Kramer, D.L. and Nowell, W. 1980. Central place foraging in the eastern chipmunk, *Tamias striatus. Anim. Behav.* **28**, 772–8.

Krebs, J.R. 1973. Behavioural aspects of predation. In *Perspectives in Ethology* (ed. P.P.G. Bateson and P.H. Klopfer), pp. 73–111. Plenum Press, New York.

Krebs, J.R. 1974. Colonial nesting and social feeding as strategies for exploiting food resources in the great blue heron (*Ardea herodias*). *Behaviour* **51**, 99–134.

Krebs, J.R. and Davies, N.B. 1978. *Behavioural Ecology, An Evolutionary Approach*. Blackwell, Oxford.

Krebs, J.R. and Davies, N.B. (ed) 1984. *Behavioural Ecology An Evolutionary Approach*, 2nd edition. Sinauer Associates, Sunderland, MA.

Krebs, J.R. and McCleery, R.H. 1984. Optimization in behavioural ecology. In *Behavioural Ecology: An Evolutionary Approach*, (eds R.J. Krebs and N.B. Davies), pp. 91–121. Sinauer Associates Inc. Publishers, Sunderland, Massachusetts.

Krebs, J.R., Erichsen, J.T., Webber, M.I. and Charnov, E.L. 1977. Optimal prey selection in the great tit (*Parus major). Anim. Behav.* **25**, 30–8.

Krebs, J.R., Kacelnik, A. and Taylor, P. 1978. Test of optimal sampling by foraging great tits. *Nature* **275**, 27–31.

Krebs, J.R., MacRoberts, M.H. and Cullen, J.M. 1972. Flocking and feeding in the great tit *Parus major* – an experimental study. *Ibis* **114**, 507–30.

Krebs, J.R., Ryan, J.C. and Charnov, E.L. 1974. Hunting by expectation or optimal foraging? a study of patch use by chickadees. *Anim. Behav.* **22**,

953–64.
Kruuk, H. 1972. *The Spotted Hyena*. University Chicago Press, Chicago.
Kruuk, H. and Sands, W.A. 1972. The aardwolf (*Proteles cristatus* Sparrman, 1783) as a predator of termites. *E. Afr. Wildl. J.* **10**, 211–27.
Kuenen, L.P.S., Baker, T.C. 1982. The effects of pheromone concentration on the flight behaviour of the oriental fruit moth, *Grapholitha molesta. Physiol. Entomol.* **7**, 423–34.
Kummer, H. 1971. *Primate Societies*. Aldine-Atherton, Chicago.
Kuẑmina, L.A. 1977. Effect of the mutation short on signaling behaviour and neurological charactersistics of the honeybee. (In Russian, English summary) *Genetika* **13**, 1552–60.
Lack, D. 1968. *Ecological Adaptations for Breeding in Birds*. Methuen, London.
Laing, J. 1937. Host-finding by insect parasites. I. Observations on the finding of hosts by *Alysia manducator, Mormoniella vitripennis* and *Trichogramma evanescens. J. Anim. Ecol.* **6**, 298–317.
Lande, R. 1976. Natural selection and random genetic drift in phenotypic evolution. *Evolution* **30**, 314–34.
Langley, W.M. 1983. Relative importance of the distance in grasshopper mouse predatory behaviour. *Anim. Behav.* **31**, 199–205.
Laverty, T.M. 1980. The flower-visiting behaviour of bumble bees: floral complexity and learning. *Can. J. Zool.* **58**, 1324–35.
Lawrence, S.W. 1982. Sexual dimorphism in between and within patch movement of a monophagous insect: *Tetraopes* (Coleoptera: Cerambycidae). *Oecologia* (Berl.) **53**, 245–50.
Lees, A.D. 1966. The control of polymorphism in aphids. *Adv. Insect Physiol.* **3**, 207–77.
Lehmann, U. 1976. Short-term and circadian rhythms in the behavior of the vole, *Microtus agrestis* (L.). *Oecologia* (Berl.) **23**, 185–99.
Leir, V. and Barlow, C.A. 1982. Effects of starvation and age on foraging efficiency and speed of consumption by larvae of a flower fly, *Metasyrphus corollae* (Syrphidae). *Can. Entomol.* **114**, 897–900.
Levin, D.A. and Anderson, W.W. 1970. Competition for pollinators between simultaneously flowering species. *Amer. Nat.* **104**, 455–67.
Lewis, A.C. 1986. Memory constraints and flower choice in *Pieris rapae. Science* **232**, 863–5.
Lewontin, R.C. 1970. The units of selection. *A. Rev. Ecol. Syst.* **1**, 1–18.
Leyhausen, P. 1973. *Verhaltensstudien an Katzen*. Paul Parey, Berlin.
Lima, S. 1984. Downy woodpecker foraging behavior: efficient sampling in simple stochastic environments. *Ecol.* **65**, 166–74.
Lima, S.L., Valone, T.J. and Caraco, T. 1985. Foraging efficiency–predation risk trade-off in the grey squirrel. *Anim. Behav.* **33**, 155–65.
Lindauer, M. 1955. Schwarmbienen auf Wohnungssuche. *Z. vergl. Physiol.* **37**, 263–324.
Lindauer, M. 1961. *Communication Among Social Bees*. Harvard University

Press, Cambridge, MA.

Lingren, P.D. Sparks, A.N., Raulston, J.R. and Wolf, W.W. 1978. Night vision equipment for studying nocturnal behavior of insects. *Bull. Entomol. Soc. Amer.* **24**, 206–12.

Linsenmair, K.E. 1972. Die Bedeutung familien spezifisher "Abzeichen" für den Familienzusammenhalt bei der sozialen Wustenassel *Hemilepistus reamuri* Audouin U. Savigny (Crustacea, Isopoda, Oniscordea). *Z. Tierpsycol.* **31**, 131–62.

Lowe, H.J.B. 1981. Resistance and susceptibility to colour forms of the aphid *Sitobion avenae* in spring and winter wheats (*Triticum aestivum*). *Ann. Appl. Biol.* **99**, 87–98.

Lowe, H.J.B. 1984. A behavioural difference amongst clones of the gra· aphid *Sitobion avenae. Ecol. Entomol.* **9**, 119–22.

Luecke, C. and O'Brien, W.J. 1981. Prey location volume of a planktivorous fish: a new measure of prey vulnerability. *Can. J. Fish. Aquat. Sci.* **38**, 1264–70.

Lyman, C.P., Willis, J.S., Malan, A. and Wang, L.C.H. (eds.) 1982. *Hibernation and Torper in Mammals and Birds*. Academic Press, New York.

McAnelly, M.L. 1985. Variation in migratory behavior and its control in the grasshopper, Melanoplus sanguinipes. *Contrib. Mar. Sci. Supple.* **27**, 687–703.

McCleery, R.H. 1983. Interaction between activities. In *Animal Behaviour: Causes and Effects*. (eds T.R. Halliday and P.J.B. Slater). pp. 134–167. Blackwell Scientific Publication, London.

McCoy, M.M. 1984. Antennal movements of the american cockroach *Periplaneta americana*. Ph.D. Thesis. University of Kansas.

McFarland, D.J. 1971. *Feedback Mechanisms in Animal Behaviour*. Academic Press, New York.

McGuire, T.R. and Hirsch, J. 1977. Behavior-genetic analysis of Phormia regina: Conditioning, reliable individual differences, and selection. *Proc. Nat. Acad. Sci.* USA. **74**, 5193–7.

McGuire, T.R. and Tully, T. 1986. Food-search behavior and its relation to the central excitatory state in the genetic analysis of the blowfly, *Phormia regina. J. Comp. Psychol.* **100**, 52–8.

Mackinnon, J. 1974. The behavior and ecology of wild orang-utans (*Pongo pygmaeus*) *Anim. Behav.* **22**, 3–74.

Manning, A. 1956. Some aspects of the foraging behaviour of bumble-bees. *Behaviour* **9**, 164–201.

Marden, J.H. and Waddington, K.D. 1981. Floral choices by honeybees in relation to the relative distances to flowers. *Physiol. Entomol.* **6**, 431–5.

Markl, H. 1972. Aggression und Beuteverhalten bei Piranhas (Serrasalminae, Characidae). *Z. Tierpsychol.* **30**, 190–216.

Markowska, A., Buresova, O. and Bures, J. 1983. An attempt to account for controversial estimates of working memory persistence in the radial maze. *Behav. Neural Biol.* **38**, 97–112.

Martindale, S. 1982. Nest defense and central place foraging: a model and experiment. *Behav. Ecol. Sociobiol.* **10**, 85–9.

Marzluff, J.M. 1988. Do pinyon jays alter nest placement based on prior experience? *Anim. Behav.* **36**, 1–10.

Mason, P.R. 1975. Chemo-klino-kinesis in planarian food location. *Anim. Behav.* **23**,460–9.

Maynard Smith, J. 1978. Optimization theory in evolution. *A. Rev. Ecol. Syst.* **9**, 31–56.

Maynard Smith, J. 1982. *Evolution and the Theory of Games.* Cambridge University Press, Cambridge.

Maynard Smith, J. 1984. Game theory and the evolution of behaviour. *Brain Behav. Sci.* **7**, 95–125.

Mayor, K.L., Aracena, J.M. and Bell, W.J. 1987. Search duration of *Drosophila melanogaster* on homogeneous sucrose patches: relative effects of starvation period, sucrose concentration and patch size, *J. Ethol.* **5**, 67–74.

Mech, L.D. 1966. The Wolves of Isle Royale. *U.S. Nat. Park Serv. Fauna. Ser. No.* 7.

Mech, L.D. 1970. The Wolf: The Ecology and Behavior of an Endangered Species. Natural History Press, New York.

Mech, L.D. and Frenzel, L.D., Jr. (eds.) 1971. Ecological studies of the Timber Wolf in North-eastern Minnesota. *US Dept Agr. Forest Ser. Res. Paper NC*, **52**, 1–62.

Mellgren, R.L. 1983. (ed.) *Animal Cognition and Behavior.* North-Holland Publishing Company, Netherlands.

Mellgren, R.L. and Roper, T.J. 1986. Spatial learning and discrimination of food patches in the European badger (*Meles meles* L.) *Anim. Behav.* **34**, 1129–34.

Menzel, E. 1978. Cognitive mapping in chimpanzees. In *Cognitive Processes in Animal Behaviour.* (eds. S. H. Hulse, H. Fowler and W.K. Honig), pp. 375–422. Lawrence Erlbaum,

Menzel, R. and Erber, J. 1978. Learning and memory in bees. *Sci. Amer.* 239, 80–7.

Metcalfe, N.B., Huntingford, F.A. and Thorpe, J.E. 1987. The influence of predation risk on the feeding motivation and foraging strategy of juvenile atlantic salmon. *Anim. Behav.* **35**, 901–11.

Metzgar, L.H. 1967. An experimental comparison of screech owl predation on resident and transient white-footed mice (*Peromyscus leucopus*). *J. Mammal.* **48**, 387–91.

Michener, C.D. 1974. *The Social Behavior of the Bees.* Harvard University Press, Cambridge, MA.

Michener, C.D., Breed, M.D. and Bell, W.J. 1979. Seasonal cycles, nests and social behavior of some Columbian Halictine bees. *Rev. Biol. Trop.* **27**, 13–34.

Mikkola, H. 1970. On the activity and food of the pygmy owl, *Glaucidium passerinum*, during breeding. *Ornis Fenn.* **47**, 10–14.

Milinski, M. 1979. An evolutionarily stable feeding strategy in sticklebacks. *Z. Tierpsychol.* **51**, 36–40.

Montgomerie, R.D. 1979. The energetics of foraging and competition in some Mexican hummingbirds. Dissertation. McGill University, Montreal, Quebec, Canada.

Moorhouse, J.E. 1971. Experimental analysis of the locomotor behaviour of *Schistocerca gregaria* induced by odour. *J. Insect Physiol.* **17**, 913–20.

Morse, D.H. 1970. Ecology aspect of some mixed-species foraging flocks of birds. *Ecol. Monogr.* **40**, 119–68.

Morse, D.H. 1980a. *Behavioral Mechanisms in Ecology*. Harvard University Press, Cambridge, M.A.

Morse, D.H. 1980b. The effect of nectar abundance on foraging patterns of bumblebees. *Ecol. Entomol.* **5**, 53–9.

Morton, M.L. 1967. The effects of insolation on the diurnal feeding pattern of white-crowned sparrows (*Zonotrichia leucophrys* gambeli). *Ecol.* **49**, 690–4.

Moss, R. 1972. Food selection by red grouse (*Lagopus lagopus scoticus* (Lath.)) in relation to chemical composition. *J. Anim. Ecol.* **41**, 411–28.

Moss, R., Miller, G.R. and Allen, S.E. 1972. Selection of heather by captive red grouse in relation to the age of the plant. *J. Appl. Ecol.* **9**, 771–81.

Mourier, H. 1964. Circling food-searching behaviour of the housefly (*Musca domestica* L.). *Dansk. Natur. Foren. Denmark Vidensk. Medde.* **127**, 181–94.

Murdie, G. and Hassell, M.P. 1973. Food distribution, searching success and predator–prey models. In *The Mathematical Theory of the Dynamics of Biological Populations* (ed. M.S. Bartlett and R.W. Hiorns), pp. 87–101. Academic Press, New York.

Murphey, R.K. and Zaretsky, M.D. 1972. Orientation to calling song by female crickets, *Scapsipedus marginatus* (Gryllidae). *J. Exp. Biol.* **56**, 335–52.

Myers, J.H. 1985. Effect of physiological condition of the host plant on the ovipositional choice of the cabbage white butterfly, *Pieris rapae*. *J. Anim. Ecol.* **54**, 193–204.

Myers, J.P., Conners, P.G. and Pitelka, F.A. 1981. Optimal territory size and the sanderling: compromises in a variable environment. In *Foraging Behavior* (ed. A.C. Kamil and T.D. Sargent), pp. 135–58. Garland and STPM Press, New York.

Nadel, L. and O'Keefe, J. 1974. The hippocampus is pieces and patchs: An essay on modes of explanation in physiological psychology. In Essay on the nervous system. A Festschrift for Prof. F. Z. Young (Ed. by R. Bellairs and E.G. Gray), pp. 367–90. Clarendon Press, Oxford.

Nagle, K.J. and Bell W.J. 1987. Genetic control of the search tactic of *Drosophila melanogaster*: an ethometric analysis of *rover/sitter* traits in adult flies. *Behav. Genet.* **17**, 385–408.

Nakamuta, K. 1985. Mechanism of the switchover from extensive to area-concentrated search behaviour of the ladybird beetle, *Coccinella septempunctata bruckii*. *J. Ins. Physiol.* **31**, 849–56.

Nakamuta, K. 1987. Diel rhythmicity of prey-search activity and its predominance over starvation in the lady beetle, *Coccinella septempunctata bruckii*. *Physiol. Entomol.* **12**, 91–8.

Narise, T. 1962. Studies on competition in plants and animals, X. Genetic variability of migratory activity in natural populations of *Drosophila melanogaster*. *Jpn. J. Genet.* **7**, 451–61.

Nelson, M.C. 1977. The blowfly's dance: role in the regulation of food intake. *J. Insect Physiol.* **23**, 603–12.

Newell, P.F. 1966. The nocturnal behavior of slugs. *Med. and Biol. Illust. part 16*, 146–59.

Newton, I. 1972. *Finches*. Collins, London.

Norton-Griffiths, M. 1969. The organisation, control and development of parental feeding in the oystercatcher (*Haematopus ostralegus*). *Behaviour* **24**, 55–114.

O'Brien, W.J. and Wright, D.I. 1985. Potential limits on the daytime planktivorous feeding depth of white crappie. *Verh. Internat. Verein. Limnol.* **22**, 2527–33.

O'Brien, W.J., Evans, B.I. and Howick, G.L. 1986. A new view of the predation cycle of a planktivorous fish, white crappie (*Pomoxis annularis*). *Can. J. Fish. Aquat. Sci.* **43**, 1894–99.

O'Brien W.J. Evans, B.I. and Browman, H.I. 1989. Flexible search tactics and efficient foraging in saltatory searching animals. *Oecologia* **80**, 100–110.

O'Keefe, J. and Nadel L. 1978. *The Hippocampus as a Cognitive Map*. Clarendon Press, Oxford.

Ollason, J.G. 1980. Learning to forage-optimally? *Theor. Popul. Biol.* **18**, 44–56.

Olton, D.S. 1979. Mazes, maps and memory. *Amer. Psychol.* **34**, 583–96.

Olton, D.S. and Samuelson, R.J. 1976. Remembrance of places passed: spatial memory in rats. *J. Exp. Psychol.: Anim. Behav. Processes* **2**, 97–116.

Opp, S.B. and Prokopy, R.J. 1986. Variation in laboratory oviposition by *Rhagoletis pomonella* (Diptera: Tephritidae) in relation to mating status. *A. Entomol. Soc. Amer.* **79**, 705–10.

Orians, G.H. 1980. Adaptations of marsh-nesting blackbirds. *Monographs in Population Biology*, no. 14. Princeton University Press, Princeton.

Orians, G.H. 1981. Foraging behavior and the evolution of discriminatory abilities. *Foraging Behavior: Ecological, Ethological and Physiological Approaches* (ed. A.C. Kamil and T.D. Sargent), pp. 389–408. Garland and STPM Press, New York.

Orians, G.H. and Horn, H.S. 1969. Overlap in foods and foraging of four species of blackbirds in the potholes of central Washington. *Ecol.* **50**, 930–8.

Orians, G.H. and Pearson, N.E. 1979. On the theory of central place foraging. In *Analysis of Ecological Systems* (ed. D.J. Horn, G.T. Stairs and R.D. Mitchell), pp. 155–77. Ohio State University Press, Columbus.

Oster, G. and Heinrich, B. 1976. Why do bumblebees major? a mathematical model. *Ecol. Monogr.* **46**, 129–33.

Owen-Smith, N. and Novellie, P. 1982. What should a clever ungulate eat? *Amer. Nat.* **119**, 151–78.

Papaj, D.R. and Rausher, M.D. 1983. Individual variation in host location by phytophagous insects. In *Herbivorous Insects, Host- Seeking Behavior and Mechanisms* (ed. S. Ahmad), pp. 77–124.

Academic Press, New York.

Parker, G.A. 1974. The reproductive behavior and the nature of sexual selection in *Scatophaga stercoraria* L. (Diptera, Scatophagidae). IX. Spatial distribution of fertilization rates and evolution of male search strategy within the reproducthve area. *Evolution* **28**, 93–108.

Parker, G.A. 1978. Searching for mates. In *Behavioral Ecology: An Evolutionary Approach.* (ed. J.R. Krebs and N.B. Davies), pp. 214–44. First edition. Blackwell, Oxford.

Parker, G.A. 1984. Evolutionarily Stable Strategies. In *Behavioural Ecology: An Evolutionary Approach.* (eds J.R. Krebs and N.B. Davis), pp. 7–29. Sinauer Associates Inc. Publishers, Sunderland, Massachusetts.

Partridge, L. 1976. Individual Difference in feeding efficiencies and feeding preferences of captive great tits. *Anim. Behav.* **24**, 230–40.

Partridge, L. and Halliday, T. 1984. Mating patterns and choice. In *Behavioral Ecology. An Evolutionary Approach.* (ed. J.R. Krebs and N.B. Davies), pp. 222–50. Blackwell, Oxford.

Payne, R.S. 1971. Acoustic location of prey by barn owls (*Tyto alba). J. exp. Biol.* **54**, 535–7.

Payne, T.L., Shorey, H.H. and Gaston, L.K. 1970. Sex pheromones of noctuid moths: Factors influencing antennal responsiveness in males of *Trichoplusia ni. J. insect Physiol.* **16**, 1043–55.

Perron, J.P. 1972. Effects of some ecological factors on populations of the onion maggot, *Hylemya antiqua* (Meigen) under field conditons in southwestern Quebec. *Ann. Entomol. Soc. Quebec*, **17**, 29–45.

Persson, L. 1986. Temperature-induced shift in foraging ability in two fish species, roach (*Rutilus rutilus*) and perch (*Perca fluviatilis*), implications for coexistince between poikilotherms. *J. Anim. Ecol.* **55**, 829–39.

Pianka, E.R. 1966. Convexity, desert lizards, and spatial heterogeneity. *Ecol.* **47**, 1055–9.

Pianka, E.R. 1988. *Evolutionary Ecology*. Harper and Row, New York.

Pienkowski, M.W. 1983. Changes in the foraging pattern of plovers in relation to environmental factors. *Anim. Behav.* **31**, 244–64.

Pierce, G.J. and Ollason, J.G. 1987. Eight reasons why optimal foraging

theory is a complete waste of time. *Oikos* **49**, 111–18.
Plasa, L. 1979. Heimfindeverhalten bie *Salamandra salamandra* (L.) *Z. Tierpsychol.* **51**, 113–25.
Powell, G.V.N. 1974. Experimental analysis of the social value of flocking by starlings (*Sturnus vulgaris*) in relation to predation and foraging. *Anim. Behav.* **22**, 501–5.
Price, S. 1978. The nutritional ecology of Coke's hartebeest (*Alcelaphus buselaphus cokei*) in Kenya. *J. Appl. Ecol.* **15**, 33–49.
Price, P.W. 1984. *Insect Ecology*. John Wiley and Sons, N.Y.
Price, P.W., Bouton, C.E., Gross, P., McPheron, B.A., Thompson, J.N. and Weis, A.E. 1980. Interactions among three trophic levels: Influence of plants on interactions between insect herbivores and natural enemies. *A. Rev. Ecol. Syst.* **11**, 41–65.
Prokopy, R.J. 1968. Sticky spheres for estimating apple maggot adult abundance. *J. Econ. Entomol.* **61**, 1082–5.
Pyke, G.H. 1978. Optimal foraging: movement patterns of bumblebees between inflorescences. *Theor. Pop. Biol.* **13**, 72–98.
Pyke, G.H. 1984. Optimal foraging theory: a critical review. *A. Rev. Ecol. Syst.* **15**, 523–75.
Pyke, G.H., Pulliam, H. and Charnov, E. 1977. Optimal foraging: a selective review of theory and tests. *Q. Rev. Biol.* **52**, 137–54.
Racey, P.A. and Swift, S.M. 1985. Feeding ecology of *Pipistrellus pipistrellus* (Chiroptera: Vespertilionidae) during pregnancy and lactation. I. foraging behaviour. *J. Anim. Ecol.* **54**, 205–15.
Raleigh, R.F. 1971. Innate control of migrations of salmon and trout fry from natal gravels to rearing areas. *Ecol.* **52**, 291–7.
Ralph, C.P. 1976. Search behavior of the large milkweed bug, *Oncopeltus fasciatus* (Hemiptera: Lygaeidae). *Ann. Entomol. Soc. Amer.* **70**, 337–42.
Ralph, C.P. 1977. Effect of host plant density on populations of a specialized seed-sucking bug, *Oncopeltus fasciatus*. *Ecol.* **58**, 799–809.
Rand, A.L. 1954. Social feeding behavior of birds. *Fieldiana Zool.* **36**, 1–71.
Rausher, M.D. 1978. Search image for leaf shape in a butterfly. *Science* **200**, 1071–3.
Rausher, M. 1979. Coevolution in a simple plant-herbivore system. Ph.D. thesis. Cornell University, Ithaca, New York.
Rausher, M.D. 1981. The effect of native vegetation on the susceptibility of *Aristolochia reticulata* (Aristolochiaceae) to herbivore attack. *Ecol.* **62**, 1187–95.
Read, D.P., Feeny, P.P. and Root, R.B. 1970. Habitat selection by the aphid parasite *Diaretiella rapae* (Hymenoptera: Broaconidae) and hyperparasite *Charips brassica* (Hymenoptera: Cynipidae). *Can. Entomol.* **102**, 1567–78.
Real, L.A. 1981. Uncertainty and pollinator–plant interactions: The foraging behavior of bees and wasps on artificial flowers. *Ecol.* **62**, 20–6.

Rechten, C., Avery, M.I. and Stevens, T.A. 1983. Optimal prey selection: why do great tits show partial preferences? *Anim. Behav.* **31**, 576–84.

Rheinlaender, J., Gerhardt, H.C., Yager, D.P. and Capranica, R.R. 1979. Accuracy of Phonotaxis by the greenfrog. (*Hyla*). *J. Comp. Physiol.* **133**, 247–55.

Ricklefs, R.E. 1979. *Ecology*, 2nd ed. Chiron Press, Portland, Oregon.

Riechert, S.E. 1986. Spider fights: a test of evolutionary game theory. *Amer. Sci.* **47**, 604–10.

Rijnsdorp, A., Daan, S. and Dijkstra, C. 1981. Hunting in the kestrel, *Falco tinnunculus*, and the adaptive significance of daily habits. *Oecologia* **50**, 391–406.

Riley, J.R. and Reynolds, D.R. 1896. Orientation at night by high-flying insects. In *Insect Flight Dispersal and Migration* (ed. W. Danthanarayana), pp. 71–87, Springer-Verlag, Heidelberg.

Risch, S.J. 1980. The population dynamics of several herbivorous beetles in a tropical agro-ecosystem: The effect of interplanting corn, beans, and squash in Costa Rica. *J. Appl. Ecol.* **17**, 593–612.

Rissing, S.W. and Wheeler, J. 1976. Foraging responses of Veromessor pergandei to changes in seed production. *Pan-Pacific Entomol.* **52**, 63–72.

Roggero, R.J. and Bell, W.J. (1989) Computer simulation of search behavior. *Entomol. Soc. Amer., San Antonio* (Abstr.).

Rohlf, F.J. and Davenport, D. 1969. Simulation of simple models of animal behavior with a digital computer. *J. Theor. Biol.* **23**, 400–24.

Roitberg, B.D. 1985. Searching dynamics in fruit-parasitic insects. *J. Insect Physiol.* **31**, 865–72.

Roitberg, B.D. and Prokopy, R.J. 1982. Influence of intertree distance on foraging behavior of *Rhagoletis pomonella* in the field. *Ecol. Entomol.* **7**, 437–42.

Roitberg, B.D., Van Lenterner, J.C., Van Alphen, J.J.M., Galis, F. and Prokopy, R.J. 1982. Foraging behaviour of *Rhagoletis pomonella*, a parasite of Hawthorn (*Crataegus viridis*), in nature. *J. Anim. Ecol.* **51**, 307–25.

Roitblat, H.L., Bever, T.G. and Terrace, H.S. (Eds.) 1984. *Animal Cognition* Lawrence Erlbaum Associates, Publishers. Hillsdale, New Jersey.

Rollo, C.D. and Wellington, W.G. 1981. Environmental orientation by terrestrial Mollusca with particular reference to homing behaviour. *Can. J. Zool.* **59**, 225–39.

Root, R.B. and Kareiva, P.M. 1984. The search for resources bycabbage butterflies (*Pieris rapae*): ecological consequencesand adaptive significance of Markovian movements in a patchyenvironment. *Ecol.* **65**, 147–65.

Rosenthal, H. and Hempel, G. 1970. Experimental studies in feeding and food requirements of herring larvae (*Clupea harengus* L.) In *Marine Food Chains* (ed. J.H. Steele), pp. 344–64. Oliver and Boyd, Edinburgh.

Royama, T. 1970. Factors governing the hunting behaviour and selection of food by the great tit (*Parus major* L.). *J. Anim. Ecol.* **39**, 619–68.

Rubenstein, D.I. 1980. On the evolution of alternative mating strategies. In *Limits to Action* (ed. J.E.R. Staddon), pp 65–100. Academic Press, New York.

Ruiter, L. de 1952. Some experiments on the camouflage of stick caterpillars. *Behaviour* **4**, 222–32.

Sabelis, M.W. and Dicke, M. 1985. Long-Range dispersal and searching behaviour. In *Spider Mites. Their Biology, Natural Enemies and Control*, Vol. 1B (eds W. Helle and M.W. Sabelis), pp. 141–60. Elsevier Sci. Publ., Amsterdam.

Sabelis, M.W., Vermaat, J.E. and Groeneveld, A. 1984. Arrestment responses of the predatory mite, *Phytoseiulus persimilis*, to steep odour gradients of a kairomone. *Physiol. Entomol.* **9**, 437–46.

Sakai, K., Narise, T. and Iyama, S. 1957. Migration studies in several wild strains of *Drosophila melanogaster. Nat. Inst. Genet. (Japan) Ann. Rep., No.* 7, pp. 73–5.

Salt, G.W. and Willard, D.E. 1971. The hunting behavior and success of forster's tern. *Ecol.* **52**, 989–99.

Sandness, J.N. and McMurty, J.A. 1972. Prey consumption behaviour of *Amblyseius largoensis* in relation to hunger. *Can. Entomol.* **104**, 461–70.

Saxena, K.N. 1969. Patterns of insect-plant relationships determining susceptibility or resistance of different plants to an insect. *Entomol. Exp. Appl.* **12**, 751–66.

Schal, C., Tobin, T.R., Surber, J.L., Vogel, G., Tourtellot, M.K., Leban, R.A., Sizemore, R. and Bell, W.J. 1983. Search strategy of sex pheromone-stimulated male German cockroaches. *J. Insect Physiol.* **27**, 575–9.

Schaller, G.B. 1967. *The Deer and the Tiger.* University of Chicago Press, Chicago.

Schaller, G.B. 1972. *The Serengeti Lion.* University of Chicago Press, Chicago.

Schmid-Hempel, P. 1984. The importance of handling time for the flight directionality in bees. *Behav. Ecol. Sociobiol.* **15**, 303–9.

Schmid-Hempel, P. and Schmid-Hempel, R. 1986. Nectar-collecting bees use distance-sensitive movement rules. *Anim. Behav.* **34**, 605–7.

Schmitz, B., Scharstein, H. and Wendler, G. 1982. Phonotaxis in *Gryllus campestris* L. (Orthoptera, Gryllidae) I. Mechanism of acoustic orientation in intact female crickets. *J. Comp. Physiol.* **148**, 431–44.

Schneirla, T.C. 1971. *The Army Ants: a Study in Social Organization.* W.H. Freeman, San Francisco.

Schoener, T.W. 1971. Theory of feeding strategies. *A. Rev. Ecol. Syst.* **2**, 369–404.

Schöne, H. 1984. *Spatial Orientation. The Spatial Control of Behavior in Animals and Man.* Princeton University Press, Princeton.

Schultz, J.C. 1983. Habitat selection and foraging tactics of caterpillars in heterogeneous trees. In *Variable Plants and Herbivores in Natural and Managed*

Systems (Ed. by R.F. Denno and M.S. McClure). pp. 61–90. Academic Press, New York.

Seeley, T.D. 1985. *Honeybee Ecology: A study of Adaptation in Social Life.* Princeton University Press, Princeton, NJ.

Seeley, T.D. 1986. Social foraging by honeybees: how colonies allocate foragers among patches of flowers. *Behav. Ecol. Sociobiol.* **19**, 343–54.

Sherry, D.F. 1985. Food storage by birds and mammals. *Adv. Study Behav.* **15**, 153–88.

Shettleworth, S.J. and Krebs, J.R. 1982. How marsh tits find their hoards: the roles of site preference and Spatial Memory. *J. Exper. Psychol. Anim. Behav. Proc.* **8**, 354–75.

Shreeve, T.G. 1986. Egg-laying by the speckled wood butterfly (*Pararge aegeria*) the role of female behaviour, host plant abundance and temperature. *Ecol. Entomol.* **11**, 229–36.

Sibly, R.M. and McFarland, D.J. 1974. A state-space approach to motivation. In *Motivational Control Systems Analysis* (ed. D.J. McFarland), pp. 213–50. Academic Press, London.

Sibly, R.M. and McFarland, D.J. 1976. On the fitness of behaviour sequences. *Amer. Nat.* **110**, 601–17.

Simons, S. and Alcock, J. 1971. Learning and the foraging persistence of white-crowned sparrows *Zonotrichia leucophrys. Ibis* **113**, 477–82.

Singer, M.C. 1982. Quantification of host preference by manipulation of oviposition behavior in the butterfly *Euphaydryas editha. Oecologia* **52**, 230–5.

Singer, M.C. 1983. Determinants of multiple host use in a phytophagous insect population. *Evolution* **37**, 389–403.

Siniff, D.B. and Jessen, C.R. 1969. A simulation model of animal movement patterns. *Adv. Ecol. Res.* **6**, 185–219.

Smith, C.C. (1977) Feeding behaviour an social organization in howling monkeys In *Primate Ecology, Studies of Feeding and Ranging Behaviour in Lemurs, Monkeys, and Apes* (ed T.H. Clutton-rock), pp. 97–126. Academic Press, New York.

Smith, E.A. 1981. The application of optimal foraging theory to the analysis of hunter-gatherer group size. In *Hunter-Gatherer Foraging Strategies Ethnographic and Archeological Analyses* (eds B. Winterhalder and E.A. Smith), pp. 36–65, University of Chicago Press, Chicago.

Smith, J.N.M. 1974a. The food searching behaviour of two European thrushes I. Description and analysis of search paths. *Behaviour* **48**, 276–302.

Smith, J.N.M. 1974b. The food searching behaviour of two European thrushes. II: The adaptiveness of the search patterns. *Behaviour* **49**, 1–61.

Smith, J.N.M. and Dawkins, R. 1971. The hunting behaviour of individual great tits in relation to spatial variations in their food density. *Anim. Behav.* **19**, 695–706.

Smith, J.N.M. and Sweatman, H.P.A. 1974. Food searching behavior of titmice in patchy environments. *Ecol.* **55**, 1216–32.

Smythe, N. 1970. Relationships between fruiting seasons and seed dispersal methods in a neotropical forest. *Amer. Nat.* **104**, 25–35.

Snow, D.W. 1965. A possible selection factor in the evolution of fruiting seasons in a tropical forest. *Oikos* **15**, 274–81.

Sokolowski, M.B. 1985. Genetics and ecology of *Drosophila melanogaster* larval foraging and pupation behaviour. *J. Insect Physiol.* **31**, 857–65.

Sokolowski, M.B. 1986. *Drosophila* larval foraging behavior and correlated behaviors. In *Evolutionary Genetics of Invertebrate Behavior* (ed. M. Huettel), pp. 197–213. Plenum Press, New York.

Sorensen, K. and Bell, W.J. 1986. Responses of isopods to temporal changes in relative humidity: simulation of a 'humid patch' in a dry habitat. *J. Insect Physiol.* **32**, 51–7.

Southwood, T.R.E. 1962. Migration of terrestrial arthropods in relation to habitat. *Biol. Rev.* **37**, 171–214.

Southwood, T.R.E. 1977. Habitat, the templet for ecological strategies? *J. Anim. Ecol.* **46**, 337–65.

Spradbery, J.P. 1969. The biology of *Pseudorhyssa sternata* Merrill (Hym., Ichneumonidae), a cleptoparasite of sircid woodwasps. *Bull. Entomol. Res.* **59**, 291–7.

Spradbery, J.P. 1970. Host finding by *Rhyssa persuasoria* (L.), an ichneumonid parasite of siricid woodwasps. *Anim. Behav.*, **18**, 103–14.

Stanton, M.L. 1984. Short-term learning and the searching accuracy of egg-laying butterflies. *Anim. Behav.* **32**, 32–40.

Stein, R.A. and Magnuson, J.J. 1976. Behavioral response of crayfish to a fish predator. *Ecol.* **57**, 751–61.

Stephens, D.W. 1982. Stochasticity in foraging theory: risk and information. Ph.D. thesis, Oxford University.

Stephens, D.W and Krebs, J.R. 1986 *Foraging Theory*. Princeton University Press. Princeton, N.J.

Stoddart, D.M. 1980. *The Ecology of Vertebrate Olfaction*. Chapman and Hall, London.

Stone, L.D. 1975. *Theory of Optimal Search*. Academic Press, New York.

Strand, M.R. and Vinson, S.B. 1982. Behavioral responses of the parasitoid *Cardiochiles nigriceps* to a kairomone. *Entomol. Exp. Appl.* **31**, 308–15.

Straw, R.M. 1972. A. Markoå model for pollinator constancy and competition. *Amer. Nat.* **106**, 597–620.

Sussman, R.W. 1977. Feeding behaviour of lemur catta and lemur fulvus. In *Primate Ecology, Studies of Feeding and Ranging Behaviour in Lemurs, Monkeys, and Apes*. (ed. T.H. Clutton-Brock), p. 1–36. Academic Press, New York.

Svendsen, G. E. 1974. Behavioral and environmental factors in the spatial distribution and population dynamics of a yellow-bellied marmot population. *Ecol.* **55**, 760–71.

Symons, D. 1978. *Play and Aggression: A Study of Rhesus Monkeys*. Columbia University Press, New York.

Tabashnik, B.E. 1980. Population structure of pierid butterflies. III. Pest populations of *Colias philodice eriphyle*. *Oecologia* **47**, 175.
Tabashnik, B.E. 1983. Host range evolution: the shift from native legume hosts to alfalfa by the butterfly, *Colias philodice eriphyle*. *Evolution* **37**, 150–62.
Tabashnik, B.E. 1986. Evolution of host plant utilization in Colias butterflies, In *Evolutionary Genetics of Invertebrate Behavior Progress and Prospects* (ed. M.D. Huettel), p. 173–184, Plenum Press, New York.
Talbot, L. and Talbot, M. 1963. The wildebeest in Western Masailand, East Africa. *Wildl. Monogr. No. 12*.
Tanimura, T., Isono, K., Takamura, T. and Shimada, I. 1982. Genetics dimorphism in the taste sensitivity to trehalose in *Drosophila melanogaster*. *J. Comp. Physiol.* **147**, 433–7.
Taylor, C.E. and Powell, J.R. 1978. Habitat choice in natural populations of *Drosophila*. *Oecologia (Berl.)* **37**, 69–75.
Taylor, L.R. and Taylor, R.A.J. 1977. Aggregation, migration and population mechanics. *Nature* **265**, 415–21.
Taylor, R.J. 1976. Value of clumping to prey and the evolutionary response of ambush predators. *Amer. Nat.* 110:13–29.
Tempel, B.L., Bonini, N., Dawson, D.R. and Quinn, W.G. 1983. Reward learning in normal and mutant *Drosophila*. *Proc. Nat. Acad. Sci., USA* **80**, 1482–6.
Thomas, G. 1974. The influence of encountering a food object on the subsequent searching behaviour in *Gasterosteus aculeatus L*. *Anim. Behav.* **22**, 941–52.
Thornhill, R. 1979. Adaptive female-mimicking behavior in a scorpionfly. *Science* **205**, 412–14.
Thornhill, R. 1981. *Panorpa* (Mecoptera, Panorpidae) Scorpionflies: systems for understanding resource-defence polygyny and alternative male reproductive efforts. *A. Rev. Ecol. Syst.* **12**, 355–86.
Thornhill, R. 1984. Alternative female choice tactics in the scorpionfly *Hylobittacus apicalis* (Mecoptera) and their implications. *Amer. Zool.* **24**, 367–83.
Tinbergen, L. 1960. The natural control of insects in pinewoods. I, Factors influencing the intensity of predation by songbirds. *Arch. Neerl. Zool.* **13**, 265–336.
Tinbergen, N., Impekoven, M. and Franck, D. 1967. An experiment on spacing out as a defence against predation. *Behaviour* **28**, 307–21.
Toates, F.M. and Birke, L.I.A. 1982. Motivation: A new perspective on some old ideas. In *Perspectives in Ethology* (Vol. 5) (eds P.P.G. Bateson and P.H. Klopfer), pp. 191–235. Plenum Press, New York.
Tobin, T.R. and Bell, W.J. 1986. Chemo-orientation of *Trogoderma variabile* in a simulated sex pheromone plume. *J. Comp. Physiol.* **158**, 729–39.
Tolman, E.C. 1948. Cognitive maps in rats and men. *Psychol. Rev.* **55**, 189–208.

Tortorici, C. and Bell, W.J. 1988. Search orientation in adult *Drosophila melanogaster*: responses of rovers and sitters to resource Dispersion in a food patch. *J. Insect Behavior*. **1**, 209–223.

Tortorici, C., Brody, A. and Bell, W.J. 1986. Influences of spatial patterning of resources on search orientation of adult *Drosophila*. *Anim. Behav*. **34**, 1569–70.

Tostowaryk, W. 1971. Relationship between parasitism and predation of diprionid sawflies. *Ann. Entomol. Soc. Amer*. **64**, 1424–7.

Tourtellot, M.K., Collins, R.D. and Bell W.J. 1990. The problem of movelength and turn definition in analysis of orientation data. *J. Theor. Biol*. (in press)

Turnbull, A.L. 1964. The searching for prey by a web-building spider. *Achaeranea tepidanonim* (C.L. Koch). *Can. Entomol*. **96**, 568–79.

Uetz, G.W., Kane, T.C., Stratton, G.E. and Benton, M.J. 1987. Environmental and genetic influences on the social grouping tendency of a communal spider. In *Evolutionary Genetics of Invertebrate Behavior* (ed. M.D. Huettel), pp. 43–53. Plenum Publishing Corporation, New York.

Van Lenteren, J.C. and Bakker, K. 1975. Discrimination between parasitised and unparasitised hosts in the parasitic wasp *Pseudocoila bochei*: a matter of learning. *Nature, Lond*. **254**, 417–19.

Van Lenteren, J.C., Nell, H.W., Sevenster-Van Der Lelie, L.A. and Woets, J. 1976. The parasite–host relationship between *Encarsia formosa* (Hymenoptera: Aphelinidae) and *Trialeurodes vaporariorum* (Hymenoptera: Aleyrodidae). 1. Host–finding by the parasite. *Entomol. Exp. Appl*. **20**, 123–30.

Vet, L.E.M. 1983. Host–habitat location through olfactory cues by *Leptopilina clavipes* (Hartig) (Hym.: Eucoilidae), a parasitoid of fungivorous *Drosophila*: the influence of conditioning. *Neth. J. Zool*. **33**, 225–48.

Via, S. 1986. Quantitative genetic analysis of feeding and oviposition behavior in the polyphagous leafminer *Liriomyza sativae*. In *Evolutionary Genetics of Invertebrate Behavior Progress and Prospects* (ed. M.D. Huettel), pp. 185–96, Plenum Press, New York.

Via, S. and Lande, R. 1985. Genotype–environment interaction and the evolution of phenotypic plasticity. *Evolution* **39**, 505–22.

Vines, G. 1980. Spatial consequences of aggressive behaviour in flocks of oystercatchers, *Haematopus ostralegus* L. *Anim. Behav*. **28**, 1175–83.

Vinson, S.B. 1975. Source of material in the tobacco budworm in host recognition by the egg–larval parasitoid, *Chelonus texanus*. *Ann. Entomol. Soc. Am*. **68**, 381–4.

Vinson, S.B. 1977. Behavioral chemicals in the augmentation of natural enemies. In *Biological Control by Augmentation of Natural Enemies* (eds by R.L. Ridgeway and S.B. Vinson), pp. 237–79. Plenum Press, New York.

Vinyard, G.L. and O'Brien, W.J. 1976. Effects of light and turbidity on the reactive distance of bluegill *Lepomis macrochirus* *J. Fish. Res. Board Can*. **33**, 2845–9.

Visscher, P.K. and Seeley, T.D. 1982. Foraging strategy of honeybee colonies in a temperate deciduous forest. *Ecol.* **63**, 1790–1801.

Vivas, H.J. and Saëther, B.-E. 1987. Interactions between a generalist herbivore the moose *Alces alces* and its food resources: an experimental study of winter foraging behaviour in relation to browse availability. *J. Anim. Ecol.* **56**, 509–20.

Waage, J.K. 1978. Arrestment responses of the parasitoid *Nemeritis canescens* to a contact chemical produced by its host, *Plodia interpunctella. Physiol. Entomol.* **3**, 135–46.

Waage, J.K. 1979. Foraging for patchily-distributed hosts by the parasitoid *Nemeritis canescens. J. Anim. Ecol.* **48**, 353–71.

Waddington, K.D. (1983a) Floral-visitation-sequences by bees:
models and experiments. In *Handbook of Experimental Pollination Biology* (eds E. Jones and R. Little), pp. 461–73. Van Nostrand-Reinhold, New York.

Waddington, K.D. (1983b) Foraging behavior of pollinators. In *Pollination Biology* (ed. L.A. Real), pp. 213–239. Academic Press, New York.

Waddington, K.D. and Holden, L.R. 1979. Optimal foraging: on flower selection by bees. *Amer. Nat.* **114**, 179–96.

Waddington, K.D., Allen, T. and Heinrich, B. 1981. Floral preferences of bumblebees (*Bombus edwardsii*) in relation to intermittent versus continuous rewards. *Anim. Behav.* **29**, 779–84.

Wahl, O. 1932. Neue untersuchungen ber das zeitgebachtnis der beinen. *Z. Vergl. Physiol.* **16**, 529–89.

Wall, S.B. van der 1982. An experimental analysis of cache recovery in Clark's nutcraker. *Anim. Behav.* **30**, 84–94.

Wallin, A. 1988. The genetics of foraging behaviour: artificial selection for food choice in larvae of the fruitfly, *Drosophila melanogaster. Anim. Behav.* **36**, 106–14.

Wallraff, H.G. 1981. Clock-controlled orientation in space. In *Handbook of Behavioral Neurobiology* (ed. J. Aschoff), p. 299–307. Plenum Press, New York.

Walsberg, G.E. 1977. Ecology and energetics of contrasting social systems in *Phainopepla nitens* (Aves, Ptilogonatidae). *Univ. Calif. Publ. Zool.* **108**, 1–63.

Waltz, E.C. 1987. A test of the information-center hypothesis in two colonies of common terns, *Sterna hirundo. Anim. Behav.* **35**, 48–59.

Ward, P. 1965. The breeding biology of the black-faced dioch *Quelea quelea* in Nigeria. *Ibis* **107**, 326–49.

Ward, P. and Zahavi, A. 1973. The importance of certain assemblages of birds as "information centers" for food finding. *Ibis* **119**, 517–34.

Waser, N. 1978. Competition for hummingbird pollination and sequential flowering in two Colorado wildflowers. *Ecol.* **59**, 934–44.

Watt, A.D. and Dixon, A.F.G. 1981. The role of cereal growth stages and crowding in the induction of alatae in Sitobion avenae and its consequences for population growth. *Ecol. Entomol.* **6**, 441–7.

Weaver, N. 1957. The foraging behavior of honeybees on hairy vetch. II. The foraging area and foraging speed. *Insectes Soc.* **4**, 43–57.

Weber, T., Thorson, J. and Huber, F. 1981. Auditory behavior of the cricket. I. Dynamics of compensated walking and discrimination paradigms on the Kramer treadmill. *J. Comp. Physiol.* **141**, 215–32.

Wecker, S.C. 1963. The role of early experience in habitat selection by the prairie deer mouse, *Peromyscus maniculatus bairdi. Ecol. Monogr.* **33**, 307–25.

Wehner, R. 1982. Himmelsnavigation bei Insekten. Neurophysiologie und Verhalten. *Neuj. Bl. Naturf. Gest. Zrich*, **184**, 1–132.

Wehner, R., Harkness, R.D. and Schmid-Hempel, P. 1983. foraging strategies in individually searching ants *Cataglyphis bicolor* (Hymenoptera: Formicidae). In *Information Processing in Animals*, Vol. 1. (ed. M. Lindauer). pp. 1–79. Mainz: Akademie der Wissenschaften und der Literatur/ Stuttgart and New York: Gustaå Fischer Verlag.

Welker, W.I. 1961. An analysis of exploratory and play behavior in animals In *The Functions of Varied Experience*, (eds D.W. Fiske and S.R. Maddi), pp. 175–226. Dorsey Press, Homewood, Illinois.

Wellington, W.G. 1965. Some maternal influence on progeny quality in the western tent caterpillar, *Malacosoma pluviale* (Dyar). *Can. Entomol.* **97**, 1–14.

Wells, M.C. and Lehner, P.N. 1978. The relative importance of the distance senses in coyote predatory behavior. *Anim. Behav.* **26**, 251–8.

Werner, E.E. and Hall, D.J. 1974. Optimal foraging and the size selection of prey by bluegill sunfish. *Ecol.* **55**, 1237–12.

White, J., Tobin, T.R. and Bell, W.J. 1984. Local search in the housefly, *Musca domestica* after feeding on sucrose. *J. Insect Physiol.* **30**, 477–88.

White, R.R. and Singer, M.C. 1974. Geographical distribution of host plant choice in *Euphydryas editha* (Nymphalidae). *J. Lepid. Soc.* **28**, 103–7.

Wiens, J.A. 1970. Effects of early experience on substrate pattern selection in Rana aurora tadpoles. *Copeia* 1970, 543–8.

Wiens, J.A. 1972. Anuran habaitat selection: early experience and substrate selection in *Rana cascadae* tadpoles. *Anim. Behav.* **20**, 218–20.

Wiens, J.A. 1973. Interterritorial habitat variation in grasshopper and savanah sparrows. *Ecol.* **54**, 877–84.

Wiens, J.A. 1976. Population responses to patchy environments. *A. Rev. Ecol. Syst.* **7**, 81–120.

Wiklund, C. 1981. Generalist vs. specialist oviposition behaviour in *Papilio machaon* (Lepidoptera) and functional aspects on the hierarchy of oviposition preferences. *Oikos* **36**, 163–70.

Williams, E. 1981. Thermal influences on oviposition in the montane butterfly *Euphydryas gillettii. Oecologia* **50**, 342–6.

Williams, K.S. and Gilbert, L.E. 1981. Insects as selective agents on plant vegetative morphology: egg mimicry reduces egg laying by butterflies. *Science* **212**, 467–9.

Willmer, P.G. 1983. Thermal constraints on activity patterns in nectar-

feeding insects. *Ecol. Entomol.* **8**, 455–69.
Willmer, P.G. 1985. Size effect on the hygrothermal balance and foraging patterns of a sphecid wasp, *Cerceris arenaria. Ecol. Entomol.* **10**, 469–79.
Wilson, E.O. 1971. *The Insect Societies.* Harvard University Press, Cambridge, MA.
Wiltschko, R. and Wiltschko, W. 1984. The development of the navigational system in young homing pigeon. In *Localization and Orientation in Biology and Engineering* (eds Varju, D. and Schnitzler, H.U.) pp. 337–43. Springer, Berlin.
Winstanley, D., Spencer, R. and Williamson, K. 1974. Where have all the whitethroats gone? *Bird Study* **21**, 1–14.
Winterhalder, B. 1981. Foraging strategies in the boreal forest: an analysis of Cree hunting and gathering. In *Hunter-Gatherer Foraging Strategies Ethnographic and Archeological Analyses* (eds by B. Winterhalder and E.A. Smith), pp. 66–98. University of Chicago Press, Chicago.
Winterhalder, B. 1981. Optimal foraging strategies and hunter– gatherer research in anthropology: theory and models. In *Hunter-Gatherer Foraging Strategies Ethnographic and Archeological Analyses* (Ed. by B. Winterhalder and E.A. Smith) pp. 13–35. University of Chicago Press, Chicago.
Wittenberger, J.F. 1981. *Animal Social Behavior.* Duxbury press, Boston.
Wolf, L.L. and Hainsworth, F.R. 1977. Temporal pattern of feeding by hummingbirds. *Anim. Behav.* **25**, 976–88.
Workman, R.B. 1958. The biology of the onion maggot *Hylemya antiqua* (Meigen) under field and greenhouse conditions. Ph.D. Thesis, Oregon State University, Corvallis.
Yano, E. 1978. A simulation model of searching behaviour of a parasite. *Res. Pop. Ecol.* **20**, 105–22.
Yu, Q., Colot, H.V., Kyriacou, C.P., Hall, J.C. and Rosbash, M. 1987. Behaviour modification by in vitro mutagenesis of a variable region within the period gene of *Drosophila. Nature* **326**, 765–9.
Zach, R. and Falls, J.B. 1976a. Ovenbirds (Aves: Parulidae) hunting behavior in a patchy environment: an experimental study. Can. J. Zool. 54, 1863–1879.
Zach, R. and Falls. J.B. 1976b. Foraging behavior, learning, and exploration by captive ovenbirds (Aves: Parulidae). *Can. J. Zool.* **54**, 1880–93.
Zach, R. and Falls, J.B. 1976c. Do ovenbirds (Aves: Parulidae) hunt by expectation? *Can. J. Zool.* **54**, 1894–1903.
Zach, R. and Falls, J.B. 1978. Prey selection by captive ovenbirds (Aves, Parulidae). *J. Anim. Ecol.* **47**, 929–43.
Zach, R. and Smith, J.N.M. 1981. Optimal foraging in wild birds? In *Foraging Behavior* Ecological, Ethological, and psychological approaches. (eds by A.C. Kamil and T.D. Sargent), pp. 95–113. Garland and STPM, New York.
Zahavi, A. 1971. The social behaviour of the white wagtail *Motacilla alba alba*

wintering in Israel. *Ibis* **113**, 203–11.
Zalucki, M.P. and Kitching, R.L. 1982. Component analysis and modelling of the movement process, the simulation of simple tracks. *Res. Pop. Ecol.* (Kyoto) **24**, 239–49.
Zaretsky, M.D. 1972. Specificity of the calling song and short term changes in the phonotactic response by female crickets (*Scapsipedus marginatus*, Gryllidae). *J. Comp. Physiol.* **79**, 153–72.
Zielinski, W.J. 1988. The influence of daily variation in foraging cost on the activity of small carnivores. *Anim. Behav.* **36**, 239–49.
Zimbardo, P.G. and Montgomery, K.C. 1957. The relative strengths of consummatory responses in hunger, thirst and exploratory drive. *J. Comp. Physiol. Psychol.* **50**, 504–8.
Zimmer-Faust, R.K. and Case, J.F. 1983. A proposed dual role of odor in foraging by the California spiny lobster, *Panulirus interruptus* (Randall). *Biol. Bull.* **164**, 341–53.
Zimmerman, M. 1979. Optimal foraging: a case for random movement. *Oecologia (Berl.)* **43**, 261–7.
Zwölfer, H. and Preiss, M. 1983. Host selection and oviposition behaviour in west European ecotypes of *Rhinocyllus conicus* Froel (Col. Curculionidae). *Z. Angew. Entomol.* **95**, 113–22.
Zohren, E. 1968. Laboruntersuchungen zur Massenzucht, Lebensweise, Eiablage und Eiablageverhalten der Kohlfliege *Chortophila brassicae* Bouche (Diptera, Anthomyidae) *Z. Angew. Entomol.* **62**, 139–88.

Index